SCIENCE AND TECHNOLOGY OF POLYMERS AND ADVANCED MATERIALS

Applied Research Methods

SCIENCE AND TECHNOLOGY OF POLYMERS AND ADVANCED MATERIALS

Applied Research Methods

Edited by

Omari V. Mukbaniani, DSc
Tamara N. Tatrishvili, PhD
Marc J. M. Abadie, DSc

APPLE ACADEMIC PRESS

Apple Academic Press Inc.
3333 Mistwell Crescent
Oakville, ON L6L 0A2
Canada

Apple Academic Press Inc.
1265 Goldenrod Circle NE
Palm Bay, Florida 32905
USA

© 2020 by Apple Academic Press, Inc.

First issued in paperback 2021

Exclusive worldwide distribution by CRC Press, a member of Taylor & Francis Group
No claim to original U.S. Government works

ISBN 13: 978-1-77463-438-7 (pbk)
ISBN 13: 978-1-77188-753-3 (hbk)

Library and Archives Canada Cataloguing in Publication

Title: Science and technology of polymers and advanced materials : applied research methods / edited by Omari V. Mukbaniani, DSc, Tamara N. Tatrishvili, PhD, Marc J. M. Abadie, DSc.

Names: Mukbaniani, O. V. (Omar V.), editor. | Tatrishvili, Tamara, editor. | Abadie, Marc J. M., editor.

Description: Includes bibliographical references and index.

Identifiers: Canadiana (print) 20190079932 | Canadiana (ebook) 20190080019 | ISBN 9781771887533 (hardcover) | ISBN 9780429425301 (ebook)

Subjects: LCSH: Polymers. | LCSH: Composite materials. | LCSH: Nanostructured materials. | LCSH: Polymerization.

Classification: LCC TA455.P58 S35 2019 | DDC 668.9—dc23

Library of Congress Cataloging-in-Publication Data

Names: Mukbaniani, O. V. (Omar V.), editor. | Tatrishvili, Tamara, editor. | Abadie, Marc J. M., editor.

Title: Science and technology of polymers and advanced materials : applied research methods / editors, Omari V. Mukbaniani, DSc, Tamara N. Tatrishvili, PhD, Marc J. M. Abadie, DSc.

Other titles: Science and technology of polymers and advanced materials (Apple Academic Press)

Description: Oakville, ON, Canada ; Palm Bay, Florida, USA : Apple Academic Press, 2019. | Includes bibliographical references and index.

Identifiers: LCCN 2019010956 (print) | LCCN 2019015746 (ebook) | ISBN 9780429425301 (ebook) | ISBN 9781771887533 (hardcover : alk. paper)

Subjects: LCSH: Polymer engineering--Research. | Polymers--Industrial applications--Research.

Classification: LCC TA455.P58 (ebook) | LCC TA455.P58 S379 2019 (print) | DDC 620.1/92--dc23

LC record available at https://lccn.loc.gov/2019010956

Apple Academic Press also publishes its books in a variety of electronic formats. Some content that appears in print may not be available in electronic format. For information about Apple Academic Press products, visit our website at **www.appleacademicpress.com** and the CRC Press website at **www.crcpress.com**

About the Editors

Omari V. Mukbaniani, DSc

Omari Vasilii Mukbaniani, DSc, is Professor and Chair of the Macromolecular Chemistry Department of Iv. Javakhishvili Tbilisi State University, Tbilisi, Georgia. He is also the Director of the Institute of Macromolecular Chemistry at Iv. Javakhishvili Tbilisi State University and a member of the Academy of Natural Sciences of Georgia. For several years, he was a member of advisory board and editorial board of the *Journal Proceedings of Iv. Javakhishvili Tbilisi State University* (Chemical Series) and contributing editor of the journals *Polymer News, Polymers Research Journal*, and *Chemistry and Chemical Technology*. His research interests include polymer chemistry, polymeric materials, and chemistry of organosilicon compounds, as well as methods of precision synthesis to build blocks and the development of graft and comb-type structure. He also researches the mechanisms of reactions leading to these polymers and the synthesis of various types of functionalized silicon polymers, copolymers, and block copolymers.

Tamara N. Tatrishvili, PhD

Tamara Tatrishvili, PhD, is Senior Specialist at the Unite of Academic Process Management (Faculty of Exact and Natural Sciences) at Iv. Javakhishvili Tbilisi State University as well as Senior Researcher at the Institute of Macromolecular Chemistry and Polymeric Materials in Tbilisi, Georgia.

Marc J. M. Abadie, DSc

Marc J. M. Abadie, DSc, is Professor Emeritus at the University of Montpellier, Institute Charles Gerhardt of Montpelier, Aggregates, Interfaces, and Materials Energy (ICGM-AIME, UMR CNRS 5253), France. He was head of the Laboratory of Polymer Science and Advanced Organic Materials–LEMP/MAO. He is currently Michael Fam Visiting Professor at the School of Materials Sciences and Engineering, Nanyang Technological University NTU, Singapore. His present activity concerns high-performance polymers for PEMFCs, composites and nanocomposites, UV/EB coatings, and biomaterials. He has published 11 books and 11 patents. He has advised nearly 95 MS and 52 PhD students with whom he has published over 402 papers. He has more than 40 years of experience in polymer science with 10 years in the industry (IBM, USA–MOD, UK & SNPA/Total, France). He created in the 1980s the International Symposium on Polyimides & High-Temperature Polymers, a.k.a. STEPI, which takes place every three years in Montpellier, France.

Contents

Contributors

M. J. M. Abadie
Institute Charles Gerhardt of Montpellier Aggregates, Interfaces, and Materials for Energy (ICGM-AIME, UMR CNRS 5253), University of Montpellier, Place Bataillon, 34095 Montpellier Cedex 5, France, E-mail: marc.abadie@umontpellier.fr

O. Alekseeva
G.A. Krestov Institute of Solution Chemistry, Russian Academy of Sciences, Akademicheskaya str., 1, Ivanovo, 153045, Russia

N. A. Alimirzoyeva
Institute of Polymer Materials of Azerbaijan National Academy of Science, S. Vurgun str., 124, Sumgait Az5004, Azerbaijan

A. T. Aliyev
Institute of Polymer Materials of Azerbaijan National Academy of Science, S. Vurgun str., 124, Sumgait Az5004, Azerbaijan

A. M. Aliyeva
Institute of Polymer Materials of Azerbaijan National Academy of Sciences, S. Vurgun str.124, Sumgayit AZ5004, Azerbaijan

N. Ananiashvili
R. Agladze Institute of Inorganic Chemistry and Electrochemistry of the Javakhishvili Tbilisi State University, Mindeli str. #11, 0186, Tbilisi, Georgia

J. N. Aneli
Institute of Machine Mechanics, 10 Mindeli Str. Tbilisi 0186, Ivane Javakhishvili Tbilisi State University, I. Chavchavadze Ave.3, Tbilisi 0179, Georgia

M. G. Areshidze
Georgian Technical University, Vl. Chavchanidze' Institute of Cybernetics, 5 Z. Anjaparidze St., 0186, Tbilisi, Georgia

N. B. Arzumanova
Institute of Polymer Materials of Azerbaijan NAS, Azerbaijan

M. Atakay
Faculty of Science, Department of Chemistry, Hacettepe University, 06800 Ankara, Turkey

R. Sh. Bakuradze
Georgian Technical University, Vl. Chavchanidze' Institute of Cybernetics, 5 Z. Anjaparidze St., 0186, Tbilisi, Georgia

I. V. Bayramova
Institute of Polymer Materials of Azerbaijan NAS, Azerbaijan

B.G. Bendeliani
Department of Cryogenic Technique and Technologies, Ilia Vekua Sukhumi Institute of Physics and Technology 0186 Tbilisi, Georgia

E. Çatıker
Faculty of Art & Science, Department of Chemistry, Ordu University, 52200, Ordu, Turkey,
E-mail: ecatiker@gmail.com

R. Chedia
Iv. Javakhishvili Tbilisi State University, Petre Melikishvili Institute of Physical and Organic
Chemistry, 31 Politkovskaya St., 0186, Tbilisi, Georgia, E-mail: chediageo@yahoo.com

I. Chikvaidze
Iv. Javakhishvili Tbilisi State University Tbilisi, Georgia, E-mail: iosebc@yahoo.com

K. Chubinidze
Tbilisi State University, 1 Ilia Chavchavadze Ave., Tbilisi 0179, Georgia/Georgian
Technical University, Institute of Cybernetics, S. Euli 5, Tbilisi 0186, Georgia

M. Chubinidze
Tbilisi State Medical University, 7 Mikheil Asatiani St, Tbilisi 0186, Georgia

E. Çil
Faculty of Education, Department of Math and Science, Ordu University, 52200, Ordu, Turkey

G. N. Dgebuadze
Department of Cryogenic Technique and Technologies, Ilia Vekua Sukhumi Institute of
Physics and Technology 0186 Tbilisi, Georgia

R. D. Dzhafarov
Institute of Polymer Materials of Azerbaijan National Academy of Sciences, S. Vurgun str.124,
Sumgayit AZ5004, Azerbaijan

D. Eliezer
Department of Materials Engineering, Ben-Gurion University of the Negev, Beer-Sheva, Israel

T. Filik
Faculty of Art & Science, Department of Chemistry, Ordu University, 52200, Ordu, Turkey

V. Gabunia
Iv. Javakhishvili Tbilisi State University, Petre Melikishvili Institute of Physical and Organic
Chemistry, 31 Politkovskaya St., 0186, Tbilisi, Georgia

V. M. Gabunia
Department of Cryogenic Technique and Technologies, Ilia Vekua Sukhumi Institute of Physics and
Technology 0186 Tbilisi, Petre Melikishvili Institute of Physical and Organic Chemistry of the Iv.
Javakhishvili Tbilisi State University, Jikia str 5, 0186, Tbilisi, Georgia

M. Gachechiladze
R. Agladze Institute of Inorganic Chemistry and Electrochemistry of the Javakhishvili Tbilisi State
University, Mindeli str. #11, 0186, Tbilisi, Georgia

J. N. Gahramanly
Azerbaijan State University of Oil and Industry

Shabnam Garayeva
Institute of Polymer Materials of Azerbaijan National Academy of Sciences, S. Vurgun str.124,
Sumgayit AZ5004, Azerbaijan

T. Gegechkori
Ivane Javakhishvili Tbilisi State University, E. Andronikashvili Institute of Physics,
Tamarashvili str., 6, 0177, Tbilisi, Georgia, E-mail: tatagegechkori@yahoo.com

M. Ghamami
Department of Mechanical Engineering, Isfahan University of Technology, Isfahan, Iran,
E-mail: mghamami@yahoo.com

S. Ghammamy
Faculty of Science, Chemistry Department, Imam Khomeini International University, Qazvin, Iran

B. Godibadze
G. Tsulukidze Mining Institute, Mindeli str., 7, 0186, Tbilisi, Georgia, E-mail: bgodibadze@gmail.com

S. Grabska
Department of Chemistry of Biomaterials and Cosmetics, Faculty of Chemistry, Nicolaus Copernicus University in Torun, Gagarin 7, 87–100 Torun, Poland, E-mail: sylwiagrabska91@gmail.com

G. K. Grigoryan
Scientific and Technological Center for Organic and Pharmaceutical Chemistry of the National Academy of Sciences of Armenia, 0014, Yerevan, Azatutyan Ave., 26

N. G. Grigoryan
Scientific and Technological Center for Organic and Pharmaceutical Chemistry of the National Academy of Sciences of Armenia, 0014, Yerevan, Azatutyan Ave., 26

I. A. Gritskova
Moscow Technological University of Fine Chemical Technologies, 117571 Moscow, Vernadsky Ave., 86

L. T. Gugulashvili
Department of Cryogenic Technique and Technologies, Ilia Vekua Sukhumi Institute of Physics and Technology 0186 Tbilisi, Georgia

A. Guliyev
Institute of Polymer Materials of Azerbaijan National Academy of Sciences, S. Vurgun str.124, Sumgayit AZ5004, Azerbaijan, E-mail: abasgulu@yandex.ru

K. G. Guliyev
Institute of Polymer Materials of Azerbaijan National Academy of Sciences, S. Vurgun str.124, Sumgayit AZ5004, Azerbaijan

O. Güven
Faculty of Science, Department of Chemistry, Hacettepe University, 06800 Ankara, Turkey

D. Gventsadze
R. Dvali Institute of Mechanics of Machines, Mindeli str. #10, 0186, Tbilisi, Georgia

M. S. Hayrapetyan
Yerevan State University, 1 Alek Manoukian Str., 0025 Yerevan, Armenia

S. S. Hayrapetyan
Yerevan State University, 1 Alek Manoukian Str., 0025 Yerevan, Armenia

A. A. Hovhannisyan
Scientific and Technological Center for Organic and Pharmaceutical Chemistry of the National Academy of Sciences of Armenia, 0014, Yerevan, Azatutyan Ave., 26, E-mail: hovarnos@gmail.com

N. Ya. Ishenko
Institute of Polymer Materials of Azerbaijan National Academy of Science, S. Vurgun str., 124, Sumgait Az5004, Azerbaijan

N. Jalabadze
Georgian Technical University, Republic Center for Structure Researches, 77 Kostava St., 0186, Tbilisi, Georgia

H. Janik
Gdansk University of Technology, Chemical Faculty, Polymer Technological Department, 11/12 Narutowicza Street, 80–232 Gdansk, Poland, E-mail: helena.janik@pg.edu.pl

M. Jastrzębska
Department of Industrial Commodity Science & Chemistry, Faculty of Entrepreneurship and Quality Science, Gdynia Maritime University, 83 Morska Str., 81–225 Gdynia, Poland, E-mail: m.jastrzebska@wpit.am.gdynia.pl

T.K. Jumadilov
JSC Institute of Chemical Sciences After A.B. Bekturov, Almaty, Republic of Kazakhstan, E-mail: jumadilov@mail.ru

B. Kaczmarek
Department of Chemistry of Biomaterials and Cosmetics, Faculty of Chemistry, Nicolaus Copernicus University, Gagarin 7, 87–100 Torun, Poland, E-mail: beatakaczmarek8@gmail.com

N. T. Kahramanov
Institute of Polymer Materials of Azerbaijan NAS, E-mail:najaf1946@rambler.ru

H. G. Khachatryan
Yerevan State University, 1 Alek Manoukian Str., 0025 Yerevan, Armenia

M. Khaddaj
Peoples' Friendship University of Russia (RUDN), 117209 Moscow, Mikloukho-Maklaya str., 6, Moscow Technological University of Fine Chemical Technologies, 117571 Moscow, Vernadsky Ave., 86

G. Khitiri
P. Melikishvili Institute of Physical & Organic Chemistry, Iv. Javakhishvili Tbilisi State University Tbilisi, Georgia

A. Khuskivadze
Tbilisi State Medical University, 7 Mikheil Asatiani St, Tbilisi 0186, Georgia

N. Kiknadze
Chemistry Department, Batumi Shota Rustaveli State University (BSU), Ninoshvili/Rustaveli str. 35/32, 6010 Batumi, Georgia, E-mail: nino-kiknadze@mail.ru

N. A. Koiava
Institute of Macromolecular Chemistry and Polymeric Materials, Ivane Javakhishvili Tbilisi State University, Faculty of Exact and Natural Sciences, I. Chavchavadze Ave.13, Tbilisi 0179, Georgia

R. Kokilashvili
Georgian Technical University Tbilisi, Georgia

R. G. Kondaurov
JSC Institute of Chemical Sciences After A.B. Bekturov," Almaty, Republic of Kazakhstan

T. Korkia
Iv. Javakhishvili Tbilisi State University, Petre Melikishvili Institute of Physical and Organic Chemistry, 31 Politkovskaya St., 0186, Tbilisi, Georgia

J. Kucinska-Lipka
Gdansk University of Technology, Chemical Faculty, Polymer Technological Department, 11/12 Narutowicza Street, 80–232 Gdansk, Poland, E-mail: juskucin@pg.edu.pl

A. M. Kuliyev
Institute of Polymer Materials of Azerbaijan National Academy of Science, S. Vurgun str., 124, Sumgait Az5004, Azerbaijan

N. I. Kurbanova
Institute of Polymer Materials of Azerbaijan National Academy of Science, S. Vurgun str., 124, Sumgait Az5004, Azerbaijan

G. Kvartskhava
Georgian Technical University, Republic Center for Structure Researches, 77 Kostava St., 0186, Tbilisi, Georgia

T. E. Lobzhanidze
Department of Chemistry, Faculty of Exact and Natural Sciences, Ivane Javakhishvili Tbilisi State University, 0179 Tbilisi, Georgia

M. Machavariani
R. Agladze Institute of Inorganic Chemistry and Electrochemistry of the Javakhishvili Tbilisi State University, Mindeli str. #11, 0186, Tbilisi, Georgia

T. Makharadze
Rafiel Agladze Institute of Inorganic Chemistry and Electrochemistry, E. Mindeli Street, 11, Tbilisi 0186, Georgia, E-mail: makharadze_tako@yahoo.com

G. Mamniashvili
Ivane Javakhishvili Tbilisi State University, E. Andronikashvili Institute of Physics, Tamarashvili str., 6, 0177, Tbilisi, Georgia, E-mail: mgrigor@rocketmail.com

E. Markarashvili
Ivane Javakhishvili Tbilisi State University, I. Chavchavadze Ave.3, Institute of Macromolecular Chemistry and Polymeric Materials, Ivane Javakhishvili Tbilisi State University, Faculty of Exact and Natural Sciences, I. Chavchavadze Ave.13, Tbilisi 0179, Georgia

T. Marsagishvili
R. Agladze Institute of Inorganic Chemistry and Electrochemistry of the Javakhishvili Tbilisi State University, Mindeli str. #11, 0186, Tbilisi, Georgia, E-mail: tamaz.marsagishvili@gmail.com

N. Megrelidze
BSU Agrarian and Membrane Technologies Institute, Grishsashvili str. 5, 601 Batumi, Georgia

J. Metreveli
R. Agladze Institute of Inorganic Chemistry and Electrochemistry of the Javakhishvili Tbilisi State University, Mindeli str. #11, 0186, Tbilisi, Georgia

I. R. Metskhvarishvili
Department of Cryogenic Technique and Technologies, Ilia Vekua Sukhumi Institute of Physics and Technology 0186 Tbilisi, Georgia, E-mail: metskhv@yahoo.com

M. R. Metskhvarishvili
Department of Engineering Physics, Georgian Technical University, 0175 Tbilisi, Georgia

O. V. Mukbaniani
Ivane Javakhishvili Tbilisi State University, I. Chavchavadze Ave.3, Institute of Macromolecular Chemistry and Polymeric Materials, Ivane Javakhishvili Tbilisi State University, Faculty of Exact and Natural Sciences, I. Chavchavadze Ave.13, Tbilisi 0179, Georgia, E-mail: omar.mukbaniani@tsu.ge

L. Nadaraia
Georgian Technical University, Republic Center for Structure Researches, 77 Kostava St., 0186, Tbilisi, Georgia

L. I. Nadareishvili
Georgian Technical University, Vl. Chavchanidze' Institute of Cybernetics, 5 Z. Anjaparidze St., 0186, Tbilisi, Georgia, E-mail: levannadar@yahoo.com

A. G. Nadaryan
Scientific and Technological Center for Organic and Pharmaceutical Chemistry of the
National Academy of Sciences of Armenia, 0014, Yerevan, Azatutyan Ave., 26, Armenia

H. Nahvi
Department of Mechanical Engineering, Isfahan University of Technology, Isfahan, Iran

N. Nonikashvili
Iv. Javakhishvili Tbilisi State University, Petre Melikishvili Institute of Physical and Organic
Chemistry, 31 Politkovskaya St., 0186, Tbilisi, Georgia

A. Noskov
G.A. Krestov Institute of Solution Chemistry, Russian Academy of Sciences, Akademicheskaya str.,
1, Ivanovo, 153045, Russia, E-mail: avn@isc-ras.ru

D. R. Nurullayeva
Institute of Polymer Materials of Azerbaijan National Academy of Science, S. Vurgun str., 124,
Sumgait Az5004, Azerbaijan

M. Nutsubidze
Iv. Javakhishvili Tbilisi State University, Department of Chemistry, Ilia Chavchavadze Ave.,
0128 Tbilisi, Georgia

B. Partsvania
Georgian Technical University, Institute of Cybernetics, S. Euli 5, Tbilisi 0186, Georgia

I. I. Pavlenishvili
Georgian Technical University, Vl. Chavchanidze' Institute of Cybernetics, 5 Z. Anjaparidze St.,
0186, Tbilisi, Georgia

A. Peikrishvili
F. Tavadze Institute of Metallurgy and Materials Science, E. Mindeli str., 10, 0186, Tbilisi, Georgia,
E-mail: apeikrishvili@yahoo.com

V. Peikrishvili
F. Tavadze Institute of Metallurgy and Materials Science, E. Mindeli str., 10, 0186, Tbilisi,
Georgia, E-mail: vaxoo3@gmail.com

G. Petriashvili
Georgian Technical University, Institute of Cybernetics, S. Euli 5, Tbilisi 0186, Georgia

G. P. Pirumyan
Yerevan State University, 1 Alek Manoukian Str., 0025 Yerevan, Armenia

G. Z. Ponomareva
Institute of Polymer Materials of Azerbaijan National Academy of Sciences, S. Vurgun str.124,
Sumgayit AZ5004, Azerbaijan

A. Przybytek
Gdansk University of Technology, Chemical Faculty, Polymer Technological Department,
11/12 Narutowicza Street, 80–232 Gdansk, Poland

M. Rutkowska
Department of Industrial Commodity Science & Chemistry, Faculty of Entrepreneurship and
Quality Science, Gdynia Maritime University, 83 Morska Str., 81–225 Gdynia, Poland

A. E. Rzayeva
Institute of Polymer Materials of Azerbaijan National Academy of Sciences, S. Vurgun str.124,
Sumgayit AZ5004, Azerbaijan

S. Saberi
Department of Mechanical Engineering, Isfahan University of Technology, Isfahan, Iran

B. Salih
Faculty of Science, Department of Chemistry, Hacettepe University, 06800 Ankara, Turkey

K. Sarajishvili
Iv. Javakhishvili Tbilisi State University, Petre Melikishvili Institute of Physical and Organic Chemistry, 31 Politkovskaya St., 0186, Tbilisi, Georgia

R. Shahnazarli
Institute of Polymer Materials of Azerbaijan National Academy of Sciences, S. Vurgun str.124, Sumgayit AZ5004, Azerbaijan

L. G. Shamanauri
Institute of Machine Mechanics, 10 Mindeli Str. Tbilisi 0186 Georgia

L. K. Sharashidze
Georgian Technical University, Vl. Chavchanidze' Institute of Cybernetics, 5 Z. Anjaparidze St., 0186, Tbilisi, Georgia

N. Sidamonidze
Iv. Javakhishvili Tbilisi State University, Department of Chemistry, Ilia Chavchavadze Ave., 0128 Tbilisi, Georgia

R. Silverstein
Department of Materials Engineering, Ben-Gurion University of the Negev, Beer-Sheva, Israel, E-mail: barrav@post.bgu.ac.il

A. Sionkowska
Department of Chemistry of Biomaterials and Cosmetics, Faculty of Chemistry, Nicolaus Copernicus University, Gagarin 7, 87–100 Torun, Poland

A. Sulowska
Gdansk University of Technology, Chemical Faculty, Polymer Technological Department, 11/12 Narutowicza Street, 80–232 Gdansk, Poland

P. Szarlej
Gdansk University of Technology, Chemical Faculty, Polymer Technological Department, 11/12 Narutowicza Street, 80–232 Gdansk, Poland

G. Tatishvili
R. Agladze Institute of Inorganic Chemistry and Electrochemistry of the Javakhishvili Tbilisi State University, Mindeli str. #11, 0186, Tbilisi, Georgia

T. Tatrishvili
Ivane Javakhishvili Tbilisi State University, I. Chavchavadze Ave.3, Institute of Macromolecular Chemistry and Polymeric Materials, Ivane Javakhishvili Tbilisi State University, Faculty of Exact and Natural Sciences, I. Chavchavadze Ave.13, Tbilisi 0179, Georgia

E. Tskhakaia
R. Agladze Institute of Inorganic Chemistry and Electrochemistry of the Javakhishvili Tbilisi State University, Mindeli str. #11, 0186, Tbilisi, Georgia

R. Vardiashvili
Iv. Javakhishvili Tbilisi State University, Department of Chemistry, Ilia Chavchavadze Ave., 0128 Tbilisi, Georgia

Abbreviations

ADA	aminododecanoic acid
AFM	atomic force microscope
AIBN	azobisisobutyronitrile
BA	buthylacrylate
BDI	1,4-butanediisocyanate
BDO	1,4-butanediol
CLSM	confocal laser scanning microscope
CNTs	carbon nanotubes
$COCH_2$	carbonyl
CS	chondroitin sulfate
DBTDL	dibutyl tin dilaurate
DCP	dicumenyl peroxide
DHB	2,5-dihydroxybenzoic acid
DMB	1,9-dimethylmethylene blue
DMF	dimethylformamide
DMSO	dimethyl sulfoxide
DSC	differential scanning calorimetry
DSS	duplex stainless steels
DTBP	ditertiary butyl peroxide
EDS	energy dispersive spectrometry
EPR	epoxy resins
ES	ethyl silicate
FAs	fulvic acids
FC	field-cooled
FIBS	focus ion beam scanning electron microscopes
FRPCs	fiber reinforced polymer composites
GAGs	glycosaminoglycans
GCO	castor oil glycerides
GNPs	gold nanoparticles
GRDF	Georgian Research and Development Foundation
HA	hyaluronic acid
HA	hydroxyapatite
HAC	hydrogen assisted cracking
HDI	hexamethylene diisocyanate

HEC	hot explosive compaction
HEL	hugoniot elastic limit
HLB	hydrophilic-lipophilic balance
HMDI	4,4'-dicyclohexylmethane diisocyanate
HTP	hydrogen transfer polymerization
IKIU	Imam Khomeini International University
IPDI	isophorone diisocyanate
IR	infrared
LDS	lean duplex stainless steel
LEO	electron microscopy Ltd
MA	maleic anhydride
MBCs	minimum bactericidal concentrations
MHA	Mueller Hinton agar
MHB	Mueller Hinton Broth
MIC	minimum inhibitory concentration
MS	micro-spherical
NCN	national science center
NF	nanofiller
NFRPNCs	nanofiller reinforced polymer nanocomposites
NMP	N-methyl-2-pyrrolidone
OCD	obsessive-compulsive disorders
ODMA	octadecyl methacrylate
OM	optical microscopy
PAA	poly(acrylic acid)
PAcHP	poly(acrylic acid-co-3-hydroxypropionate)
PBS	phosphate-buffer saline
PCL	poly(ε-caprolactone)
PEG	poly(ethylene glycol)
PMCs	polymer matrix composites
PMMA	poly(methyl methacrylate)
PMNC	polymer matrix nanocomposites
PP	potassium persulfate
PSMA	prostate-specific membrane antigen
PU	polyurethanes
PVA	polyvinyl alcohol
RP-HPLC	reversed-phase high-performance liquid chromatography
RT	room temperature
SC/PL	solvent casting/particulate leaching
SDSS	super duplex stainless steel
SEM	scanning electron microscope

SF	silk fibroin
SG	sol-gel method
SMSS	supermartensitic stainless steels
SRNSF	Shota Rustaveli National Science Foundation
SSR	solid-state reaction method
TBI	traumatic brain injury
TDA	thermal desorption analysis
TDS	thermal desorption spectrometry
TEM	transmission electron microscopy
TFA	trifluoroacetic acid
TGA	thermogravimetric analysis
THF	tetrahydrofuran
TPP	triphenylphosphine
UP	unsaturated polyester
UV	ultraviolet
VA	vinyl acetate
VER	vinylester
VISAR	velocity interferometer for any reflector
XRD	x-ray diffraction
ZFC	zero-field-cooled

Preface

Selected papers presented at the 5th International Caucasian Symposium on Polymers and Advanced Materials (ICSP&AM 5) are collected in this volume. This conference took place in Tbilisi, Georgia, on July 2–5, 2017 at the Ivane Javakhishvili Tbilisi State University. This book assembles a collection of interdisciplinary papers on the state of knowledge on various topics under consideration through a combination of overviews and original unpublished recent research.

The basic purpose of this book is to review the present status of several domains of polymer science, such as advanced polymers, composites, and nanocomposites. The book is organized along four parts: Part I: Composites and Nanomaterials; Part II: Polymer Synthesis and Application; Part III: Materials and Properties; and Part IV: Constitutional Systems for Medicine

Part I reviews research on several nanoparticles to reinforce polymer matrices, thermoplastics, and thermosets as well. Diverse composite-based epoxy, polyester, ethylene/propylene, polyethylene, PMMA, even natural polymer are presented for the different types of applications. Interface/interphase is not ignored to reinforce composites and nanocomposites. Special attention is brought to medical applications, such as tissue engineering, prostate cancer biomarker, or dealing with nanomedicine.

Part II treats some new syntheses by classical techniques and photochemistry for different applications such as health care cosmetics, water-proofing materials, sorbents for different nature ions sorption, and antibacterial activity.

Part III is related to hydrogels, intergel systems, sol-gel, and zeolites. And Part IV is related to the constitutional systems for medicine.

This book presents interdisciplinary papers on the state of knowledge of each topic under consideration through a combination of overviews and original unpublished research.

This book is addressed to all those working in the field of polymers and composites, i.e., academics, institutes, research centers, as well as engineers working in the industry.

—Omari V. Mukbaniani, DSc
Tamara N. Tatrishvili, PhD
Marc J. M. Abadie, DSc

PART I
Composites and Nanomaterials

CHAPTER 1

Chemistry of the Interfaces/Interphases in Composites and Nanocomposites

MARC J. M. ABADIE

Institute Charles Gerhardt of Montpellier Aggregates, Interfaces, and Materials for Energy (ICGM-AIME, UMR CNRS 5253), University of Montpellier, Place Bataillon, 34095 Montpellier Cedex 5, France, E-mail: marc.abadie@umontpellier.fr

ABSTRACT

Fiber reinforced polymer composites (FRPCs) are still an emerging class of engineering materials as well as nanofiller (NF) reinforced polymer nanocomposites (NFRPNCs). In polymer composites or nanocomposites, the interface/interphase is a key issue of which depends on the performance of the materials, but not for the same reason.

For the composites, the interface/interphase is a key issue since it guarantees the necessary stress transfer from the matrix (weak part) to the reinforcement (strong part). In the nanocomposites, the main issue is the homogeneous dispersion of the NF into the matrix due to its nano size and low wt.% used (1–5 wt.%); therefore, a good control of the interface/interphase will strongly help in the homogeneity and guarantee performance of these materials.

After defining interface/interphase in such materials, we will describe the recent progress in chemical treatments of inorganic fillers (glass, clay, POSS) as well as organic fillers (aramid, GFs, CFs & NFs: CNFs, CNTs, graphite, GOs).

SEM for composites, TEM, and XRD analysis for nanocomposites will be discussed in order to control the performance properties of both composites and nanocomposites.

1.1 INTRODUCTION

Chemistry has been the driven force for the development of high-performance polymer composites and new materials [1, 2]. Polymeric materials have been developed to fulfill the requirements imposed by constraints of high tech products such as lightweight, toughness, high-temperature resistance, oxidative thermal stability, low shrinkage, high modulus, etc. [3, 4].

If polymers were developed since the beginning of last century, composites appeared to emerge in the 60s mainly for the aerospace conquest first and then applied to all sectors of advanced technologies. Since the discovery made by the engineers of the carmaker Toyota loading polyamide with nano-fillers (NFs) as clay, the drastic improvement observed in the mechanical and barrier properties. This new field is considered as emergent and so-called nanocomposites [5].

1.2 COUPLING AGENT [6–8]

Glass fibers are the only fibers that can commonly be modified on the surface by chemical treatment, so-called coupling agent. This surface treatment is done online during the process of fabrication of the fibers. Generally, it is a water emulsion of a coupling agent that is spread on the surface of the fibers.

The role of the coupling agent is to link the reinforcing agent and matrix by covalent bonds that give higher mechanical properties to the composite (Figure 1.1).

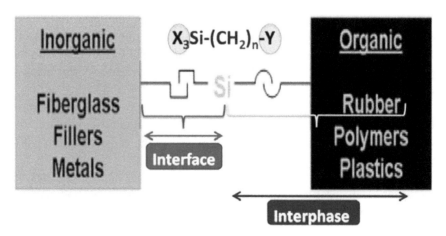

FIGURE 1.1   Interface/Interphase in a composite.

It is important to underline that the coupling agent depends on the matrix used and will be different if we use unsaturated polyester (UP) or vinylester (VER) and epoxy (EPR) resins.

The coupling agent is principally constituted by two parts: one polar side (for glass fiber) in contact with reinforcement agent and having at the other end a chemical function that matches with the matrix considered. A general formula is $X_3Si (CH_2)_nY$, where X is a methoxy or ethoxy group and Y the chemical function adapted to the matrix used; i.e., acrylate for UP and VER, epoxy or amine for EPR (Figure 1.2).

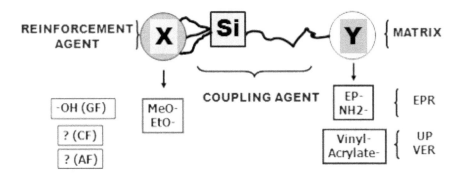

FIGURE 1.2 **(See color insert.)** Interaction of the coupling agent with matrix and reinforcement agent.

Table 1.1 describes some coupling agents currently used in the industry under spray solution during manufacturing continuous filaments.

TABLE 1.1 Industrial Coupling Agents and Compatibility with Resins: Unsaturated Polyester Resins UPR, Vinylester Resins VER, and Epoxy Resins EPR

Structure	Name	Commercial Name	Compatibility with Resin
NH_2-$(CH_2)_2$-NH- $(CH_2)_2$ -$Si(OCH_3)_3$	N-β-aminoethyl-γ-aminopropyl trimethoxy silane	Z 6020 A 1120	EPR
CH_2-CH(O)-CH_2-O- $(CH_2)_3$-$Si(OCH_3)_3$	γ-glycidoxy propyl	Z 6040	EPR
Cl-$(CH_2)_3$-$Si(OCH_3)_3$	γ-chloro propyl trimethoxy silane	Z 076	EPR
NH_2-$(CH_2)_3$-$Si(OCH_3)_3$	γ-aminopropyl trimethoxy silane	A 110	EPR
$CH_2 = C(CH_3)$-CO_2 -$(CH_2)_3$-$Si(OCH_3)_3$	methacroyl trimethoxy silane	Z 6030	UPR/VER
$CH_2 = CH$-$Si(OCH_3)_3$	Vinyl trimethoxy silane	A 171	UPR/VER

1.3 HOW TO DEFINE INTERFACES/INTERPHASES [9]

A composite is mainly constituted by three parts: the matrix which is the weak part of the composite, the reinforcement agent which is the strong part of the composite, and the interface/interphase of the composite on which depends the toughness and the integrity of the material.

For nanocomposite, the problem is the same in term of composition (nanofiller dispersed in an organic phase, interface/interphase) but due to the nanosize of the filler and its low percentage (1 to 5 wt.%) into the organic phase, the homogeneous dispersion of the nanofiller is crucial with regards to the mechanical performance.

When a composite is submitted to any force, the first part receiving that force is the matrix-envelope of the composite. As the matrix is the weak part of the composite, forces have to be transferred to the strong part of the composite that is the reinforcement agent.

This transfer will be done in two steps (Figure 1.3) [10, 11]:

- First step via the network part of the matrix to the "vicinage" of the reinforcement agent properly treated by a coupling agent and which represents the interphase.
- Second step via the immediate phase of the network directly in contact with the reinforcement agent that represents the interface.

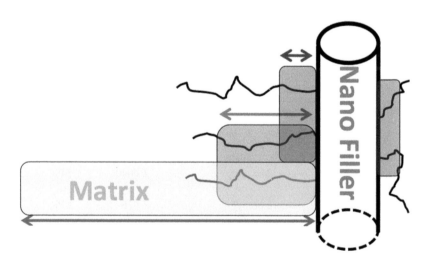

FIGURE 1.3 (See color insert.) Interface (1 to 100 Å)–blue part, Interphase (up to 100 µm)–red part and the integration of the filler into the composite–yellow part. Filaments represent the coupling agent.

Through the coupling agent we can consider the interface, i.e., the part of the matrix which is directly in contact with the reinforcement agent and which represent ~ 1 to 10 Å and the interphase which concerns the integration of the coupling agent within the network of the matrix and represent ~ 100 µm.

The control of the interface allows good links between matrix and reinforcements, to transfer strain from the matrix to reinforcements and maintain cohesion of the composite. The control of the interphase permits to assure no discontinuity within the matrix and have a three-dimensional network [12].

1.4 EXPERIMENTAL

Materials: Unsaturated polyester and VER resins with styrene reinforced glass fibers (45 wt.%) was cross-linked by conventional procedures *viz.* peroxide MEKP (3 wt.%) accelerated by cobalt salt (0.5 wt.%) at RT. EPR Epiclon HP7200 was crosslink in the presence of amino-5, amino methyl-1, trimethyl-1,3,3 cyclohexane.

Measurement: Scanning electronic microscopy (SEM) is used for composites whereas transmission electronic microscopy (TEM) spectroscopy and XRD are used for nanocomposites to confirm exfoliated structures.

1.5 RESULTS AND DISCUSSION

1.5.1 COMPATIBILIZATION MATRIX/REINFORCEMENT FOR COMPOSITES

1.5.1.1 UPR/GLASS FIBERS [12]

We prepared different samples of composites and studied the interface/interphase after splitting the composite into two parts by disembowelment and pulling out.

It is clearly seen that the glass fiber after appropriate chemical treatment is fully integrated into the UPR matrix. We observe (Figure 1.4a) that the glass fiber after tearing is correctly sized by the appearance of peak resin and residue all along the fiberglass. In the absence of sizing and due to the polarity of the glass fiber we clearly observe (Figure 1.4b) a repulsion of the resin. Such a composite, under the effect of constraints, cannot transfer the stress on the glass fiber and will give poor mechanical properties.

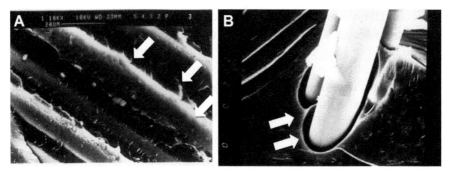

FIGURE 1.4   (a) UPR/GF with a coupling agent and (b) without a coupling agent.

1.5.1.2   *SYNTACTIC FOAM BASED EPR/HOLLOW SPHERES [13]*

We have studied the effect of a hollow glass sphere on syntactic foam composite based on EPR [5]. The coupling agent Z 6040 (Table 1.1 and Figure 1.5), is used to improve the integrity of the epoxy foam with hollow microspheres (Figure 1.6) [14].

FIGURE 1.5   (3-Glycidyloxypropyl)trimethoxysilane compatible with epoxy resin.

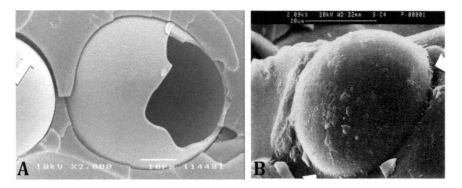

FIGURE 1.6   (a) Hollow glass microsphere without chemical treatment, (b) with (3-glycidyloxypropyl)trimethoxysilane Z 6040.

Hollow glass microspheres K15 and H50 from 3M Scotchlite, of density 0.15 and 0.5 g/cm^3, respectively, have been used for making syntactic foam. Without treatment of the hollow microspheres by (3-glycidyloxypropyl) trimethoxysilane there is still discontinuity between the matrix and the hollow glass (Figure 1.6a), whereas once bonded to the coupling agent (Figure 1.6b), better mechanical properties are observed (Table 1.2).

TABLE 1.2 Mechanical Properties of Foam Based Epoxy [14]

System	d	Void (%)	Compression Strength (MPa)	Compression Modulus (GPa)	Hardness (MPa)
Neat Epoxy	1.17	0.16	99.16	0.67	119.11
K15 10% wo.	0.97	5.18	64.60	0.47	76.28
K15 10% w.	0.95	7.04	65.15	0.58	87.83
H50 10% w.	0.96	9.52	69.97	0.44	93.71
K15 50% wo.	0.58	8.83	40.78	0.31	46.69
K15 50% w.	0.60	4.65	46.69	0.39	50.23
H50 50% w.	0.77	4.32	71.39	0.54	76.93

1.5.1.3 MATRICES/CARBON FIBERS [15] OR MATRICES/ARAMID FIBERS [16]

If it is easy to treat glass fibers, aramid (Kevlar®), carbon fibers or polyethylene such as UHMWPE (Spectra®) cannot be treated chemically in industrial scale without losing their mechanical properties. However, some chemical transformation has been tested with success (Figure 1.7), for polyaramid AFs and carbon fibers CFs. Generally, these fibers are treated by a physical process such as Corona or Plasma effect that makes the surface rough, facilitate impregnation and allows the resin to be hanged; however, it is difficult to control the physical effect that makes the fiber less resistant.

1.5.2 COMPATIBILIZATION MATRIX/NANOFILLER FOR NANOCOMPOSITES

In nanocomposites, the compatibility of the nanofillers is crucial and conditions their correct homogeneously distribution within the polymer matrix. Moreover, agglomeration of nanoparticles should be prevented at all costs to maximize mechanical properties.

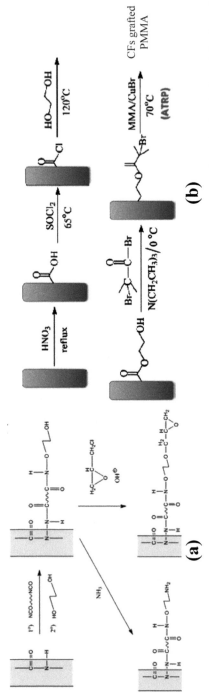

FIGURE 1.7 (a) Chemical modification of aramid surface for aramid fiber/epoxy resin AFs/EPR composite, (b) Chemical modification of carbon surface for CFs/PMMA composite.

1.5.2.1   MONTMORILLONITE CLAY [17]

Montmorillonite clay MMT was the first nanofiller used by carmaker Toyota to reinforce polyamide 6. Clay is hydrophilic in nature and incompatible with the hydrophobic polymer matrix. Therefore, the surface of the clay is modified with an organic modifier to convert the hydrophilic surface into the hydrophobic surface. The objective is to improve clay dispersion and form exfoliated morphology vs. maximum reinforcement.

By modifying the surface of the clay–hydrophilic, making it organophilic vs. compatible with the organic matrix by exchanging Na^+ in the clay with an amino acid such as 12-aminododecanoic acid (ADA) (Figure 1.8), according to the reaction:

$$HO_2C\text{-}(CH_2)_{11}\text{-}N(CH_3)_3^+Cl^- + Na^+CLAY \rightarrow HO_2C\text{-}(CH_2)_{11}$$
$$\text{-}N(CH_3)_3^{+-}CLAY + NaCl$$

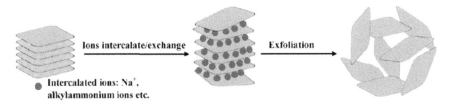

FIGURE 1.8   Ammonium salt of 12-aminododecanoic acid.

Thanks to this chemical modification of the clay, as the exfoliated structures could be reached (Figure 1.9) and spectacular improvements of mechanical properties obtained.

Ions intercalate/exchange   Exfoliation

● Intercalated ions: Na⁺,
alkylammonium ions etc.

FIGURE 1.9   **(See color insert.)** Intercalated *vs.* exfoliated structure.

The characteristic of the interfacial bond is also critical to the long-term stability of the PMC, playing a key role in fatigue properties, environmental behavior, and resistance to hot/wet conditions. Therefore, relatively small amounts (2–5 wt.%) of nanometer-sized clay particles can provide large improvements in mechanical and thermal properties, as well as the gas barrier and flame resistance.

1.5.2.2 THE TNO CONCEPT [18]

An alternative approach to compatibilizing clays with polymers has been introduced by TNO (The Netherlands), based on the use of block or graft copolymers where one component of the copolymer is compatible with the clay and the other with the polymer matrix.

A typical block copolymer would consist of a clay-compatible hydrophilic block-ethylene oxide and a polymer-compatible hydrophobic block-polystyrene such as:

$$HO - \left(CH_2 - CH_2 - O \right)_n \left(CH_2 - CH \right)_m$$

The block length must be controlled and must not be too long. High degrees of exfoliation are claimed using this approach.

For high-temperature thermoplastics in clay-based nanocomposites, imidazolium salts (see structure below), more thermally stable than the ammonium salts, are used.

1.5.2.3 CARBON NANOTUBES (CNTS) [19]

Carbon nanotubes (CNTs) have extraordinary mechanical properties, 100 times stronger than stainless steel and six times lighter. CNTs are the ultimate reinforcement material for advanced composites, owing to their very small diameters, very high aspect ratios, and exceptional strength.

As of now, carbon nanotubes reinforced composites are not produced at industrial levels of production, due to fundamental barriers based on dispersion and or interfacial shear strength issues:

- *Dispersion*: CNTs must be uniformly dispersed in order to achieve efficient load transfer, uniform stress distribution and to minimize stress-concentration centers.
- *Interfacial stress transfer*: The most important requirement is that external stress applied to the matrix is efficiently transferred to the CNTs.

There has been an immense effort to establish the most suitable conditions for the transfer of either mechanical load or electrical charge to individual nanotubes in a polymer composite components.

A prerequisite for such an endeavor is the efficient dispersion of individual nanotubes and the establishment of a strong chemical affinity (covalent or non-covalent) with the surrounding polymer matrix. Various methods of CNT chemical modification have been proved quite successful in introducing functional moieties which contribute to better nanotube dispersion and eventually to efficient thermodynamic wetting of nanotubes with polymer matrices [20].

Another area of intense research is the grafting of macromolecules onto the nanotube surface. Indeed, the addition of a whole polymer chain is expected to have a greater influence on the nanotube properties and their affinity to polymer matrices as compared to the addition of low molecular mass functionalities. The modification of CNTs by polymers is separated into two main categories, based on whether the bonding to the nanotube surface is covalent or not. The covalent modification itself involves either "grafting to" or "grafting from" strategies

For carbon nanostructure, i.e., CNTs, an excellent review of chemical surface modifications has been recently published [21].

We may cite two specific types of reaction which we have used successfully to modify the surface of CNTs that make them compatible with thermoplastics either by ATRP process (Figure 1.10a) [22], or ROP for biodegradable polymers (Figure 1.10b) [23].

1.5.2.4 FUNCTIONALIZED GRAPHENE SHEETS (FGSS)

Nowadays, graphene oxide GO-polymer composites have attracted much attention due to their unique organic-inorganic hybrid structure and

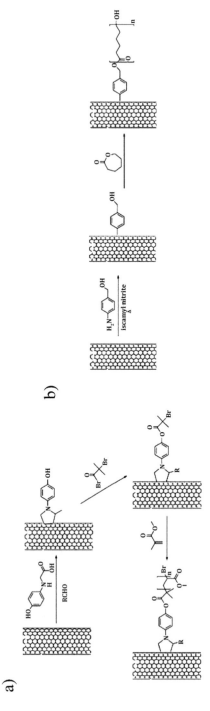

FIGURE 1.10 (a) ATRP modification approach of CNTs, (b) Graft by ROP of ε-caprolactone.

exceptional properties. One of the major interest is the surface of graphene oxide as it contains epoxy, hydroxyl, and carboxyl groups, which can interact with matrices of composites (Figure 1.11).

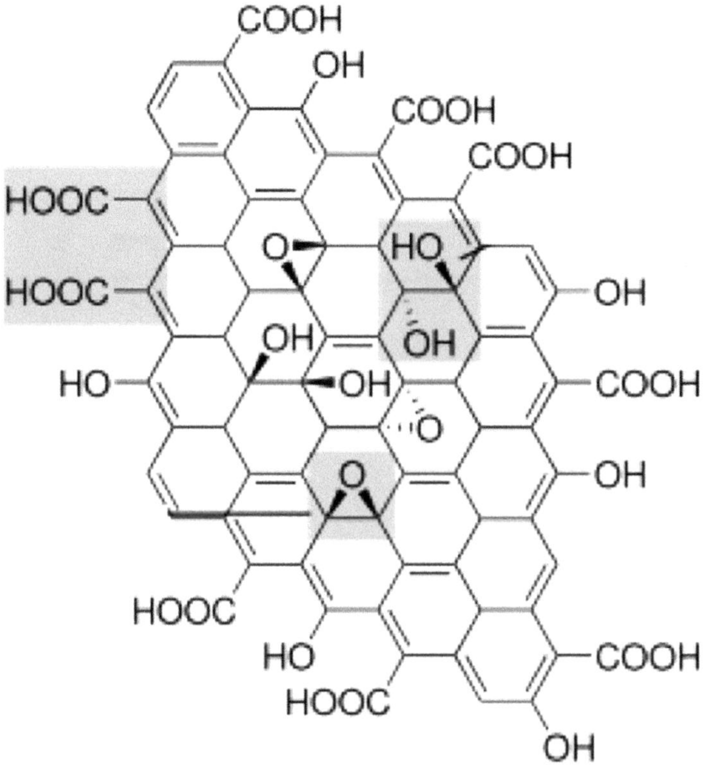

FIGURE 1.11 GOs functions.

We have been tested, for an oil services and equipment company three different elastomers: one saturated nitrile rubber SNR and two fluoroelastomers FEA & FEB, which might be used for drilling under Antarctic conditions [22].

The neat resins were homogenized with GO, oxidizer, surfactant, and vulcanizer at three defined temperatures for a specific mixing period. Vulcanization process was subsequently performed after homogenizing the elastomers with NF. The mixed samples were cured by hot pressing at 177°C for 15 min followed by post-curing at 200–220°C at the variable time.

As it can be seen in Figure 1.12, the glass transition Tg is fairly dependent of the percentage in wt.% of GO, whereas the temperature of decomposition Td is affected by the wt.% of GO (Figure 1.13).

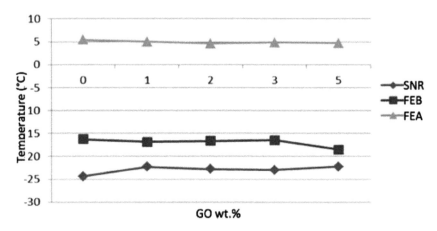

FIGURE 1.12   Comparison of Tg of different composites cured at 50°C.

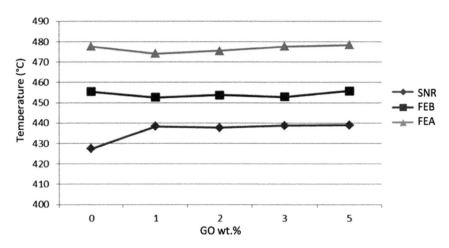

FIGURE 1.13   Comparison of Td of different composites cure at 50°C.

An expectation was made that the E-modulus of the composites would increase with increasing wt.% addition of NF. This was based on the fact of mechanical properties transformation from stronger NF to the softer polymer matrix, which has been studied and reported for the polymer/CNT composites researches. Through the experiment data analysis, we found the

results match well with this expectation, and the most remarkable improvement was found with 5 wt.% graphene in the composites. Detailed analyzes
are given in Figure 1.14.

Comparisons of the E-modulus of the three elastomer/GO-composites
was further performed and discussed below.

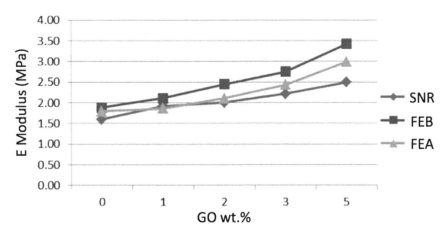

FIGURE 1.14   Comparison of E-modulus of different composites cured at 50°C.

Figure 1.14 is the comparison of E-modulus of the composites cured at
50°C, from which we found that FEB composite showed the best mechanical
property among the three while SNR/GO composite had the lowest value.
Similar conclusions were also reached for composites cured at 75 and 95°C.
In addition, FEB/GO composites achieved the most impressive improvement
of 106% increase of E-modulus with 5 wt.% graphene oxide cured at 95°C.

From the above analysis of the mechanical characterization results,
we may expect that the E-modulus of the elastomer/GO composite would
generally increase with increasing wt.% graphene oxide incorporated in
the composite. Since the dog bone samples used for mechanical tests were
much larger than the amount of sample used for thermal tests; therefore, the
mechanical properties characterization results would be less sensitive to NF
dispersion compared to thermal properties characterization results. Based
on the researches reported for polymer/CNT composite, we may also expect
it to be attributed to the mechanical property transferred from the graphene
oxide to the elastomer phase in the polymer/graphene oxide composite under
load. And with the increasing wt.% GO, there are more interfacial interactions between the polymer matrix and the graphene. Therefore more load

would be transferred from the softer polymer matrix to the graphene oxide filler, the stronger the composite will be.

In conclusion, the composite FEB/GO was the better compromise for low Tg, high Td, and increase e-modulus and qualified to serve in arctic conditions.

1.5.2.5   *POLYHEDRAL OLIGOMERIC SILSESQUIOXANE (POSS) [23, 24]*

Polyhedral oligomeric silsesquioxanes, now commonly referred to under the trade name POSS, often referred to as "nano silica," were discovered in the 1940s. The cage-like shape of the molecule vs. cage-like hydrocarbons (Figure 1.15), is considered as a hybrid structure combining organic such as C and inorganic such as Si. The functionalization of POSS allows it to be fully integrated into the matrix providing a good interface/interphase and therefore enhanced mechanical properties.

Note that for high-performance polymers there is adapted POSS structure depending on the matrix used vs. rigid POSS imidazolium salt to reinforce the composites or flexible POSS imidazolium salt to enhance the soft structure (Figure 1.16) [25, 26].

1.6   TEM AND XRD ANALYSIS

The dispersion and homogenization of the NF within the polymer matrix is difficult to achieve and depend on the efficient chemical treatment of the NF. Spectroscopy analysis may confirm the obtention of well-dispersed nanocomposites. Three techniques can give us convincing results of the formation of a nanocomposite with optimal properties. SEM will give the first information; i.e., if you observe something, an agglomeration of NFs, your dispersion is not effective, but if you see nothing you may expect to have succeeded in the homogenization of your material as nanoscale cannot be seen by SEM. TEM will allow to see if you have intercalated or exfoliated structure, but will also show if you have aggregates (Figure 1.17).

XRD (WAXS) will confirm if you have exfoliated dispersed nanofiller (Figure 1.18). Using TEM images and XRD patterns, one can confirm the nanocomposite with optimized mechanical properties.

In the case of clay incorporated blend such as LPA/LLDPE [27], mixing techniques and the number of mixing above all play a great role in the homogenization of the nanofiller within the matrix. WAXS patterns confirm the better compatibilization for two steps mixing than for one step (Figure 1.19).

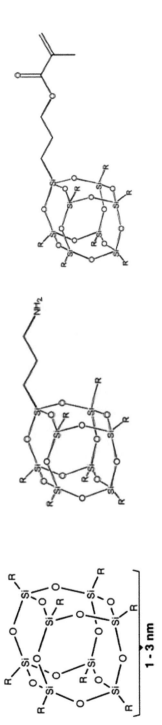

FIGURE 1.15 Structures of octahedral cage POSS with R = functionalized organic groups–NH$_2$ for epoxy resins, acrylate for unsaturated polyester, vinylester or vinyl resins.

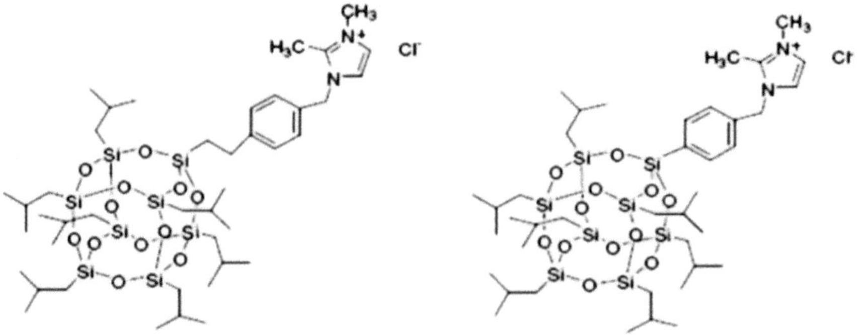

FIGURE 1.16 Chemical structure of POSS imidazolium salt: (a) flexible POSS imidazolium salt [25], (b) rigid POSS imidazolium salt [26].

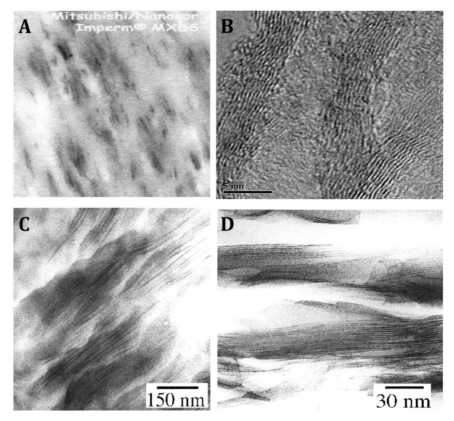

FIGURE 1.17 Intercalated structures: (a) exfoliated structure MMT clay modified, (b) CNTs, (c) and (d) MMT clay modified.

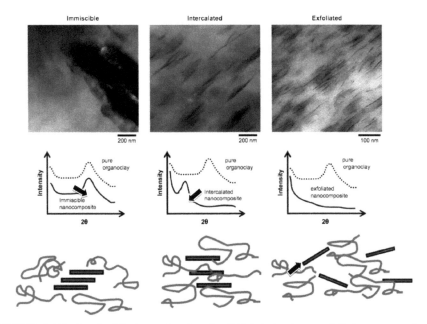

FIGURE 1.18 Immiscible, intercalated, and exfoliated structures and their corresponding XRD patterns.

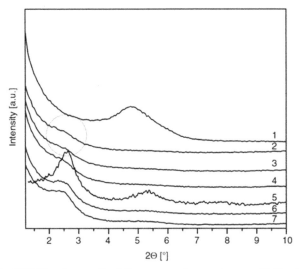

FIGURE 1.19 WAXS patterns of pristine organoclays and compatibilized PLA/LLDPE nanocomposites [27].

Curve 1 – MMT clay modified.

Curves 2, 3, 4 – Two steps mixing for 3, 4.5 and 6 wt.% of clay, respectively.

Curves 5, 6, 7 – One step mixing for 3, 4.5 and 6 wt.% of clay, respectively.

1.7 CONCLUSIONS

The different types of chemistry presented above give rise to modern technologies and access to new classes of organic matrices for polymer matrix composites (PMCs) and polymer matrix nanocomposites (PMNCs) where the homogenization, the distribution and the compatibility of the treated nanofillers within the matrix allow the obtention of performant composites or nanocomposites.

KEYWORDS

- **composites and nanocomposites**
- **interfaces/interphases**
- **SEM**
- **TEM**
- **XRD analysis**

REFERENCES

1. Peters, S. T., (2017). *Handbook of Composites*, Published by Springer Science & Business Media, pp. 1–1043.
2. www.bris.ac.uk/composites (accessed on 11 January 2019).
3. Voytekunas, V. Y., Jones, D., Pratomo, L. W., & Abadie, M. J. M., (2010). Curing kinetics of 3d structures for marine and aerospace applications. *Proceedings of the Second International Conference on Science and Engineering* (Vol. 3, pp. 8–14). Sedona Hotel, Yangon, Myanmar.
4. Carey, J. P., (2018). *Handbook of Advances in Braided Composite Materials – Theory, Production, Testing, and Applications*, Published by Elsevier WP, pp. 1–496.
5. Okpala, C. C., (2014). The benefits and applications of nanocomposites, *Int. J. Adv. Engg. Tech.*, 2(4), 12–18.
6. www.seichemical.com/products/coupling.html (accessed on 11 January 2019).
7. www.specialchem4polymers.com/tc/silanes (accessed on 11 January 2019).
8. www.potterseurope.org/documents/Coupling_Agents (accessed on 11 January 2019).
9. Pegoretti, A., & Karger-Kocsis, J., (2015). *eXPRESS Polymer Letters* (Vol. 9, pp. 10, 838).
10. Voytekunas, V. Y., Pratomo, L. W., & Abadie, M. J. M., (2010). "Controlling interface/interphase in composites & nanocomposites." *Proceedings of the Second International Conference on Science and Engineering* (Vol. 3, pp. 2–7). Sedona Hotel, Yangon, Myanmar.

11. Karger-Kocsis, J., Mahmood, H., & Pegoretti, A., (2015). Recent advances in fiber/ matrix interphase engineering for polymer composites, *Progress in Materials Science, 73*, 1–43.

12. Thakur, V. K., Thakur, M. K., & Kessler, M. R., (2017). *Handbook of Composites From Renewable Materials, Functionalization* (Vol. 4, p. 608).

13. Wouterson, E. W., Boey, F. Y. C., Hu, X., & Wong, S. C., (2007). Effect of fiber reinforcement on the tensile, fracture and thermal properties of the syntactic foam. *Science Direct (Polymer), 48*(11), 3183–3191.

14. Chua, W. T., & Abadie, M. J. M., (2012). *High-Performance Syntactic Foam,* Final Year Project. Nanyang Technological University, Singapore, MSE-Library Ref: 2012-55, 1–62.

15. Beggs, K. M., Perus, M. D., Servinis, L., O'Dell, L. A., Fox, B. L., Gengenbach, T. R., & Henderson, L. C., (2016). *Rapid Surface Functionalization of Carbon Fibers Using Microwave Irradiation in An Ionic Liquid,* RSC Advances, *39*(7), 26658–26664.

16. Yatvin, J., Sherman, S. A., Filocamo, S. F., & Locklin, J., (2015). Direct functionalization of Kevlar® with copolymers containing sulfonyl nitrenes. *Polym. Chem., 6*, 3090–3097.

17. Chow, W. S., & Mohd, I. Z. A., (2015). Polyamide blend-based nanocomposites: A review, *eXPRESS Polymer Letters, 9*(3), 211–232.

18. TNO founded by law in 1932, is a Dutch company, to enable business and government to apply knowledge. It is an organization regulated by public law and independent – not part of any government, university or company.

19. Saeed, K., & Khan, I., (2013). Carbon nanotubes-properties and applications: A review, *Carbon letters,* 14, 3. doi: 10. 5714/CL.2013.14.3.131.

20. Schwarz A. J., Contescu C. I., Putyera K., Dekker Encyclopedia of Nanoscience and Nanotechnology, Stanislaus, S. W., & Sarbajit, B. *Functionalization of Nanotube Surfaces,* Vol. 2, pp. 1251–1268. doi: 10.1081/E-ENN 120013711.

21. Zdenko, S., Dimitrios, T., Konstantinos, P., & Costas, G., (2010). Carbon nanotube–polymer composites: Chemistry, processing, mechanical and electrical properties, *Progress in Polymer Science, 35*, 357–401.

22. Li, Y., & Abadie, M. J. M., (2012). *Thermal & Mechanical Properties of Graphene Oxide Reinforced Elastomers,* Final Year Project, Nanyang Technological University, Singapore, MSE-Library Ref: 2012-21, pp. 1–55.

23. Ro, H. W., & Soles, C. L., (2011). Silsesquioxanes in nanoscale patterning applications, *Materials Today, 14*(1 & 2), 20–33.

24. Kawakami, Y., Kakihana, Y., Miyazato, A., Tateyama, S., & Hoque, M. A., (2011). Polyhedral oligomeric silsesquioxanes with the controlled structure: Formation and application in new Si-based polymer systems. *Advances in Polymer Science, 235*, 185–228.

25. Taraia, A., & Baruah, J. B., (2017). Competing phenol-imidazole and phenol-phenol interactions in the flexible supramolecular environment of *N,N'*-bis(3-imidazole–1-cyclopropyl)naphthalenediimide causing domain expansion, *New Journal of Chemistry, 41*, 10750–10760.

26. Paladugu, S., Chatla, N. B., & Ganesan, P., (2015). Semi-rigid imidazolium carboxylate controlled structural topologies in zwitterionic coordination networks, *Polyhedron, 89*, pp. 322–329.

27. Ashabi, L., Jafari, S. H., Khonakdar, H. A., Boldt, R., Wagenknecht, U., & Heinrich, G., (2013). Tuning the processability, morphology, and biodegradability of clay incorporated PLA/LLDPE blends via selective localization of nanoclay induced by melt mixing sequence, *eXPRESS Polymer Letters, 7*(1), 21–39.

CHAPTER 2

Composites Based on Epoxy Resin with Chemically Modified Mineral Fillers

L. G. SHAMANAURI[1], E. MARKARASHVILI[2,3], T. TATRISHVILI[2,3], N. A. KOIAVA[3], J. N. ANELI[1,2], and O. V. MUKBANIANI[2,3]

[1]Institute of Machine Mechanics, 10 Mindeli Str. Tbilisi 0186, Georgia

[2]Ivane Javakhishvili Tbilisi State University, I. Chavchavadze Ave. 3, Tbilisi 0179, Georgia, E-mail: omar.mukbaniani@tsu.ge

[3]Institute of Macromolecular Chemistry and Polymeric Materials, Ivane Javakhishvili Tbilisi State University, Faculty of Exact and Natural Sciences, I. Chavchavadze Ave. 13, Tbilisi 0179, Georgia

ABSTRACT

Ultimate strength, softening temperature, and water absorption of the polymer composites based on epoxy resin (type ED–20) with unmodified and modified by ethyl silicate minerals diatomite and andesite are described in this chapter. Experimental results obtained for investigated composites show that one's containing modified filler have the better technical parameters mentioned above than composites with unmodified filler at corresponding loading. Experimentally is shown that the composites containing binary fillers diatomite and andesite at a definite ratio of them possess the optimal characteristics–so-called synergistic effect. The experimental results are explained in terms of structural peculiarities of polymer composites.

2.1 INTRODUCTION

In recent times, the mineral fillers attract attention as active filling agents in polymer composites [1, 2]. Thanks to these fillers, as many properties of the composites, are improved–increases the durability and rigidity, decrease the shrinkage during hardening process and water absorption, improves thermal stability, fireproof and dielectric properties, and finally, the price of composites becomes cheaper [3–5]. At the same time, it must be noted that the mineral fillers at high content lead to some impair of different physical properties of composites. Therefore, the attention of the scientists is attracted to substances, which would be remove mentioned leaks. It is known that silicon organic substances (both low and high molecular) reveal hydrophobic properties, high elasticity, and durability in a wide range of filling and temperatures [6, 7].

The purpose of presented work is the investigation of the effect of modifying by ethyl silicate (ES) of the mineral-diatomite as main filler and same mineral with andesite (binary filler) on some physical properties of composites based on epoxy resin.

2.2   EXPERIMENTAL

Mineral diatomite as a filler was used. The organic solvents were purified by drying and distillation. The purity of starting compounds was controlled by an LKhM-8-MD gas-liquid chromatography; phase SKTF–100 (10%, the NAW chromosorb, carrier gas He, 2 m column). FTIR spectra were recorded on a Jasco FTIR–4200 device.

The silanization reaction of diatomite surface with ES was carried out by means of three-necked flask supplied with a mechanical mixer, thermometer, and dropping funnel. For obtaining of modified by 3 mass % diatomite to a solution of 50 g grind finely diatomite in 80 mL anhydrous toluene the toluene solution of 1.5 g (0.0072 moles) ES in 5 ml, toluene was added. The reaction mixture was heated at the boiling temperature of used solvent toluene. Then the solid reaction product was filtrated, the solvents (toluene and ethyl alcohol) were eliminated, and the reaction product was dried up to constant mass in vacuum. Another product modified by 5% tetraethoxysilane was produced via the same method.

Following parameters were defined for obtained composites: ultimate strength (on the stretching apparatus of type "Instron"), softening temperature (Vica method), density, and water absorption (at saving of the corresponding standards).

2.3 RESULTS AND DISCUSSION

The masses of ES were 3 and 5% from the mass of filler. The reaction systems were heated at the solvent boiling temperature (~110°C) during 5–6 hours by stirring.

The direction of reaction defined by FTIR spectra analysis shown that after reaction between mineral surface hydroxyl, -OSi(OEt)$_3$ and the -OSi(OEt)$_2$O- groups are formed on the mineral particles surface.

In the FTIR spectra of modified diatomite, one can observe absorption bands characteristic for asymmetric valence oscillation for linear ≡Si-O-Si≡ bonds at 1030 cm^{-1}. In the spectra, one can see absorption bands characteristic for valence oscillation of ≡Si-O-C≡ bonds at 1150 cm^{-1} and for ≡C-H bonds at 2950–3000 cm^{-1}. One can see also broadened absorption bands characteristic for unassociated hydroxyl groups.

On the basis of modified diatomite and epoxy resin (of type ED–20), the polymer composites with different content of filler were obtained after careful wet mixing of components in the mixer. The blends with hardening agent (polyethylene-polyamine) were placed to the cylindrical forms (in accordance with standards ISO) for hardening, at room temperature, during 24 h. The samples hardened later were exposed to temperature treatment at 120°C during 4 h.

The concentration of powder diatomite (average diameter up to 50 microns) was changed in the range 10–60 mass %.

The curves on Figure 2.1 show that at increasing of filler (diatomite) concentration in the composites the density of materials essentially depends on both of diatomite contain and on the degree of concentration of modify agent (ES). Naturally, the decreasing of the density of composites at increasing of filler concentration is due to increasing of micro empties because of one's localized in the filler particles (Figure 2.1, curve 1). The composites with modified by ES diatomite contain less amount of empties as they are filled with modify agent (Figure 2.1, curves 2 and 3).

The dependence of ultimate strength on the content of diatomite (modified and unmodified) presented in Figure 2.2 shows that it has an extreme character. However, the positions of corresponding curves maximums essentially depend on the amount of modified agent ES. The general view of these dependencies is in full conformity with well-known dependence of σ–C [8]. The sharing of the maximum of the curve for composites containing 5% of modified diatomite from the maximum for the analogous composites containing 3% modifier to some extent is due to increasing of the amount of the bonds between filler particles and macromolecules at increasing of the concentration of the filler.

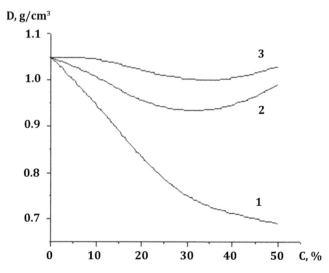

FIGURE 2.1　Dependence of the density of the composites based on epoxy resin on the concentration of unmodified (1), modified by 3% (2) and 5 mass % (3) ES diatomite.

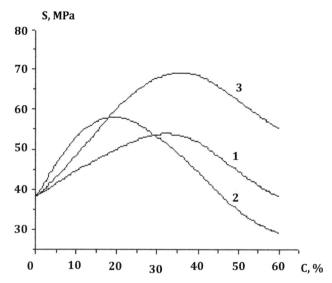

FIGURE 2.2　Dependence of ultimate strength of the composites based on ED–20 with unmodified (1) and modified by 3 (2), and 5 mass % (3) ES diatomite.

Investigation of composites softening temperature was carried out by the apparatus of Vica method. Figure 2.3 shows the temperature dependence of the indentor deepening to the mass of the sample for composites with fixed (20 mass %) concentration of unmodified and modified by ES.

Based on the character of curves in Figure 2.3, it may be proposed that the composites containing diatomite modified by ES possess thermo-stability higher than in case of analogous composites with unmodified filler. Probably, the presence of increased interactions between macromolecules and filler particles due to modify agent leads to increasing of thermo-stability of composites with modified diatomite.

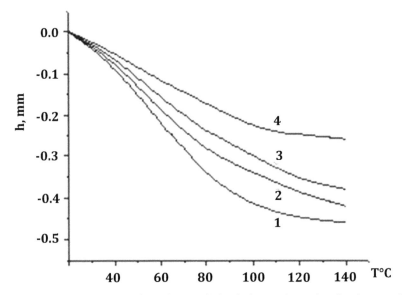

FIGURE 2.3   Temperature dependence of the indentor deepening in the sample for composites containing 0 (1), 20 mass % (2), 20 mass % modified by 3% ES (3), 20 mass % modified by 5% ES (4) diatomite.

Effect of silane modifier on the investigated polymer composites also reveals in the water absorption. In accordance with Figure 2.4, this parameter is increased at increasing of filler content. However, if the composites contain the diatomite modified by ES this dependence becomes weak.

There were conducted the investigation of binary fillers on the properties of the composites with same polymer basis (ED–20). Two types of minerals diatomite and andesite with different ratios were used as fillers. It was interesting to establish effect both of ratio of the fillers and effect of modifier ES on the same properties of the polymer composites investigated above.

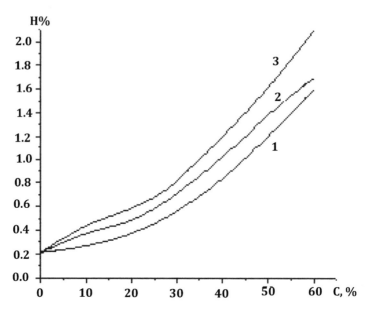

FIGURE 2.4   Dependence of the water-absorption on the concentration of filler in the composites based on epoxy resin containing diatomite modified by 5% (1) and 3% (2) ES and unmodified one (3).

The curves presented in Figure 2.5 show the effect of modify agent ES on the dependence of the density of composites containing the binary filler diatomite and andesite on the ratio of lasts when the total content of fillers is 50 mass % to which the maximal ultimate strength corresponds. The maximum of noted effect corresponds to composite, filler ratio diatomite/andesite in which is about 20/30. Probably microstructure of such composite corresponds to the optimal distribution of filler particles in the polymer matrix at minimal inner energy of statistical equilibration, at which the concentration of empties is minimal because of the dense disposition of the composite components. It is known that such structures consist minimal amount both of micro and macro structural defects [8].

Such an approach to the microstructure of composites with an optimal ratio of the composite ingredients allows supposing that these composites would be possessed high mechanical properties, thermo-stability, and low water absorption. Moreover, the composites with same concentrations of the fillers modified by ES possess all the noted above properties better than ones for composites with unmodified by ES binary fillers, which may be proposed early (Figures 2.6–2.8). Indeed, the curves on the Figures 2.6–2.8 show that the maximal ultimate strength, thermo-stability, and simultaneously hydrophobic properties correspond to composites with the same ratio of fillers to which the maximal density corresponds.

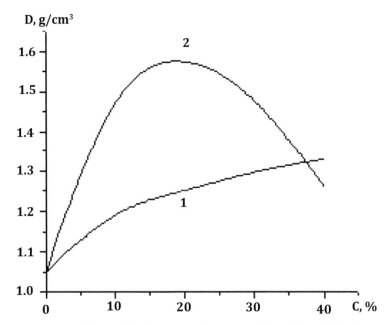

FIGURE 2.5   Dependence of the density on the concentration of diatomite in binary fillers with andesite. (1) Unmodified and modified by 5% ES (2) fillers for composites based on epoxy resin. Full concentration of binary filler in composites 50 mass %.

The obtained experimental results may be explained in terms of composite structure peculiarities. Silane molecules displaced on the surface of diatomite and andesite particles lead to activation of them and participate in chemical reactions between active groups of ES (hydroxyl) and homopolymer (epoxy group). Silane molecules create the "buffer" zones between the filler and the homopolymer. This phenomenon may be one of the reasons for increasing of strengthening of composites in comparison with composites containing unmodified fillers. The composites with modified diatomite display more high compatibility of the components than in the case of same composites with unmodified filler. The modified filler has more strong contact with polymer matrix (thanks to silane modifier) than unmodified diatomite. Therefore, mechanical stresses formed in composites by stretching or compressing forces absorb effectively by relatively soft silane phases; i.e., the development of micro defects in carbon chain polymer matrix of composite districts and finishes in silane part of material the rigidity of which decreases.

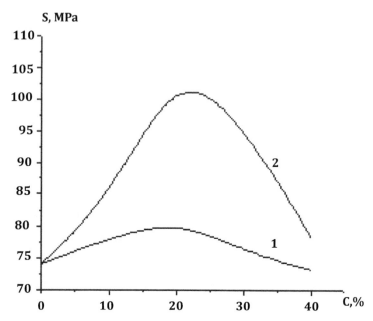

FIGURE 2.6 Dependence of ultimate strength on the concentration of diatomite in binary fillers with andesite. (1) Unmodified fillers and modified by 5% ES (2) ones for composites based on epoxy resin. Full concentration of binary filler in composites 50 mass %.

The structural peculiarities of composites also display in thermo-mechanical properties of the materials. It is clear that softening of composites with modified by ES composites begins at relatively high temperatures. This phenomenon is in good correlation with corresponding composite mechanical strength. Of course, the modified filler has more strong interactions (thanks to modifier) with epoxy polymer molecules, than unmodified filler.

The amplified competition of the filler particles with macromolecules by ES displays well also on the characteristics of water absorption. In general, loosening of microstructure because of micro empty areas is due to the increasing of filler content. Formation of such defects in the microstructure of composite promotes the water absorption processes. Water absorption of composites with modified diatomite is lower than that for one with unmodified filler to some extent. The decreasing of water absorption of composites containing silane is the result of hydrophobic properties of ones.

Composites with binary fillers possess so-called synergistic effect–non-additive increasing of technical characteristics of composites at containing fillers with a definite ratio of them, which is due to the creation of the dense distribution of ingredients in composites.

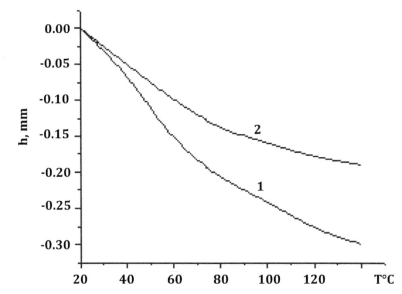

FIGURE 2.7 Thermostability of composites with binary fillers at ratio diatomite/andesite = 20/30.

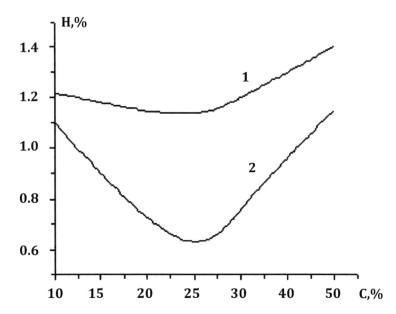

FIGURE 2.8 Dependence of the water-absorption of composites based on epoxy resin on the concentration of diatomite in binary fillers with andesite. (1) Unmodified and modified by 5% ES (2) fillers. The total concentration of binary fillers in composites is 50 mass %.

2.4 CONCLUSIONS

Comparison of the density, ultimate strength, softening temperature, and water absorption for polymer composites based on epoxy resin and unmodified and modified by ES mineral fillers diatomite and andesite lead to the conclusion that modify agent stipulates the formation of heterogeneous structures with higher compatibility of ingredients and consequently to enhancing of noted above technical characteristics.

ACKNOWLEDGMENT

This work has been fulfilled by the financial support of Georgian National Science Foundation GNSF/ST06/4–070.

KEYWORDS

- **ethyl silicate**
- **modified filler**
- **polymer composite**
- **synergistic effect**

REFERENCES

1. Katz, H. S., & Milewski, J. V., (1987). *Handbook of Fillers for Plastics*, RAPRA, Springer US, 468 p.
2. Mareri, P., Bastrole, S., Broda, N., & Crespi, A., (1998). *Composites Science and Technology*, *58*(5), 747.
3. Tolonen, H., & Sjolind, S., (1996). *Mechanics of Composite Materials*, *31*(4), 317.
4. Rothon, S., (2003). *Particulate Filled Polymer Composites*, RAPRA Technology Limited, NY, 275 p.
5. Lou, J., & Harinath, V., (2004). *Journal of Materials Processing Technology*, *152*(2), 185.
6. Khananashvili, L. M., Mukbaniani, O. V., & Zaikov, G. E., (2006). Monograph, *"New Concepts in Polymer Science, Elementorganic Monomers: Technology, Properties, Applications."* Printed in Netherlands, VSP, Utrecht, 497 p.
7. Aneli, J. N., Khananashvili, L. M., & Zaikov, G. E., (1998). *Monograph, "Structuring and Conductivity of Polymer Composites."* Nova Sci. Publ., NY, 326 p.
8. Zelenev, Y. V., & Bartenev, G. M., (1978). *Physics of Polymers, Visshaia shkola* (High school), (in Rus.), 267 p.

CHAPTER 3

Material Model of Polyester Composites with Glass Reinforced Polyester Recyclate and Nanofiller

M. JASTRZĘBSKA and M. RUTKOWSKA

Department of Industrial Commodity Science and Chemistry,
Faculty of Entrepreneurship and Quality Science,
Gdynia Maritime University, 83 Morska Str., 81–225 Gdynia, Poland,
E-mail: m.jastrzebska@wpit.umg.edu.pl

ABSTRACT

The earlier studies showed that glass reinforced polyester recyclate might be used as filler in polyester with the dolomite dust composites [1–4], but sometimes mechanical properties of the composites were decreased. In this case, the glass reinforced polyester waste was ground and added to composites. In composites, the dolomite dust was partially replaced by glass reinforced polyester recyclate. In order to improve the properties of our new composites with recyclate the nanofiller (modified montmorillonite) was added. The influence of nanofiller on compressive and flexural strengths of composites containing glass reinforced polyester recyclate have been tested, according to the content of recycled material and polyester resin.

The subject of this paper is the statistical analysis of the synergic effects and the estimation of the relative importance of components of the polyester composite with glass reinforced polyester recyclate and the nanofiller. For this aim, the material model of polyester composite has been developed on the basis of experimental results, and then the statistical significances of the elements of the model have been estimated.

In this study, the sought relationships have been determined in the form of 2nd-degree polynomials (quadratic functions). Conformity of the model to experimental data was evaluated using determination coefficient R^2.

Multi-criteria optimization, based on the statistical experimental material model of the composite, has been carried out and composition of polyester composites with glass reinforced polyester recyclate of high strength has been formulated. Results have confirmed that the addition of 2 wt.% nano-filler to the polyester composite with 20 wt.% polyester resin and no more than 12 wt.% recyclate significantly improves the mechanical properties in comparison to the properties of the composite without the nanofiller.

3.1   INTRODUCTION

Nowadays, glass reinforced polyesters are applied more extensively, especially for transporting structures, vehicle, and vessels. Glass fiber reinforced plastics dominate the composite market accounting for approx. 95% of total volume. Glass fiber used for reinforcement generally contributes between 15% and 70% of the material used in these composites depending on the manufacturing process and application. The average proportion of glass used over all segments is around 25–35%. In Europe, the production volume of glass fiber reinforced composites reached nearly 1.4 million tonnes in 2017 [5]. Unsaturated polyester resins are still by far the most commonly used. Commercial unsaturated polyesters are based on phthalic acid, maleic acid, ethylene glycol, and butanediol. Technical advantages of these composites (mainly great durability and resistance to the environment) became a serious disadvantage during tests on their utilization. The polyesters are difficult to be recycled because the material is fully cured and contains incorporated glass reinforcement. The curing process of composites is irreversible. The high melting temperature of glass ($\sim 1550°C$) makes energy intensity the major environmental issue. The Landfill Directive (1999/31/EC) encourages to waste minimization, recovery, and recycling initiatives. Under many national legislations, the landfilling of organic waste is prohibited. Current and unavoidable waste management legislation will put more pressure on industry to address the option available for dealing with composite waste. An efficient system for recycling composite materials would markedly improve their impact on the environment during their complete lifetime. Several techniques of glass reinforced polyesters recycling do exist, but they are not yet commercially available. Unlike monolithic material, the heterogeneous nature of composite materials makes recycling very challenging. Composite recycling technologies are categorized into thermal, material, chemical, and co-processing in cement plants.

The preferred option is material recycling. After suitable size reduction, the material is ground in a hammer mill and graded into different fractions. The use of ground composite materials can have two purposes: filler or reinforcement. The powdered products recovered after sorting can be used as filler, but this is not considered commercially viable because of the very low cost of virgin fillers such as calcium carbonate or silica. The incorporation of filler material in new materials is limited to typically less than 10 wt.% because of the deterioration in mechanical properties and increased processing problems at higher contents as a result of the higher viscosity of the compound. For glass reinforced polyester waste, product-specific development is needed to incorporate regrind as a reinforcing filler, e.g., in new building materials. As a result of these degraded mechanical properties, glass reinforced polyester recyclates are usually used in low-end applications such as construction fillers. This case study was carried out on the material recycling of glass reinforced polyester waste which is shredded into smaller fragments and can be added into polymer composites. Our earlier studies showed that glass reinforced polyester recyclate might be used as filler in new polyester composites [1–4], but mechanical properties of the composites were decreased. In these composite materials, the empty spaces have been observed, and they have an influence on weakening and increasing of composite absorption. In the field of polymers, a wide variety of materials are used as fillers; much research is focused on the nanoscale. The unique properties obtained in by a nanocomposite may be attributed to a well-dispersed reinforcing phase creating a large interfacial surface area. Nanoclays, in addition to their primary function as high aspect ratio reinforcements, also have the important functions such as thermal and barrier properties and synergistic flame retardancy. Many nanoclays are built of the smectites clay known as montmorillonite, a hydrated sodium calcium aluminum magnesium silicate hydroxide, $(Na, Ca)(Al, Mg)_6(Si_4O_{10})_3 \cdot nH_2O$. Montmorillonite is most often applied because of its high availability, low price, and its specific surface and large ion exchange capacity. However, the hydrophilic character of montmorillonite is a barrier for the easy dispersion of clay platelets in most of the polymer. A modification of montmorillonite is necessary in order to decrease the surface tension, and decrease the wettability and give montmorillonite an organophilic character. Quaternary ammonium salts containing at least an organophilic n-alkyl chain are most often used to replace the cations in the galleries. Szustakiewicz et al. [6] reported that the addition of the modified montmorillonite NanoBent® ZW1 to composites had caused a significant improvement in their mechanical properties. The tensile strength of composites based on waste polyamide 6 with neat polyamide 6 and 5 wt.% nanofiller was near to the corresponding value of neat polyamide 6. Moreover, Young's modulus of the composites was 15–30% higher than for neat polyamide 6. The main diffraction maximum from the layered structure of

NanoBent® ZW1 is located at $2\theta = 5°$ which corresponds to the basal spacing of the modified silicate. Bakar [7] added 1% NanoBent® ZW1, and butadiene-acrylonitrile copolymer terminated with an amine group to epoxy resin and observed that the impact strength increased approx. by 200% in relation to neat epoxy resin. The value of the interlayer distance calculated according to Braggs's law equals 1.8 nm. The very little research has been carried out on the unsaturated polyester system [8–12]. Suh et al. [8] studied the mechanism of mixing unsaturated polyester with organophilic–treated montmorillonite. The styrene monomer moves more easily than uncured polyester chains. This may generate a higher styrene monomer concentration in the montmorillonite gallery than in any other part in a simultaneous mixing system. If polymerization occurs in these conditions, the total crosslinking density of the sample decreases due to the low concentration of the styrene in the uncured polyester linear chains. Hence, the styrene monomers, which act as a curing agent, are much more dispersed inside and outside of the silicate layers as the mixing time increases. Therefore, the crosslinking reaction takes place homogeneously inside and outside of the silicate layers, and the crosslinking density reaches the degree of crosslinking density of cured pure polyester.

In order to improve the properties of our new composites with recyclate the nanofiller (modified montmorillonite) NanoBent® ZW1 was added. The influence of nanofiller on compressive and flexural strengths of composites containing glass reinforced polyester recyclate have been tested, according to the content of recycled material and polyester resin.

For this aim, the material model of polyester composite has been developed on the basis of experimental results, and then the statistical significances of the elements of the model have been estimated.

The subject of this chapter is the statistical analysis of the synergic effects and the estimation of the relative importance of the components of the polyester composite with glass reinforced polyester recyclate and the nanofiller.

3.2   EXPERIMENTAL

3.2.1   MATERIALS

The materials used for the composites were:

- The recyclate of glass fiber reinforced cold-cured polyester laminates was a mixture of cured polyester resin particles and glass fiber. The waste was ground in a shredder manufactured in Kubala Sp. z o.o. to the maximal size of the particles not larger than 7 mm. The recyclate contains 45 wt.% of glass fiber in a polyester resin matrix.

- Nanofiller NanoBent® ZW1 manufactured in ZGM Zębiec S.A. (Poland) in cooperation with Wroclaw University of Technology, organophilized montmorillonite (cation exchange capacity CEC minimum 80 meq/100 g, d_{001} = 1.84 nm).
- Unsaturated ortophthalic polyester resin Polimal 109–32 K– manufactured in "Organika–Sarzyna" Chemical Works S.A. (Poland).
- Initiator (methyl ethyl ketone peroxide).
- Accelerator (cobalt naphthenate).
- Dolomite dust manufactured in Kambud Sp. z o.o.

In this work, polyester-dolomite with different amounts of polyester resin (15, 18, 20 wt.%) and different amount of glass-polyester recyclate (8%, 10%, 12%, 15 wt.%) and 2 wt.% of organically modified montmoril- lonite NanoBent® ZW1 were prepared and the mechanical properties have been studied. The compositions were mixed with the initiator (in the amount of 0.01 wt.%) and the accelerator (in the amount of 1 wt.%) at 22°C in our laboratory. Specimens of 40 x 40 x 160 mm were made.

3.2.2   MEASUREMENT

The mechanical properties (compressive strength and flexural strength) of the composites were measured by using a Universal Testing Machine EDB–60 according to PN-EN 1926:2007 and PN-EN 12372:2010 standards. The experimental date has enabled to elaborate the mathematical model (second-degree polynomials functions of 2 variable) of polyester composites with glass reinforced polyester recyclate and the nanofiller. The material model is a quantitative relationship composition-properties. Calculations, analysis, and presentation of the model have been carried out using STATISTICA computer program.

3.3   RESULTS AND DISCUSSION

The objective of this study was to enhance the performance of polyester composites with a glass reinforced polyester recyclate. Our earlier studies showed that glass reinforced polyester recyclate might be used as fillers in new polyester composites, but the mechanical properties of the composites were decreased [1–4]. Preparation of composites with the nanofiller Nano-Bent® ZW1 proved to be an effective method of improving the mechanical properties. When the nanofiller was added in 2 wt.%, to the composites with 20 wt.% of resin and 10 wt.% glass reinforced polyester recyclate the compressive strength increased by up to 70% and the flexural strength—by up to 40% compared to the properties of the sample without the nanofiller. The values of the compressive strength and the flexural strength were near the standard recommendation for window sills [13]. This was a result of the high surface area attained by adding nanofiller to the polyester with recyclates. The polyester interacts of the filler surface forming interphase of absorbed polymer and the overall polymer-filler adhesion increases due to the high surface area and thereby improves the strength. The addition of 2 wt.% nanofiller to composites with 20 wt.% resin and a 12 wt.% recyclate increased the compressive strength (by 96%) and the flexural strength (by 11%) compared to the properties of the sample without the filler. The addition of 2 wt.% nanofiller to composites with 15 and 18 wt.% recyclates did not improve the mechanical properties of the composites with 20 wt.% resin. At a higher recyclate amount, the positive effects of the nanofiller are not observed because a high amount of recyclate can create a restriction on obtaining high crosslinking density, thus leading to lower strength.

In this work, the mechanical properties of polyester composites with different amounts of polyester resin (15, 18, 20 wt.%) and different amount of glass-polyester recyclate (8%, 10%, 12%, 15 wt.%) and 2 wt.% of organic modified montmorillonite NanoBent® ZW1 have been studied. In Table 3.1, the material model of polyester composite has been developed on the basis of experimental results, and then the statistical significances of the elements of the model have been estimated. High values of determination coefficients ($R^2 > 0.80$) from STATISTICA computer program have confirmed good conformity of the model to the experimental data.

TABLE 3.1 Material Model of Composites with Glass-Polyester Recyclate and Nanofiller

Property	Relationship	Determination coefficient, R^2
Compressive strength	$F_c = -536.18 - 66.6\,p - 1.46\,p^2 - 0.40\,p \times r - 10.79\,r + 0.42\,r^2$	0.80
Flexural strength	$F_f = -625.70 + 87.54\,p - 2.71\,p^2 + 1.03\,p \times r - 22.94\,r - 0.02\,r^2$	0.99

p – polyester resin; r – recyclate of glass-polyester.

The compressive strengths and the flexural strengths of the polyester composites with 2 wt.% nanofiller and with different amounts of glass reinforced polyester recyclate and different amount of polyester resin are given in Figures 3.1 and 3.2.

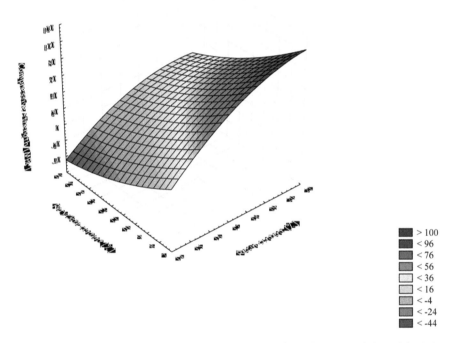

■	> 100
■	< 96
☐	< 76
☐	< 56
☐	< 36
☐	< 16
☐	< -4
■	< -24
■	< -44

FIGURE 3.1 (See color insert.) Graphical presentation of the material model of the polyester composite with glass reinforced polyester and nanofiller, showing the 'compressive strength' as a function of the component variables: Polyester resin and glass polyester recyclate.

In this study, the sought relationships have been determined in the form of 2nd-degree polynomials (quadratic functions). Conformity of the model to experimental data was evaluated using determination coefficient R^2. Multi-criteria optimization, based on the statistical experimental material model of the composite, has been carried out and composition of polyester composites with glass reinforced polyester recyclate of high strength has been formulated.

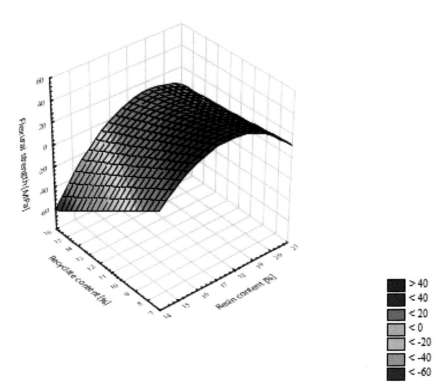

FIGURE 3.2   (See color insert.) Graphical presentation of the material model of the polyester composite with glass reinforced polyester and nanofiller, showing the 'flexural strength' as a function of the component variables: Polyester resin and glass polyester recyclate.

The measure of the coefficient's significance is the absolute value of the *t*-student parameter. For the given significance level α (i.e., the accepted risk of mistake; in the technical problems α is usually equal to 0.05) there is a critical value of parameter *t*, above which the regression coefficient is considered as statistically significant. The higher value is an absolute value of *t*; the larger is a significance of the estimated effect. The evaluation of the significance of the regression coefficients of the determined model is presented in the form of Pareto charts.

The statistical analysis of the synergic effects and estimation of the relative importance of the components of the polyester composite with glass reinforced polyester recyclate and the nanofiller has been done.

In the case of compressive strength (Figure 3.3), all factors are not statistically significant.

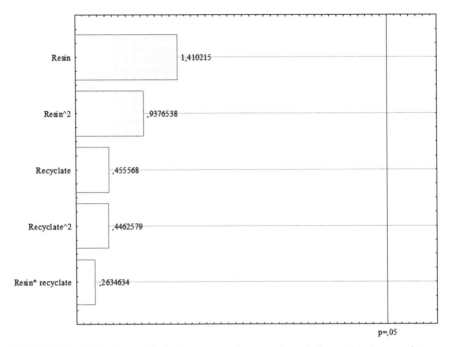

FIGURE 3.3 Estimation of the influence of polymer resin and glass-polyester recyclate on 'compressive strength' of composites with nanofiller.

In the case of flexural strength, the dominated importance of the polyester resin is particularly evident (Figure 3.4). The linear effect of the recyclate content is statistically significant, but the quadratic effect is nonsignificant. The synergic effect is also statistically significant.

3.4   CONCLUSIONS

The material model of polyester composites with glass reinforced polyester recyclate modified with 2 wt.% nanofiller NanoBent®ZW1 are presented. The positive effect the nanofiller has on the mechanical properties is observed when the composites contain only 10 or 12 wt.% recyclate. This is because styrene diffuses through the galleries of the organoclay more easily owing to its smaller molecular structure than the polymer. This reduces the styrene amount available for crosslinking in the medium which is the reason for the lower molecular weight between the crosslinking site, leading to restrictions for chain mobility and increasing the strength.

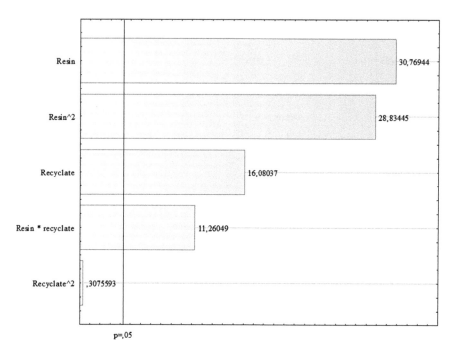

FIGURE 3.4   Estimation of the influence of polymer resin and glass-polyester recyclate on 'flexural strength' of composites with nanofiller.

The relationships between compressive or flexural strengths according to the content of recycled material and polyester resin in composites have been determined in the form of 2nd-degree polynomials (quadratic functions).

The study results suggest that the use of glass reinforced polyester waste in composite with nanofiller gives the technical and environmental benefits. Glass reinforced polyester recyclate can be applied as a filler in building materials like window sills.

Mechanical reprocessing is a good way for recycling glass reinforced polyester waste. The new composite is modeled on the basis of the properties (the compressive strength and the flexural strength) of material, which are depended on the amount of components. This simply and accuracy model, which is to be useful in theoretical considerations as well as engineering practice, showed the viable technological option to help with glass reinforced polyester waste management.

KEYWORDS

- **composite**
- **glass-polyester recyclate**
- **material recycling**
- **waste**

REFERENCES

1. Jastrzębska, M., & Jurczak, W., (2008). *Kompozyty (Composites)*, *8*(1), 59.
2. Jastrzębska, M., & Jurczak, W., (2011). *Polish Journal of Environmental Studies*, *20*(5A), 70.
3. Jastrzębska, M., & Blokus- Roszkowska, A., (2015). *Przemysł Chemiczny*, *94*(6), 1003.
4. Jastrzębska, M., & Rutkowska, M., (2007). New composites containing glass-reinforced recyclate. *Proceedings of the XXIV Reinforced Plastics, House of Technology Ltd. and Association of Composites Manufactures* (p. 159), Karlovy Vary (Carlsbad).
5. Witten, E., (2017). The GRP-market Europe 2017. *Composites Market Report 2017; Market Developments, Trends, Outlook and Challenges*, AVK.
6. Szustakiewicz, K., Gazińska, M., Kiernowski, A., & Pigłowski, J., (2011). *Polimery*, *56*, 397.
7. Bakar, M., & Szymańska, J., (2014). *Journal of Thermoplastic Composite Material*, *27*(9), 1239.
8. Suh, D. J., Lim, Y. T., & Park, O. O., (2000). *Polymer, 41*, 8557.
9. Kornmann, X., Berglund, L. A., Sterte, J., & Giannelis, E. P., (1998). *Polymer Engineering and Science, 38*, 1351.
10. Jastrzębska, M., Janik, H., & Paukszta, D., (2014). *Polimery*, *59*(9), 656.
11. Bharadwaj, R. K., Mehrabi, A. R., Hamilton, C., Tujilo, C., Murga, M., Fan, R., Chavira, A., & Thompson, A. K., (2003). *Polymer, 43*, 2033.
12. Mironi-Harpaz, I., Narkis, M., & Siegmann, A., (2005). *Polymer Engineering and Science, 45*(2), 174.
13. The Building Research Institute Technical Approval for Windowsills AT-06-0764/2005 (in Polish).

CHAPTER 4

Structure and Properties of Polymer Composites Based on Statistical Ethylene-Propylene Copolymer and Mineral Fillers

N. T. KAHRAMANOV[1], N. B. ARZUMANOVA[1], I. V. BAYRAMOVA[1], and J. N. GAHRAMANLY[2]

[1]Institute of Polymer Materials of Azerbaijan NAS, Azerbaijan, E-mail: najaf1946@rambler.ru

[2]Azerbaijan State University of Oil and Industry, Azerbaijan

ABSTRACT

The influence of the multi-functional organic structurants and mineral fillers on the structural features and physicomechanical properties of polymer composites based on statistical ethylene-propylene copolymer is reviewed. The principal possibility of improvement of strength characteristics and fluidity of polymer composites in the process of joint use of mineral fillers and ingredients is shown.

In recent years, scientists worldwide have all increasingly been attracted to researches aimed at studying the influence of the structure and composition of mineral fillers on the process of formation of crystalline per molecular organization in polymer materials [1–7]. A unique ability of finely divided particles of minerals to affect the mechanism of occurrence and growth of heterogeneous crystal nucleus in the polymer matrix has been detected in the process of mixing the polymers with the minerals. The latter fact has an essential impact on improving the properties of polymer composites filled with minerals.

The following are widely applied in order to implement a set of measures aimed at changing the mechanical properties of the original polymer matrix:

- processing additives that improve processing conditions.
- additives that modify the mechanical properties (structurants, plasticizers, reinforcing fillers, etc.).

Therewith the efficiency of simultaneous use of a number of additives has the combined action on the properties of the polymer composite. The mechanism of the existence of the components in the boundary areas of the interphase area is predetermined by the physical and physicochemical forms of interaction existing in the additive-macrochain-filler system [7–10].

There are almost no systematic studies aimed at establishing the effect of multifunctional organic structurants on the structural characteristics and properties of filled polymer composites. The urgency of this problem is that the introduction of the minimum concentrations of organic structurants contributes not only to improving the deformation-strength characteristics of polymer composites, but also significantly facilitates the miscibility of the blend components and, as a consequence, their processability [11, 12].

We have repeatedly confirmed in our works [8–12] that the use of organic structurants significantly affects the formation of finely-divided spherulitic structures that contribute to the improvement of the technological compatibility of limited-compatible polymers [11], increasing the strength properties, melt flow index and processability of composite materials [12].

Investigations on the effects of finely-divided particles of the organic structurants on the mechanism of formation of per molecular structures in polymer composites and processes occurring in the interphase area and boundary areas of the polymer-filler system still remain open. The occurrence of certain structures in the filled polymers and the consequential impact of filler on the regularity of the changing of their properties are one of the most important criteria predetermining the degree of "strengthening" of the polymer base. The increase in the elasticity coefficient and strength of the samples by dispersing the filler is considered as a form of "strengthening" of polymer composites.

Taking into account the complexity and the insufficient illumination of this problem in the literature, the purpose of this paper is to show how essential the role of multi-functional structurants in the regularities of changes of main physicomechanical and rheological properties of filled polymer composites is.

4.1 EXPERIMENTAL

The South Korean Industrial statistical ethylene-propylene copolymer (REP) (ethylene content 2–4%), trade name RP2400 has been used as the polymer base. The properties of the polymer are given below:

- Ultimate tensile stress – 28.5 MPa
- Flexural modulus – 975 MPa
- Tensile strain – 600%
- Melt flow index – 0.36 g/10 min
- Vicat softening point – 125°C
- Melting point – 138°C

Cement, chalk, and silica flour have been used as the filler of polymer.

Cement: The main components of this building material are the binding materials of inorganic origin. The particle size varies in the range of 107–400 nm.

Silica flour: Typical chemical composition of the silica flour is: SiO_2 (99.46%), Fe_2O_3 (0.048%), Al_2O_3 (0.21%), TiO_2 (0.027%), CaO (0.021%). Bulk density is 653 kg/m³ and the average particle size is 50–100 nm.

Chalk: The base of the chemical composition of the chalk is the calcium carbonate with a small amount of magnesium carbonate, but generally there is also non-carbonate part, mainly metal oxides. The average particle size is 200–400 nm.

Alizarin and zinc stearate have been used as organic structurants.

Zinc stearate $(C_{17}H_{35}COO)_2Zn$ – white amorphous powder, melting point is 403K, used concentration 0.3–1.0 wt%, used as the lubricant agent in the processing of polymers by injection molding and extrusion.

Alizarin $C_{14}H_8O_4$ – 1,2-dihydroxy anthraquinone, colorant red crystals with molecular mass is 240.2, melting point is 562 K. Below is the structural formula of alizarin:

The polymer compositions were prepared in the process of mechano-chemical modification (hot rolling) at a temperature of 463 K, the rolling time is 10 minutes.

For carrying out physicomechanical testing of the polymer composites, they are subjected to pressing at a temperature of 473K. Samples were punched from these to determine flexural modulus, ultimate tensile stress and tensile strain of filled composites. Ultimate tensile stress and tensile strain were determined in accordance with the GOST 11262–80 (State Standard). The flexural modulus was determined in accordance with the GOST 4648–71 (State Standard).

The melt index of composites was determined on the IIRT device at a temperature of 190°C and a load of 5 kg.

The crystallization temperature was determined on IIRT device fitted for dilatometric measurements [13].

4.2   RESULTS AND DISCUSSION

Taking into account the complexity and insufficient study of this problem, it seemed interesting to carry out a phased approach to the study of regularity of changes in the structure and properties of polymer composites, depending on the type of filler and structurant. It should be noted that often in the literature zinc stearate are only lubricant agent, which improves the processing of polymer composites, but do not reveal the main reasons for the improvement of the quality of the products obtained with the participation of this ingredient. At the same time, alizarin is known as a colorant and therefore practically is not perceived as a structurant in polymer composites.

For a more complete interpretation of detected regularities turn to the results of experimental studies, shown in Table 4.1. As can be seen from this table, the introduction of such fillers like cement, silica flour and chalk (without ingredients) contributes to the continuous growth of flexural modulus. As regards ultimate tensile stress, the maximum value of its values is accounted for by the samples with 5–10 wt% filler content. Further increase in the concentration of these fillers leads to a natural decrease in ultimate tensile stress, tensile strain and melt index of samples. Reducing the strength of filled composites during uniaxial tension is the evidence to complex processes in the inter-spherulite area. We do not exclude that in small concentrations of the filler particles (5–10 wt%), the latter predominantly involved in the formation of heterogeneous nucleation [14]. A further increase in the concentration of filler helps in the process of crystallization from the melt and the crystal growth of particles in excess pushed in the inter-spherulite amorphous area. If the crystallinity of the polymer base is approximately 50%, the amount of filler in the amorphous field is doubled.

In other words, if the polymer is introduced 20 wt% filler, its concentration in the amorphous field of semi-crystalline REP is approximately 40 wt%. Accumulating in this space, the filler particles reduce conformational mobility "continuous chains," increase the stiffness of the amorphous field, which immediately affects the deterioration of strength and tensile strain of polymer composites.

Attention shall be paid to the fact that the separate introduction of 1wt% alizarin and zinc stearate into the REP leads to serious changes in its qualitative characteristics. For example, in this case, it provided a significant increase in all its strength characteristics, tensile strain and MFI of REP. It is clear that such significant improvement in the quality characteristics of the REP can be clearly interpreted based on the peculiarities of the formation of fine spherulitic per molecular structures.

The important aspect is that the simultaneous introduction of alizarin and zinc stearate into the composition of REP leads to even more improvement of final properties of polymer composites. It is precisely this feature of joint participation of alizarin and zinc stearate towards improving the quality of REP that can be regarded as a "synergetic effect." Therefore, in our further investigations in all the filled composites of REP, these ingredients were introduced simultaneously in the amount of 1 wt% alizarin and 1 wt% zinc stearate.

According to the data given in Table 4.1, the introduction of the described ingredients into the composition of filled composites leads to a substantial improvement of their properties. Of pairwise comparing samples of filled composites with or without ingredients can establish that the maximum increase in the flexural modulus is 22–25%, ultimate tensile stress 15–21% and MFI 4.0–4.8 times. There is a reason to assume that the smaller spherulite size, the fewer defects in crystalline formations and the more likely uniformly dispersing of the filler particles in the inter-spherulite area and generally throughout the volume of the polymer matrix. On the other hand, we do not exclude that alizarin and zinc stearate may to some extent contribute to laying the grain of surface of filler particles, thereby creating favorable conditions for the aggregate flow of the melt and supporting maximum wall sliding in evaluating of MFI of samples. In view of the nature of the filler and polymer, and mutual dispersion, the process of choosing the optimal ratio of components in the polymer matrix may be associated with certain difficulties. In several cases, this circumstance is explained by a variety of approaches to the interpretation of the observed regularities [13, 14]. Depending on the type and concentration of filler, the latter has a significant structuring influence on polymer base both in solid and in the viscous-flow state.

TABLE 4.1 The Composition and the Physico-Mechanical Properties of the Filled Composites Based on REP Containing 1 wt% Alizarin and 1 wt% Zinc Stearate

№	Composition of polymer compound	Ultimate tensile stress, MPa	Flexural modulus, MPa	Tensile strain, %	MFI, g/10 min
1	REP	28.5	975	600	0.36
2	REP+1% alizarin	29.8	1001	750	1.56
3	REP+1% zinc stearate	29.1	988	680	2.98
4	REP + alizarin + zinc stearate	30.4	1006	780	3.11
5	REP+5% cement	31.3	1010	560	0.44
6	REP+5% cement + alizarin + zinc stearate	32.2	1028	640	1.94
7	REP+10% cement	31.5	1025	320	0.35
8	REP+10% cement + alizarin + zinc stearate	34.6	1176	510	1.98
9	REP+20% cement	24	1075	85	0.25
10	REP+20% cement + alizarin + zinc stearate	32.3	1198	250	1.85
11	REP+30% cement	22.8	1080	25	Don't flow
12	REP+30% cement + alizarin + zinc stearate	24.6	1215	125	0.95
13	REP+5% silica flour	29.8	1015	240	0.35
14	REP+5% silica flour + alizarin + zinc stearate	31.4	1022	525	2.04
15	REP+10% silica flour	27	1020	180	0.3
16	REP+10% silica flour + alizarin + zinc stearate	30.8	1132	295	1.84
17	REP+20% silica flour	22.6	1040	70	0.11
18	REP+20% silica flour + alizarin + zinc stearate	27.3	1155	110	1.92
19	REP+30% silica flour	20.8	1050	35	Don't flow
20	REP+30% silica flour + alizarin + zinc stearate	24.5	1175	70	0.73
21	REP+5% chalk	27.1	980	495	0.37
22	REP+5% chalk + alizarin + zinc stearate	30.3	1035	210	2.01
23	REP+10% chalk	25.8	1000	315	0.2
24	REP+10% chalk + alizarin + zinc stearate	27.9	1121	145	1.72
25	REP+20% chalk	19.8	1010	95	0.12
26	REP+20% chalk + alizarin + zinc stearate	24.7	1123	105	1.45
27	REP+30% chalk	18.2	1015	15	Don't flow
28	REP+30% chalk + alizarin + zinc stearate	22.6	1155	65	0.69

For example, Table 4.2 presents the results of investigations on the influence of type and concentration of the filler and organic structurants on the crystallization temperature of the polymer composites. After analyzing the data contained in this table, it can be stated that the used organic structurants have a strong impact on the value of the crystallization temperature of composites. Particularly, there is a strong influence of alizarin. Joint use of discussed organic structurants once again has confirmed the existence of "synergism" effect, which was expressed in the increase of crystallization temperature of composites 5–7°S. It is characteristic that filler itself (except chalk) exhibits a structure-forming effect expressed an increase of the temperature of crystallization of polymer composites. As shown from this table, the most intensive crystallization process occurs in those samples in which cement and silica flour are used as a filler. Changing the crystallization temperature of the polymer matrix is only possible when foreign finely dispersed particle of organic or mineral origin is capable of forming heterogeneous crystallization nucleus.

TABLE 4.2 Influence of Structurants and Mineral Fillers on Heat Resistance and Crystallization Temperature of REP

№	Composition of polymer compound	Chilling point, °C	Vicat softening point, °C
1	REP	141	125
2	REP+1% alizarin	144	125
3	REP+1% zinc stearate	142	125
4	REP+ alizarin + zinc stearate	146	125
5	REP+5% cement	142	127
6	REP+5% cement + alizarin + zinc stearate	147	128
7	REP+10% cement	143	129
8	REP+10% cement + alizarin + zinc Stearate	147	130
9	REP+5% silica flour	143	127
10	REP+5% silica flour + alizarin + zinc Stearate	148	129
11	REP+5% chalk	141	125
12	REP+5% chalk + alizarin + zinc stearate	142	126

The filled polymer composites may have two types of crystallization nucleus: homogeneous and heterogeneous. In our view, the homogeneous nucleus is the primary microcrystalline formations which emerged as a result of thermo-fluctuation changes in the melt of polymer at the level of oriented segments of macro chains. In the process of further cooling, these micro oriented areas grow into larger crystalline formations–spherulites. Heterogeneous crystallization nucleus is formed with the participation of solid foreign particles, which can orient macro segments of the polymer matrix on their surface. It is obvious that during polymer processing by extrusion or injection molding, the cooling process of products is accompanied, at first, by the formation of heterogeneous crystallization nucleus, and then homogeneous crystallization nucleus [13, 14].

Thus, on the basis of the foregoing, it can be concluded that the use of structurants with various mineral fillers has a positive impact on the improvement of physicomechanical, thermal, and rheological properties of filled polymer composites. It becomes obvious that the use of mineral fillers and structurants allows solving a number of problems related to improving the technological capabilities of processing of filled polymer composites. As a matter of fact, new polymer composites with improved operating and processing characteristics were generated in the process of modification of REP.

KEYWORDS

- **chains**
- **continuous**
- **crystallinity**
- **crystallization**
- **filler**
- **flexural modulus**
- **interfacial area**
- **spherulites**
- **tensile strength at yield**
- **ultimate tensile stress**

REFERENCES

1. Berlin, A. A., Volfson, S. A., & Oshman, V. Q., (1990). "Principles of creation composite materials." *M.: Chemistry*, pp. 240.

2. Lipatov, Y. S., (1977). "Physical chemistry of filled polymers." *M.: Chemistry*, pp. 304.
3. Keleznev, V. N., & Shershenov, V. A., (1988). *"Chemistry and Physics of Polymers."* M.: V. Sh., pp. 312.
4. Osama, Al, X., Osipchik, V. S., Petuxova, A. V., Kravchenko, T. P., & Kovalenko, V. A., (2009). Modification of filled polypropylene. *Plastic Mass, 1*, 43–46.
5. Osipchik, V. S., & Nesterenkova, A. I., (2007). Talc filled compositions based on polypropylene. *Plastic Mass, 6*, 44–46.
6. Ermakov, S. N., Kerber, M. L., & Kravchenko, T. P., (2007). Chemical modification and mixing of polymers during reactive extrusion. *Plastic Mass, 10*, 32–41.
7. Pesetsky, S. S., & Boqdanovich, S. P., (2015). Nanocomposites obtained by dispersing clay in polymer melts. *International Scientific and Technical Conference "Polymer Composites and Tribology, "* (pp. 5). Gomel.
8. Kahramanov, N. T., (1990). *The Mechanism of Modifying the Permolecular Structure of Polyolefins by Grafting Acrylic Monomers*, v.32A, *11*. C. 2399–2403.
9. Kahramanov, N. T., Baladzhanova, G. M., & Shahmaliyev, A. M., (1991). Investigation of sorption kinetics of graft copolymers on the filler surface. *Macromolecular Compounds. 32B, 5*, 325–329.
10. Kahramanov, N. T., Kahramanly, Y. N., & Faradzhev, G. M., (2007). The properties of filled crystalline polymers. *Azerbaijan Chemical Journal, 2*, 135–141.
11. Kahramanov, N. T., Meyralieva, N. A., & Kahramanly, Y. N., (2011). The technological parameters of processing of PP compositions filled by natural zeolite. *Plastic Mass, 1*, 57–59.
12. Kahramanov, N. T., Hajiyeva, R. Sh., Kuliev, A. M., & Kahramanly, Y. N., (2013). The influence of different ingredients on the properties of the polymer mixtures based on polyamide and polyurethane. *Plastic Mass, 12*, 9–13.
13. Kahramanov, N. T., Dyachkovsky, F. S., & Buniyat-zade, A. A., (1982). Volumetric properties and crystallization of polymerization-filled polyethylene. *Compilation IX. Synthesis of Polymerization-Filled Polyolefins, Institute of Chemical Physics* (pp. 130). Academy of Sciences of the USSR. Chernogolovka.
14. Kahramanov, N. T., & Arzumanova, N. B., (2015). The problematic questions of mechanochemical synthesis of polymer compositions during their processing. *International Scientific Institute "Educatio," Novosibirsk, 3*(10), 147–148.

CHAPTER 5

Preparation of Copper-Containing Nanoparticles in Polyethylene Matrix Without the Use of Solvents

N. I. KURBANOVA, A. M. KULIYEV, N. A. ALIMIRZOYEVA, A. T. ALIYEV, N. YA. ISHENKO, and D. R. NURULLAYEVA

Institute of Polymer Materials of Azerbaijan National Academy of Science, S. Vurgun Str., 124, Sumgait Az5004, Azerbaijan, E-mail: ipoma@science.az

ABSTRACT

By a method of high-speed thermal decomposition of salts of the organic acids in the conditions of high shear deformations, there have been prepared the nanoparticles of copper oxide in a matrix of polyethylene of high pressure. The phase composition and structure of the prepared nanocomposites have been investigated by RPhA and SEM methods. It has been shown the formation of a layered structure, which possesses high destruction viscosity.

5.1 INTRODUCTION

One of the perspective directions in a science of the polymers and material science of last years is the development of principles of preparation of the polymer nanocomposites, which are the newest type of functional materials and can be used in very various spheres of application [1, 2].

Recently, it is shown a considerable interest to the composition materials on the basis of polymer matrices and nanosized metal particles, which has been stipulated by a wide spectrum of their application from catalysis to nanotechnology in information technology [1].

The metal-polymer composition materials are mainly in electronic and radio-technical industry and also in avia- and rocket production [2].

Use of metal nanoparticles of variable valency (copper, cobalt, nickel, etc.) in the polymers allows to prepare principally the new materials, which find wide application in radio- and optoelectronics as the magnetic, electro-conducting, and optical media [1, 3].

One of the methods of formation of the metal polymers is the high-speed thermal decomposition of precursors in a solution of the polymer melt. A short-range order of structure of the original polymer remains in the melt, and the existing voids become available for localization of the forming particles. At first, they are implemented in the interpherulitic field of the polymer matrix in the space between lamels and in the centers of spherulites. In this, the strong interaction between nanoparticles and polymer chains is observed [4].

Thermal method of preparation of the nanoparticles based on decomposition of the organic salts or metal-organic compounds (method "klaspol") [5] is a simple, it can be carried out in usual heat-resistant glass flask at $t = 300–350°C$; however, it is used a large number of solvent, a yield of nanocomposite is low. The vacuum oil (VM–1) is used as a solvent. The prepared samples are separated from oil with multiple washing by benzene [6].

In connection with rigid demand of the ecologists to an increase of safety of the polymer materials and obligatory utilization of the production wastes, we offer the ecological mechano-chemical method of preparation of the metals nanoparticles without the use of the organic solvents in the extruder–mixer of the closed type [7].

The mechano-chemical approach (implementation of the nanotechnology "upwards") to the preparation of nanocomposites provides the possibility of creation of the effective and ecologically safe and resource-saving technologies, as these technological processes are based on chemical reactions in the solid phase, i.e., in the absence of solvents and technological operations connected with their use. In addition, the mechano-chemical method is most suitable for industrial application.

In connection with above-mentioned one, an application of the mechano-chemical method for preparation of the nanocomposites is the actual problem.

This work has been devoted to the ecological method of preparation of the metal-containing nanoparticles in polyethylene matrix without the use of solvents by decomposition of salts of the organic acids in a medium of polymer in the conditions of the high shear deformations and to the investigation of properties of the prepared nanocomposite.

5.2 EXPERIMENTAL

In the work that have been used, polyethylene of high pressure of mark 10803–020 (PE) having the following characteristics: content of crystalline phase – 60÷70%, density – 0.94 g/cm^3, melting index – 1.3, melting temperature determined by a method DTA, 100°C; as metal-containing compounds (precursors) – copper formate dehydrate $Cu(HCOO)_2 \cdot 2H_2O$.

By a method of high-speed thermal decomposition of salts of organic acids in the conditions of the high shear deformations, there have been prepared the metal nanoparticles in a matrix of polyethylene of high pressure in two stages. At the first stage at temperature 130–140C it was made the binary mixture of polymer and precursor on laboratory rolls. At the second stage the mixture was heated in micro-extruder "Brabender" in a medium of nitrogen at temperature 170–190°C for 10–12 min.

The phase composition and structure of the prepared nanocomposites have been investigated by RPhA and SEM methods. For carrying out of RPhA and SEM investigations the samples of the initial PE and prepared nanocomposites as films by thickness 0.5–1.0 mm by pressing at temperature 130–135°C and pressure 10 atm have been prepared.

The phase composition of the prepared nanocomposites has been investigated on X-ray diffractometer of wide purpose "PANalytical Empyrean" (The Netherlands). For carrying out of RPhA and SEM investigations, there have been made the samples of the initial PE and prepared nanocomposites as the films by thickness 0.5–1.00 mm by pressing at temperature 130–135°C and pressure 10 atm.

SEM investigations have been carried out on apparatus "SEM HITACHIS 3400N" (Japan). Thermostability of the studied samples of nanocomposites was studied on derivatograph of mark Q–1500D of firm MOM, Hungary. The tests have been carried out in an air atmosphere in the dynamic regime at heating 5 deg.·min^{-1} from 20 to 500°C, sample – 100 mg, the sensitivity of channels–DTA – 250mcV, TG–100, DTG – 1 mV.

5.3 RESULTS AND DISCUSSION

The nanocomposite polymer materials on the basis of PE with copper-containing nanoparticles have been prepared. The composition and structure of the prepared nanocomposites have been investigated.

The phase composition of the prepared nanocomposites was investigated by X-ray structure analysis. The phase identification was carried out according to the interplanar distances, using card file ASTM. It has been shown that in the investigated nanocomposites the reflections from planes of metal crystal lattice corresponding on card file ASTM in a series of d_{hkl} of copper oxide I (Cu_2O) were observed.

In Figures 5.1–5.3, the diffractograms of the initial PE and also PE with copper-containing nanoparticles are presented.

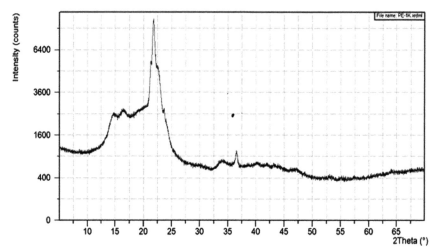

FIGURE 5.1   Diffractometer of the initial PE.

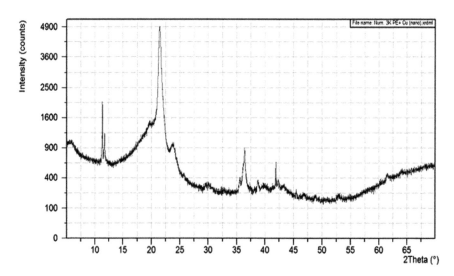

FIGURE 5.2   Diffractogram of PE with copper-containing nanoparticles.

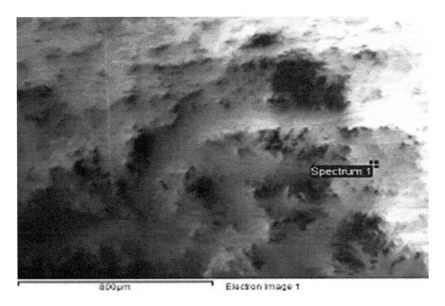

FIGURE 5.3   SEM-image of the initial PE ПЭ.

In Figures 5.3–5.5, SEM-images of PE of the initial PE and PE with copper-containing nanoparticles are presented.

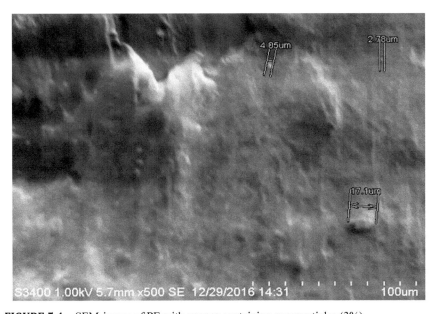

FIGURE 5.4   SEM-image of PE with copper-containing nanoparticles (3%).

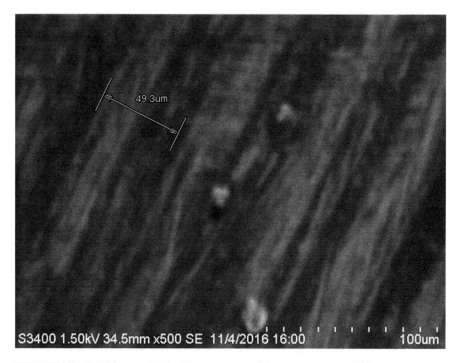

FIGURE 5.5   SEM-image of PE with copper-containing nanoparticles (5%).

As seen from Figure 5.3, microphotography of the initial PE is a loose, shapeless structure.

It is seen from Figure 5.4 that on the background of polymer matrix there are light formations–metal nanoparticles, in the form of spherical or close to spherical. A distribution of particles on sizes is not very wide (2.78, 4.05, 17.1) nm.

As is seen from Figures 5.4 and 5.5 at a low content of nanoparticles (3%) they do not interact between themselves, as they have been divided by polymer matrix. In the increase of the concentration of nanoparticles (5%), it arises percolation–charge exchange between particles. In this case, an agglomeration of particles and also their interaction with polymer matrix on nanolevel occurs, as a result of which it is per molecular structure. The microphotographs of the prepared nanocomposites evidence about the formation of layered structure, which possesses high destruction viscosity [8, 9], is not characteristic for the initial PE. The distribution of particles on sizes is not wide (49.3 nm).

The thermo-oxidative properties of metal-containing nanocomposites have been studied. Thermostability of the prepared samples of copper-containing nanocomposites was estimated on activation energy (E_a) of decay of thermo-oxidative destruction calculated by a method of double logarithm TG [10], on temperature of 10% (T_{10}), 20% (T_{20}), and 50% (T_{50}) decay of the studied samples and also on their half-decay time–$\tau_{1/2}$. The data prepared as a result of derivatographic investigations are presented in Table 5.1.

As is seen from data of Table 5.1, an introduction of nanoparticles of copper oxide in composition of polyethylene composition favors temperature rise of decay of the samples: T_{10} and T_{20} by 20°C, T_{50} by 30°C; half-decay time $\tau_{1/2,}$ is increased from 72 to 78 min., an activation energy (E_a) of decay of thermo-oxidative destruction of the prepared nanocomposites is increased to 16.2 kJ/mol.

TABLE 5.1 Thermal Properties of Studied Samples

Composition	T_{10}, °C	T_{20}, °C	T_{50}, °C	$\tau_{1/2}$, min.	E_a, kJ/mol
PE	290	340	380	72	129.35
PE + np Cu$_2$O (3%)	300	350	400	76	140.86
PE + np Cu$_2$O (5%)	310	360	410	78	145.57

It has been shown that an introduction of the copper nanoparticles in the composition of PE stabilizes the composition increases the beginning temperature of thermo-oxidative destruction by 30°, in this a melting temperature is not practically changed and equal to 100°C.

The derivatographic investigations showed that an introduction of nanoparticles of the copper oxide in the composition of PE favors improvement of thermo-oxidative stability of the prepared nanocomposites.

The nanocomposites containing metals or semiconductors attract attention, first of all, clusters with unique properties formed by a different number of metal atoms or semiconductor–from ten to several thousand incomings in their composition. The nanoparticles also show the superparamagnetism and catalytic properties. The developed surface of the composition nanomaterials stipulates their wide application as the adsorbents and catalysts [11].

5.4 CONCLUSIONS

By the ecological mechano-chemical method, the nanocomposite polymer materials on the basis of PE with copper-containing nanoparticles have been prepared. The composition and structure of the prepared nanocomposites have been investigated.

It has been shown that in the investigated nanocomposites the reflections from planes of metal crystal lattice corresponding on card file ASTM in a series of d_{hkl} of copper oxide I (Cu_2O) and zinc oxide (ZnO) were observed. The distribution of particles on sizes is not wide.

The microphotographs of the prepared nanocomposites evidence about the formation of layered structure possess high destruction viscosity.

The derivatographic investigations showed that an introduction of copper nanoparticles in the composition of PE stabilizes the composition increases the beginning temperature of thermo-oxidative destruction by 30°.

The prepared composites containing nanoparticles of the metal oxides can be used as the catalysts of the organic and petrochemical synthesis, adsorbents for water purification and also modifying agents of the industrial polymers with the aim of improvement of their operational properties.

KEYWORDS

- **copper-containing nanoparticles**
- **ecological method**
- **polyethylene matrix**
- **polymer melt**
- **RPhA and SEM methods**

REFERENCES

1. Gubin, S. P., Yurkov, G. Y., & Kosobudsky, I. D., (2005). *International Journal of Materials and Product Technology, 23*(1 & 2), 2.
2. Pomogaylo, A. D., Rozenberg, A. S., & Uflyand, I. E., (2000). *Nanoparticles of Metals in Polymers* (pp. 672). Khimiya, Moscow.
3. Mikhaylin, Y. A., (2009). *Polymer Materials, 7*, 10.
4. Gubin, S. P., (2000). *Russian Chemistry, J., XLI*(6), 23.
5. Kosobudskii, I. D., Kashkina, L. V., Gubin, S. P., et al., (1985). *Polymer Science, U.S.S.R., 27*(4), 768.
6. Yurkov, G. Y., Kozinkin, A. V., Nedoseikina, T. I., et al., (2001). *Inorganic Materials, 37*(10), 997.

7. Tarasova, N. N. P. O. M., & Lunin, V. V., (2010). *Russian Chemical Reviews, 79*(6), 439.

8. Berlin, A. L., (2010). *Polymer Science Series A, 52*(9), 875.

9. Chvalun, S. N., (2000). *Priroda, 7*, 1.

10. Kurenkova, V. F. K., (1990). *Practical Work on Chemistry and Physics of Polymers* (p. 299). Moscow.

11. Suzdalev, I. P., & Suzdalev, P. I., (2001). *Russian Chemical Reviews, 70*(3), 177.

CHAPTER 6

Electrodeposition and Properties of Copper Coatings Modified by Carbon Nanomaterial Obtained From Secondary Raw Materials

T. MARSAGISHVILI[1], G. TATISHVILI[1], N. ANANIASHVILI[1], M. GACHECHILADZE[1], J. METREVELI[1], E. TSKHAKAIA[1], M. MACHAVARIANI[1], and D. GVENTSADZE[2]

[1]R. Agladze Institute of Inorganic Chemistry and Electrochemistry of the Javakhishvili Tbilisi State University, Mindeli str. #11, 0186, Tbilisi, Georgia, E-mail: tamaz.marsagishvili@gmail.com

[2]R. Dvali Institute of Mechanics of Machines, Mindeli Str. #10, 0186, Tbilisi, Georgia

ABSTRACT

With the development of electroplating, composite coatings of CEP, which are deposited from electrolyte suspensions, are increasingly being used. Composite coatings are obtained in those cases when it is envisaged to modify the surface of metal products to give them new properties. Perspective dispersed material can be carbon particles, because of their large specific surface area, high porosity, and subminiature structure. The obtaining of such material is possible from cheap secondary raw materials (car tires, nutshells, sawdust, bamboo).

The purpose of this work is to create a CEP-based on copper with a carbon-dispersed phase. For the experiment, we used a carbon material with a particle size 40 nm, obtained by us, by pyrolysis from secondary raw materials (used tires of machines).

To study the effect of the dispersed phase on copper electrodeposition, the optimum conditions for obtaining the CEP Cu-C were determined.

Morphology and coefficients of friction-sliding of coatings of copper and CEP are studied.

Incorporating into the base metal matrix, the carbon particles, embedded in the deposit, determine its further growth. The change in the structure of the electrolytic deposit affects its functional properties.

The test of the obtained samples showed that the best tribological properties were found in CEP Cu-C, where the concentration of carbon material was 15 g/L. For CEP Cu-C, the friction coefficient (f) decreases three times compared to copper deposit, and wear by 12 times. Probably this is due to the fact that the carbon material performs the function of dry lubrication due to its layered structure.

6.1 INTRODUCTION

The rapid development of modern technology increases the need for the creation of new materials with specific and unique properties.

With the development of electroplating, composition coatings (CEPs) are increasingly being used. CEPs are applied from electrolyte suspensions, i.e., electrolytes modified with additives of nanomaterials or nanofibers, when nanoparticles during electrodeposition are covered with metal, being fixed on the surface of the article in a metal matrix.

Composite coatings are obtained in those cases when it is envisaged to modify the surface of metal products to give them new properties (increasing corrosion resistance, reducing friction and wear, increasing hardness, etc.) [1].

The process of CEP formation is influenced by many factors [2, 3], one of which is the nature of the material that is used for the modification. A promising material may be nanocarbon particles, because of their large specific surface area, high porosity, and subminiature structure. As is known from the literature [4], due to these properties, nanocarbon materials are used in various industries, including as a solid lubricant in antifriction coatings, which significantly reduce the coefficient of friction and wear. Reducing the coefficient of friction by means of such coatings, it is possible to achieve several very positive results at once: increasing the service life of parts and saving on additional lubricants. On this basis, it is possible to create modified, composite metal coatings of multifunctional purpose with unique properties. Conventional traditional coatings of such properties cannot be achieved.

The production of carbon nanomaterials is possible from cheap secondary raw materials–household waste of agriculture, used car tires, etc. Their rational and complete use acquires substantial scientific, economic, and ecological significance.

Our institute has experience in obtaining carbon nanomaterial from secondary raw materials (car tires, nutshells, sawdust, bamboo, etc.).

The aim of this work is to create a CEP based on copper with nano-carbon. For the experiment, a carbon nanomaterial was used, with a particle size of ~40 nm obtained in our laboratory by pyrolysis from secondary raw materials (machine tires).

During electrolysis, carbon nanoparticles are included in the matrix of the base metal, which significantly changes the properties of the galvanic coating (tribological, magnetic, etc.).

6.2 EXPERIMENTAL

6.2.1 MATERIALS

The studies were carried out using the standard copper coating electrolyte of the following composition, g/L: $CuSO_4 \cdot 5H_2O$–200; H_2SO_4 50; pH = 0.35, with constant stirring with a magnetic stirrer. The concentration of carbon nanoparticles in the electrolyte was from 1.0 to 25.0 g/L.

For better wetting and uniform distribution of the dispersed nanomaterial the carbon particles were treated with ethyl alcohol, filtered, mixed with the electrolyte and thoroughly mixed.

Coatings were applied to the steel plates and to the ends of the bushings with an area of 3 cm². The copper plate served as the anode. The thickness of the coatings was approximately ~ 25–40 μm.

6.2.2 MEASUREMENT

The morphology of the surface of copper coatings was studied using a Euromex microscope. Tribotechnical tests were carried out on an inertial friction test machine by a particular method at a load of 0.1 MPa, a sliding speed of 0.25 m/s and an ambient temperature of 25°C. The material of the counter body was steel 40X, hardness 60 HRC. Running-in was carried out at this same load until complete contact was established over the entire friction surface. The coefficient of friction was determined for the steady-state friction regime without lubrication.

6.3   RESULTS AND DISCUSSION

To study the effect of carbon nanomaterial on copper electrodeposition, optimal conditions for obtaining CEP Cu-C were determined. The current yield (W) of copper was studied and was determined for the following values of cathode current densities (ik): 0.05; 1.0; 1.5; 2.0; 3.0; (A/dm^2). The experimental data are presented in Figure 6.1.

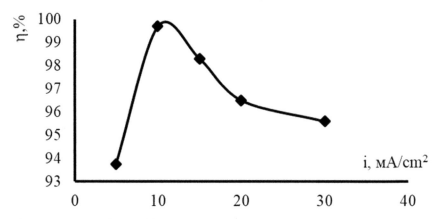

FIGURE 6.1   Dependence of the copper current output on the current density at the content in the electrolyte of carbon nanomaterial 1 g/L.

Figure 6.1 shows that with a current density of 10.0 mA/cm^2, the current output is maximal and is ~100.0%. At a given current density, qualitative coatings, uniform in color and without dendrites, are formed. At current densities below and above 10.0 mA/cm^2, the current output of copper drops. In further studies on the formation of CEP from this electrolyte were carried out at a current density of 10.0 mA/cm^2, because to this corresponds the maximum output of copper current and high-quality coatings for all concentrations of the carbon nanomaterial.

When the transition from the copper coating (Figure 6.2a) to the CEP Cu-C (Figure 6.2b), the micrograph of the surface of the electrolytic deposition changes. The composite coating has a coarse-grained surface, due to the overgrowth of carbon nanoparticles by metal. Consequently, the carbon particles, embedded in the sediment, determine its further growth.

The inclusion of the carbon nanomaterial in coating leads to structural changes in the metal matrix, which affects the exploitation properties of the electrolytic deposit.

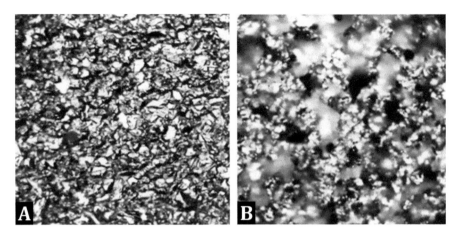

FIGURE 6.2 (See color insert.) Microstructure of the surface of electrolytic copper (a) and CEP Cu-C (b). Current density $i = 10.0$ mA/cm^2.

Of considerable interest in practical terms is the definition of the tribological properties of the Cu-C CEP obtained by us, in particular, the coefficient of sliding friction (f). As noted above, during the electrolysis for obtaining CEP Cu-C, the concentration of carbon nanomaterial in the electrolyte changed from 1.0 to 25.0 g/L. The test of the obtained samples showed that the best tribological properties were found in CEP Cu-C, where the carbon nanomaterial concentration was 15 g/L. Table 6.1 shows the test results.

TABLE 6.1 Tribological Properties of Copper Coatings at a Dispersed Phase Concentration of 15.0 g/L

Name of coating material	Speed of friction, V m/s	Friction temperature, T°C	Coefficient of sliding friction, f	Wear, mg/h
Copper	0.125	55	1.0	102
Copper + carbon nanomaterial (substrate – steel 45)	0.125	32	0.32	8
Copper + carbon nanomaterial (substrate – stainless steel)	0.125	28	0.29	6

As can be seen from the Table 6.1, in the case of CEP Cu-C, the values of the coefficient of friction (f) decreases by a factor of three in comparison with the copper deposit, and the wear by 12 times. This is due to the fact that carbon nanomaterial at electrodeposition is included and redistributed in the structure of copper coatings and performs the function of dry solid lubrication during friction.

6.4   CONCLUSIONS

Thus, in the production of CEP Cu-C, using a carbon nanomaterial (obtained in our laboratory from used tires of machines) with a concentration of 15 g/L, the coefficient of friction (f) decreases three times compared to copper deposit, and the wear by 12 times.

KEYWORDS

- carbon nanomaterial
- composite electrochemical coatings
- secondary raw materials
- tribological properties

REFERENCES

1. Mingazova, G. G., Fomina, R. E., Vodopianova, S. V., & Saifulin, R. S., (2012). *Kazan, Bulletin KTTU, 20*(2), 81–84.
2. Saifulin, R. S., (1983). *Inorganic Composition Materials* (p. 304). Moscow. "Khimia."
3. Tarasevich, M. P., (1984). *Electrochemistry of Carbonic-Materials* (p. 253). Moscow. "Nauka."
4. Saifulin, R. S., (1977). *Composition Coatings and Materials* (p. 227). Moscow. "Khimia."

CHAPTER 7

Impregnation of Iron and Magnetite Phases in Wood and Partial Pyrolized Wood

K. SARAJISHVILI[1], N. JALABADZE[2], L. NADARAIA[2], G. KVARTSKHAVA[2], T. KORKIA[1], N. NONIKASHVILI[1], V. GABUNIA[1], and R. CHEDIA[1]

[1]Iv. Javakhishvili Tbilisi State University, Petre Melikishvili Institute of Physical and Organic Chemistry, 31 Politkovskaya St., 0186, Tbilisi, Georgia, E-mail: chediageo@yahoo.com

[2]Georgian Technical University, Republic Center for Structure Researches, 77 Kostava St., 0186, Tbilisi, Georgia

ABSTRACT

Nanoiron-containing adsorbents have been widely used for cleaning of waters contaminated with organic and inorganic pollutants. Nanoiron can be spread over an organic and inorganic liner. Using renewable bioresources (timber, agricultural wastes, waste of biotechnological processes, etc.) to obtain sorbents containing nanoiron and iron oxide is quite promising. Iron-containing sorbents were obtained using *Alnus incana* wood. To avoid the emission of organic components from the wood into water, their partial pyrolysis was carried out with the formation of 1–3 mm coal layer. 13–17% of iron was impregnated in pyrolyzed and non-pyrolyzed samples. Impregnation was performed by 0.2 M $FeCL_3 \cdot 6H_2O$ and 0.2 M $Fe(NO)_3 \cdot 9H_2O$ solutions. Reduction of impregnated in the wood Fe^{+3} ions to nanoiron was carried out by $NaBH_4$ in an inert atmosphere. By this method, polyfunctional sorbents like Fe/wood and Fe^0/C/wood were obtained. Fe_3O_4/C/Wood type sorbents were obtained by pyrolysis of $Fe(NO_3)_3$–wood complexes. The nanoiron was spread on wood linen by passing steam of iron (0) pentacarbonyl at 150–200°. Wood–$Fe(CO)_5$ system fibers composed from spatial grains are formed in an autoclave during heating. Magnetite partially impregnated into the wood. The obtained sorbents easily cause degradation of halogenated organic pollutants (1,4-dichlorbenzol and 4-bromanaline) and completely remove Cu^{+2} ion from model solutions.

7.1 INTRODUCTION

Various organic and inorganic substances dramatically increase pollution of soil, water, and air due to anthropogenic activity. One of the biggest problems of the modern world is water pollution by industrial, urban, and agricultural waste flows. The most of enterprises operating in Georgia do not remove organic and inorganic pollutants from wastewaters. Application of methods used in foreign countries is limited due to their high cost. However, these problems can be partially solved through the preparation of mono- and multifunctional sorbents in Georgia. The existing environmental situation, especially in terms of water pollution requires the installation of effective purification systems. Obtaining effective multifunctional sorbents is foreseen based on modern civilization experience. Application of unique properties of nanozerovalent iron (nZVI) is considered as one of the upcoming trends in technology for cleaning of wastewaters. It has been established that the nZVI in water undergoes some transformations by the following scheme:

$$2Fe^0_{(s)}+4H^+_{(aq)}+O_{2(aq)}\rightarrow 2Fe^{2+}_{(aq)}+2H_2O_{(l)}\ Fe^0_{(s)}+2H_2O_{(l)}\rightarrow 2Fe^{2+}_{(aq)}+2H_{2(g)}+2OH^-$$
$$4Fe^{2+}_{(aq)}+4H^+_{(aq)}+O_{2(aq)}\rightarrow 4Fe^{3+}_{(aq)}+2H_2O_{(l)}\ 4Fe^{2+}_{(aq)}+2H_2O_{(l)}+O_{2(aq)}\rightarrow$$
$$2Fe^{3+}_{(aq)}+2H_{2(g)}+2OH^-_{(aq)}$$

where nZVI acts as a reducing agent because of its standard reduction potential (E^o_h = 0.44 V). Standard reduction potential for dehalogenation half-reaction of various alkyl halides ranges from +0.5 to +1.25 V at pH = 7, so the reaction goes to the following direction:

$$Fe^0 + RX + H^+\rightarrow Fe^{2+} + RH + X^+$$

As a result, chloro-organic compounds are converted into hydrocarbons, which are easily degraded by biochemical processes or removed from water by physical procedures. Iron nanoparticles with the surface of 500–2000 m^2/g represent reducing agents for heavy metal ions and easily participate in redox reactions. For example, reduction of $Cr_2O^{2-}_{7-}$ or CrO^{2-}_4 is implemented by the following scheme [1–8]:

$$Cr^{6+} + Fe^0 \rightarrow Cr^{3+} + Fe^{3+}(1-x)Fe^{3+} + xCr^{3+}$$
$$+ 3H_2O \rightarrow CrxFe_{(1}-x)(OH)_{3\downarrow} + 3H^+$$

Part of these issues is a well-known worldwide practice, but to achieve their final practical application in our conditions (and also in our neighborhood regions) it is necessary to conduct comprehensive scientific research for the development of new technological processes.

The content of inorganic and organic pollutants in the environment depends on different types of production and the quality of the water pollution. Therefore, in each case, appropriate sorbents and purification technologies must be adopted. It is possible to remove inorganic and organic pollutants from wastewaters separately or simultaneously through mono- and multi-functional sorbents, which will be hybrid systems containing organic-inorganic or inorganic components. It is well known that biosorbents obtained from renewable bio-resources can remove ions of heavy metals from polluted water [9–21]. Biosorbents obtained by us from wastes of oak (*Quercus*), hornbeam (*Carpinus*), poplar (*Populus italic, populous pyramidalis*), plane tree (*Platanus*) and beech (*Fagus*) wood processing) have been tested for removing of Crions from waters [22–24]. Removal of chromium from waste-waters of the tannery is a topical challenge: chromium sulfate (basic) is used in the leather production technology of Georgian tanneries. Only 60–70% of chromium is used in leather, all the rest appear in the wastewaters. From the local inorganic mineral resources, it is possible to use aluminosilicates (zeolite tuffs, clays, etc.); renewed bioresources often are used as heavy metal adsorbents, but they have less degradation ability for organic pollutants. For assigning of multi-functional properties to natural biosorbents, the necessary components should be added, which will increase their ability to absorb heavy metal ions and degraded organic pollutants. It is necessary to impregnate nZVI or iron-containing compounds (oxides, hydroxides) into biosorbents and as a result, inorganic-organic sorbents will be created. Mild chemical modifica-tion of biosorbents (including partial pyrolysis of wood) is also necessary with the aim of making lining inert towards active component–nanoiron and to avoid its deactivation. This article applies to the development of methods for impregnation of ultradispersed powders of iron and magnetite into the wood and partially pyrolyzed wood.

7.2 EXPERIMENTAL

$FeCl_3 6H_2O$, $Fe(NO_3)_3 9H_2O$, $NaBH_4$, $Fe(CO)_5$ purchased from Sigma Aldrich were used. The microstructure of the samples was studied by optical and scanning electron microscopes (Nikon ECLIPSE LV 150, LEITZ WETZLAR and Jeol JSM–6510 LV-SEM). Samples X-ray diffraction (XRD) patterns were obtained with a DRON–3M diffractometer (Cu-Kα, Ni filter, 2°/min). Fe content has been established by ISO–11047–1998 standard with atomic-absorption spectrophotometer AA 350. Water sampling and pre-processing was carried out in accordance with the ISO 11466 standard.

1. **Partial pyrolysis of wood**: Partial pyrolysis of wood was carried out by the flame and in a high-temperature Kejia tube furnace at 300–1000°C. Duration of flame pyrolysis was 5–30 sec. Pyrolysis in the inert area was carried out in a previously heated furnace for 10–30 min. A volume of sample was 1–3 cm^3 and charcoal deep-ness–0.5–3.4 mm.

2. **Impregnation of Iron (III) compounds in the wood:** Impregnation of Iron (III) compounds in the wood, and charcoaled wood samples were carried out by using 0.2 M solution of $FeCl_3 6H_2O$ and $Fe(NO_3)_3 9H_2O$ during 48 h. After filtration, samples were dried at room temperature. Crystallization of salts on the surface of the sample was not noticed.

3. **Reduction of impregnated iron compounds using sodium boro-hydride:** Reduction was carried out in desiccator like a reactor, with three-necked removable lid. Samples were placed in a glass reactor with a 50% Ethanol-water solution. Samples were sinking in solution by using a hollow plastic plate. The reactor was cooled with ice water, and 0.5 M $NaBH_4$ were added. Mole ratio $Fe^{+3}:NaBH_4 = 1:15$. Reduction time 8–10 h. Samples were separated from the solution, washed with ethanol and dried in vacuum at 50°C 8 hours. Ultra-dispersed iron powder separated from samples was washed with ethanol and dried in vacuum for 2 h.

4. **$Fe(NO_3) 9H_2O$-wood, and $Fe(NO_3) 9H_2O$-partially pyrolyzed wood samples reduction in hydrogen flow**: Samples reduced in hydrogen flow were carried out into a high-temperature vacuum furnace (Kejia Furnace). Samples were placed in the furnace and 5 min after argon, and then argon/hydrogen mixture (50:50) were inputted. The temperature was increased up to 250–450°C with a heating rate of 10°C min^{-1}. Reduction of samples continued for 30 min at the highest temperature, and rapid cooling was achieved by moving samples in a cool part of the furnace.

5. **Impregnation of nanoiron in supports by using iron (0) pentacarbonyl:** Impregnation of supports with iron was carried

out at low temperature (150–200°C) by passing a vapor of $Fe(CO)_5$. 5 ml of $Fe(CO)_5$ was placed in the gas bubbler and heated to 50–55°C in the area of argon. Carbonyl vapor flows through a quartz pipe where samples of supports are placed. Impregnation of supports with nanoiron continues for 60 min. During this process, partial pyrolysis of supports and decomposition of iron (0) pentacarbonyl take place.

6. **Pressure treatment of $Fe(CO)_5$:** Wood samples were placed into flask and 5 ml of iron (0) pentacarbonyl was added. The mixture was stirred on a magnetic stirrer for 12 h at room temperature in the area of argon. Samples without drying were transferred to 0.5 L high-pressure reactor (autoclave, the inner surface is covered with Teflon) in the area of argon. Operative conditions were: pressure 5 atm, temperature 200°C, time 2 h. Obtained Fe_3O_4 was deposited on wood and reactor walls.

7.3 RESULTS AND DISCUSSIONS

As a research object was used samples of *Alnus incana* wood (Georgia, Gonio). After drying at 105°C, samples were treated with 0.2 M $FeCl_3 \cdot 6H_2O$ and $Fe(NO_3)_3 \cdot 9H_2O$ at ambient temperature during 48 h. Impregnated samples were dried at room temperature. The deepness of vertically impregnated with salts wood capillaries was 3–5 mm, while horizontal migration of iron ions was just 0.5–1 mm. Results are visible on a slice of wood sample. Limit of impregnation is more contrast when the surface of the wood slice is treated with a dilute solution of KSCN or NH_4SCN. SNC^- ions are used for qualitative analysis of Fe^{+3} ions. Reduction of wood samples size or mechanical destruction of structure leads to complete impregnation. XRD and SEM analysis show that wood retains structure, but iron content varies in a large range. When Iron (III) nitrate was used, the content of iron was 6.5–13.6% m/m, in average 9.47% (Figures 7.1–7.4).

It should be noted that the atomic ratio of N:Fe and Cl:Fe is more than 1, quite different from stoichiometry ratio in salts (3:1). We can conclude, that impregnation causes interaction between salts and supports. EDX analysis shows that iron content and ratio (Fe:N and Fe:Cl) is different in different part of the wood. XRD analysis shows—that there are no separate phases of salts—products of hydrolysis or products obtained by the interaction of functional groups (hydroxyl, phenolic OH, carboxyl, amide) presented in wood. The same results were obtained in case of iron (III) chloride. The content of iron varies 13.8–22.2%, average – 17.44%, which is obviously more than in case of impregnation by using iron (III) nitrate.

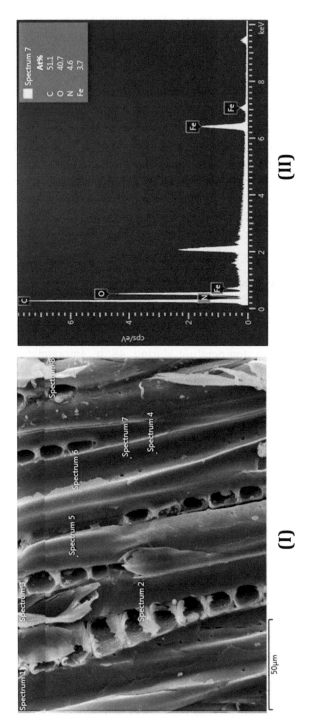

FIGURE 7.1   SEM micrograph (I) and EDX spectrum (II) of $Fe(NO_3)_3$-Wood systems.

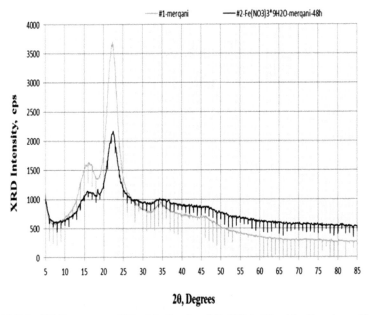

FIGURE 7.2   XRD patterns of Wood (yellow) and Fe(NO$_3$)$_3$-Wood (red) systems (I).

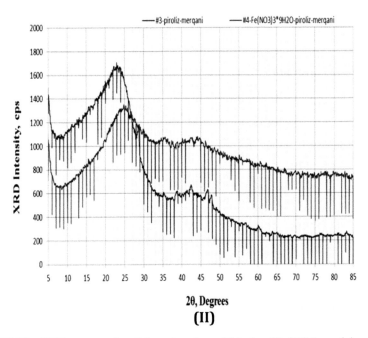

FIGURE 7.3   XRD patterns of partial pyrolyzed wood (blue) and Fe(NO$_3$)$_3$-partial pyrolyzed wood systems (450°C, 30 min.),(II, red).

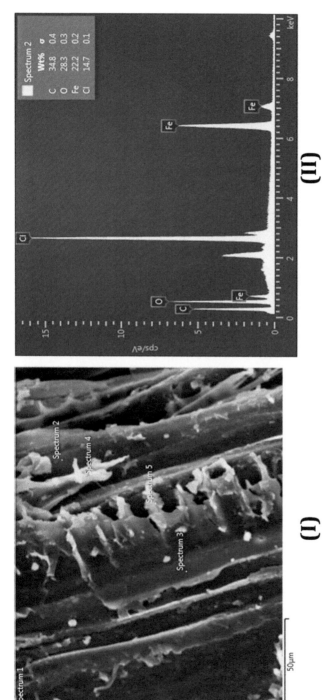

FIGURE 7.4 SEM micrograph (I) and EDX spectrum (II) of FeCl$_3$-wood systems.

Certain changes in structure and chemical composition take place during partial flame pyrolysis or pyrolysis in the inert area. At the same time, the morphology of the charcoal layer is similar to initial wood (Figure 7.5). During rapid flame pyrolysis (10–30 sec) the only surface forms charcoal, the lower surface remains unchanged, the formation of charcoal on the lateral surface is not significant. Fe^{+3} ions migrate in sample volume from any side during impregnation. Reduction of these ions leads to obtaining nanoiron impregnated sorbents. One part of this surface is Fe^0/wood, but another part –Fe^0/C. In common, Fe^0/wood surface adsorbs heavy metals, Fe^0/C adsorbs heavy metals and organic pollutants, but they easily degraded by nanoiron. Thus, polyfunctional sorbent for inorganic (radionuclides) and organic pollutants are obtained, by using this method.

In inert area, partial pyrolysis occurs on the surface, and charcoal layers are obtained. Its impregnation with Fe^{+3} and further reduction with $NaBH_4$ gives adsorbents with surfaces Fe^0/C.

Thus, partial pyrolysis makes it possible to receive polyfunctional sorbents. It was established that rapidly pyrolyzed wood sample (700°C, 30 sec) adsorbs 12.26% (mass) of iron from $FeCl_3$ solution. These results are less than for unpyrolyzed samples (17.44%). It may be explained by the hydrophobicity of the charcoal layer. Pyrolysis of wood-$Fe(NO_3)_3$ samples gives sorbents impregnated with iron oxides. Such sorbents are perspective for removing different pollutants from wastewaters.

Iron (0) pentacarbonyl was used for low-temperature nanoiron impregnation of supports. Ultrasonic decomposition of iron (0) pentacarbonyl is one of the easiest methods for deposition of iron on the supports [25]. Iron was deposited on the surface of wood samples by passing iron (0) pentacarbonyl through a quartz pipe (Figure 7.6). 6.3% of iron was deposited when pentacarbonyl/Ar mixture was used (200°C). Composite Fe^0/wood particles are paramagnetic and were easily moved through the water by the effect of the magnetic field. The same process was used when cotton fiber was impregnated with iron (0.3%).

Iron pentacarbonyl was easily decomposed, and iron, iron oxides, and carbides were formed. In a closed system in the presence of wood samples iron (0) pentacarbonyl gives iron (II, III) oxide (magnetite), because it undergoes oxidation by organic compounds and water released from the wood (Figure 7.6). Partial pyrolysis of organic components of the wood takes place, so magnetite powder contains carbon (7% m/m). Magnetite is formed in the whole volume of the autoclave. Just 8% iron is impregnated in wood. Most part of Fe_3O_4 has fiber structure and contains spherical grains (2–4 μm); in turn, it contains nanosize (50–200 nm) primary crystallites (Figures 7.7 and 7.8).

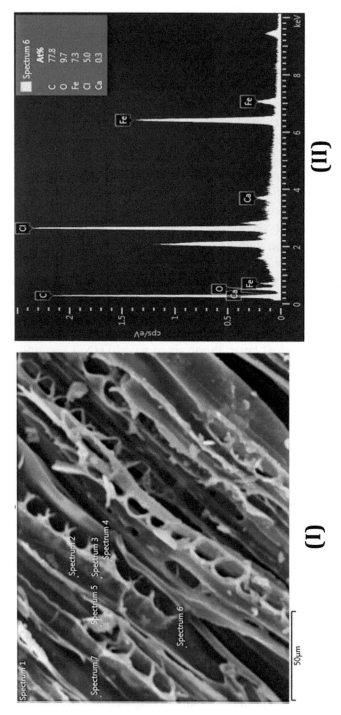

FIGURE 7.5   SEM micrograph (I) and EDX spectrum (II) of FeCl$_3$- partial pyrolyzed wood (450°C, 30 min) systems.

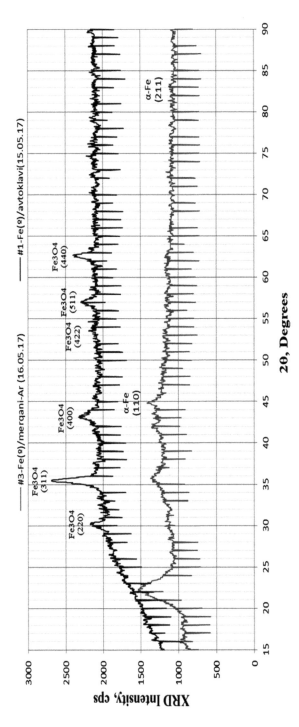

FIGURE 7.6 XRD patterns of Fe_3O_4 obtained from $Fe(CO)_5$ in an autoclave (200°C, 2 hr, red) and deposited ultra-dispersive iron (green) on wood with $Fe(CO)_5$ vapor flows into the quartz pipe.

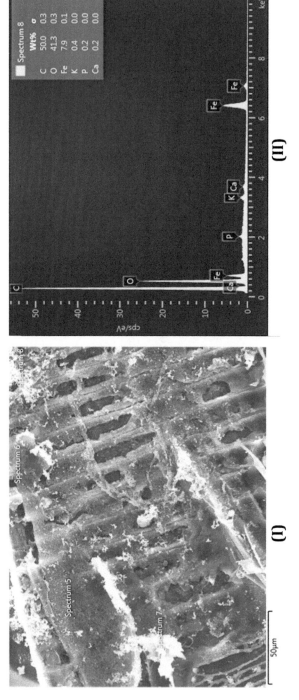

FIGURE 7.7 SEM micrograph (I) and EDX spectrum (II) of Fe_3O_4-wood systems (autoclave, 200°C, 2 hr).

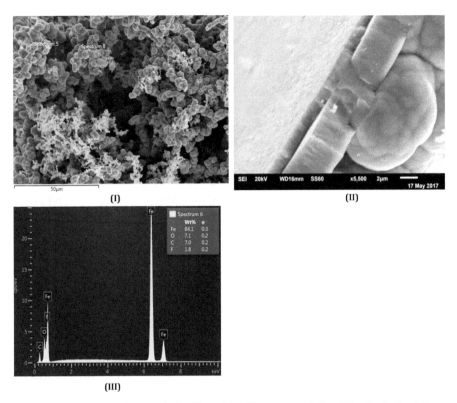

FIGURE 7.8 SEM micrograph (I, II) and EDX spectrum (III) of Fe_3O_4 obtained from $Fe(CO)_5$ in an autoclave (200°C, 2 hr).

The density of magnetite powders is 22–26 mg/ml. Iron content in powders attained to 81.55% m/m, but the content of iron in Fe_3O_4 is less (73%). XRD analysis shows that the formation of crystalline carbides from carbonyl is not noticeable. It is possible that magnetite powder contains an amorphous iron or other iron-containing phases. The obtained sorbents easily cause degradation of halogenated organic pollutants (1,4 dichlorbenzol and 4-bromanaline) and completely remove Cu^{+2} ion from model solutions.

7.4 CONCLUSIONS

Iron (III) nitrate and chloride are impregnated into the wood, and it's partially pyrolyzed samples. XRD and SEM analysis show that wood retains its structure, but iron content varies in wide ranges. When iron (III) nitrate has used the content of Fe was 6.5–13.6% (average 9.47%). In the case of iron (III) chloride the content of Fe varies from 13.8 to 22.2% (average 17.44%). It was established that rapidly pyrolyzed wood sample (700°C, 30 sec) adsorbs 12.26% % of Fe^{+3} from $FeCl_3$ solution. These results are lower than for unpyrolyzed samples. This can be explained by the hydrophobicity of the charcoal layer. Pyrolysis of wood–$Fe(NO_3)_3$ samples gives sorbents impregnated with iron oxides. Iron was deposited on the surface of wood samples by passing iron (0) pentacarbonyl vapor through a quartz pipe. 6.3% of nanoiron was deposited when pentacarbonyl/Ar mixture was used (200°C). In the closed system and in the presence of wood samples $Fe(CO)_5$ gives iron (II, III) oxide (magnetite). Most of Fe_3O_4 has fiber structure and contains spherical grains (2–4 µm); moreover, it contains nano-size (50–200 nm) primary crystallites.

KEYWORDS

- **biosorbents**
- **impregnation**
- **iron**
- **iron oxides**
- **multifunctional sorbents**
- **organic and inorganic pollutants**
- **partial pyrolysis**
- **wood**

REFERENCES

1. Yang, S. C., Lei, M., Chen, T. B., Li, X. Y., Lang, Q., & Ma, C., (2010). Application of zerovalent iron (Fe(O)) to enhance degradation of HCHs and DDX in soil from a former organochlorine pesticides manufacturing plant. *Chemosphere, 79*(7), 727–732.
2. Muller, N. C., Braun, J., Bruns, J., Cernik, M., Rissing, P., Rickerby, D., & Nowack, B., (2012). Application of nanoscale zerovalent Iron (NZVI) for groundwater remediation in Europe, *Environ. Sci. Pollut. Res., 19*, 550–558.
3. Machado, S., Pinto, S. L., Grosso, J. P., Nouws, H. P., & Delerue-Matos, C., (2013). Green production of zerovalent iron nanoparticles using tree leaf extracts. *Sci. Total Environ., 1–8*, 445–446.

4. Cao, M., Wang, L., Chen, J., & Lu, X., (2013). Remediation of DDTs contaminated soil in a novel Fenton-like System with zerovalent iron, *Chemosphere, 90*(8), 2303.

5. Chen, J. L., Souhail, R., Ryan, J. A., & Li, Z., (2001). Effects of pH on dechlorination of trichloroethylene by zerovalent iron. *Journal of Hazardous Materials, B83*, 243–254.

6. Cirwertny, D. M., Bransford, S. L., & Roberts, A. L., (2007). Influence of the oxidizing species on the reactivity of iron-based bimetallic reductants. *Environ. Sci. Technol., 41*, 3734–3740.

7. Suwannee, J., (2005). Use of zerovalent iron for wastewater treatment. *KMITL Sci. Tech. J., 5*, 587–595.

8. He, F., & Zhao, D., (2008). Manipulating the size and dispersibility of zerovalent iron nanoparticles by use of carboxymethyl cellulose stabilizers. *Environ. Sci. Thechnol., 1*, 3479.

9. Bhatnagar. A., & Silanpaa, M., (2010). Utilization of agro-industrial and municipal waste materials as potential adsorbents for water treatment. *Chemical Engineering Journal, 157*(2 & 3), 277–296.

10. Memon, R., Memon, S. Q., Bhanger, M. I., & Khuhawar, M. Y., (2008). Banana peel: A green and economical sorbent for Cr (III) removal. *Pak. J. Anal. Environ. Chem., 9*(1), 20–25.

11. Botelho, C. M. S., & Boaventura, R. A. R., (2007). Chromium and zinc uptake by algae *gelidium* and agar extraction algal waste: Kinetics and equilibrium. *J. Hazard. Mat., 149*(3), 643–649.

12. Sarin, V., & Pant, K. K., (2006). Removal of chromium from industrial waste by using eucalyptus bark. *Bioresource Technology, 97*, 15–20.

13. Dhankhar, R., & Hooda, A., (2011). Fungal biosorption--an alternative to meet the challenges of heavy metal pollution in aqueous solutions. *Environ. Technol., 32*(5 & 6), 467–491.

14. Ashraf, M. A., Wajid, A., Mahmood, K., Maah, M. J., & Yusoff, I., (2011). Low-cost biosorbent banana peel (*Musa sapientum*) for the removal of heavy metals, *Scientific Research, and Essays, 6*(19), 4055–4064.

15. Qaiser, S., Saleemi, A. R., & Umar, M., (2009). Biosorption of lead (II) and chromium (VI) on groundnut hull: Equilibrium, kinetics, and thermodynamics study. *Electronic Journal of Biotechnology, 12*(4). 15–20.

16. Wang, J., & Chen, C., (2009). Biosorbents for heavy metals removal and their future, *Biotechnol. Adv., 27*(2), 195–226.

17. Apiratikul, R., & Pavasant, P., (2008). Batch and column studies of biosorption of heavy metals by *Caulerpalentillifera*. *Bioresource Technology, 99*(8), 2766–2777.

18. Sangi, M. R., Shahmoradi, A., Zolgharnein, J. A., Gholam, H., & Ghorbandoost, M., (2008). Removal and recovery of heavy metals from aqueous solution using *Ulmuscarpinifoliaand Fraxinus excelsior* tree leaves, *J. Hazard. Mat., 155*, 513–522.

19. Seyed, N. A., Abasalt, H. C., & Seyede, M. H., (2012). Removal of Cd (II) from the aquatic system using *oscillatoria sp.* biosorbent. *The Scientific World Journal, ID 347053*, p. 7.

20. Kvartskhava, G., Goletiani, A., Sarajishvili, Q., Korkia, T., Jinikashvili, I., & Chedia, R., (2013). Removal of heavy metal ions wastewaters by using renewable bio sources of Georgia. *Third Intern Caucasian Symposium on Polymers & Advanced Materials* (p. 95). Tbilisi, Georgia.

21. Jalagonia, N. T., Korkia, T. V., Sarajishvili, K. G., Kvartskhava, G. R., & Chedia, R. V., (2014). Immobilization of zero-valent iron nanoparticles in biopolymers. *Nanotech.,* (p. 84), Italy, Venice, Italy.
22. Jalagonia, N. T., Kuchukhidze, T. V., Kvartskhava, G. R., Sanaia, E. E., & Chedia, R. V., (2016). Impregnation of zerovalent iron in biopolymers for remediation of wastewater. Advanced materials and multifunctional materials, *Physics, 9,* 109–120.
23. Peters, D., (1996). Ultrasound in materials chemistry. *J. Mater. Chem., 6*(10), 1605–1618.

CHAPTER 8

Effect of the Material Manufacturer Factor on the Deformation States of Bistable Composite Plates

M. GHAMAMI, H. NAHVI, and S. SABERI

Department of Mechanical Engineering, Isfahan University of Technology, Isfahan, Iran, E-mail: mghamami@yahoo.com

ABSTRACT

Nowadays, composites have many applications in various industries. A novel type of composites, which are called bistable morphing structures, is generally used in cases where the structure is undergoing frequent deformities. The practical application of these structures requires their analysis and testing in different conditions. Bitable composite plates have two stable states and can be transformed under continuous force between these two stable states. In this chapter, firstly, a summary of the types of commonly used fibers and resins in the composite industry is introduced, and then is given some details on the structures of morphing and bistable composite plates. In the following are investigated the effect of materials and the layers arrangement on the deformed states of the bistable composite plates. For this purpose, bistable composite plates with different layers and bistable hybrid composite plates are studied, and their equilibrium and stable states are assessed. The Rayleigh-Ritz method and the principle of potential energy minimization are analyzed to determine the equations governing the structure. After determining the governing equations of structure and solving the nonlinear equations system, the equilibrium and stable states of the structure are obtained. The curvature value will be calculated in each stable state for a bistable composite plate with the above-mentioned layers arrangements. In this research, the results are validated using finite element method and ABAQUS software which shows that the results have acceptable accuracy.

8.1 INTRODUCTION

In engineering applications, it is often inevitable to integrate materials properties. In the pioneer industries, there is no neat material that provides all the requirements. For example, in the aerial industry, materials are required that have high strength and abrasion resistance, lightweight, etc. Since a neat material can't be found to have all the desired properties, it should be preferred using composite materials. The use of these materials throughout history was also common, the most important of which is the use of adobe made of clay and it's strengthening with straw by Egyptians, a prototype of composites. The use of composites due to many advantages such as high strength, hardness, corrosion resistance, abrasion resistance, thermal resistance, impact load resistance, sound insulation, and lightweight are most important factors of industrial progress in three past decades. Examples of composite materials applications in various industries are shown in Figure 8.1.

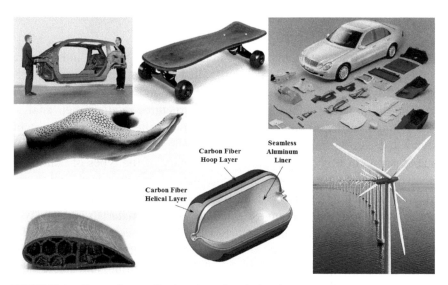

FIGURE 8.1   Composites applications in various industries.

Composite material refers to a substance that consists of two or more components that are mixed in the macroscopic level and are not solved in together. The composite properties are significantly different from the properties of the constituent materials, as composite properties are usually better than constituent properties. The constituents of each composite materials are included:

Matrix: This constituent acts as maintaining reinforcing phase and load transfer environment between the reinforcing materials.

Reinforcement: This constituent is functioned for strengthening the material and improving its properties, and is distributed in the matrix phase. The reinforcement have fibers, particles, and strands states. Examples of a composite material and structures are shown in Figure 8.2.

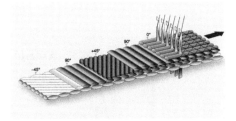

FIGURE 8.2   Examples of a composite material and structures.

In the third decade of the twentieth century, glass fibers were used to reinforce resins. The manufactured material from this composite is called fiberglass. Since 1970, the advent of new fibers such as carbon, boron, and aramid fibers and the application of metal and ceramic matrix, the use of composites has made significant developments.

Composite material due to advantages such as: high strength to weight ratio, high corrosion, abrasion, and fatigue resistance and appropriate thermal insulation, etc., have wide applications [1].

8.2   TYPES OF COMPOSITES

The components are divided into four general components in terms of component geometry:

- particles composites;
- flake composites (piecewise);
- fiber composites; and
- nanocomposites.

The most common form of the composite in the industry is the fiber composite materials with the polymer matrix and various fibers. The used fibers in this category of materials can be divided into four general categories:

- mineral fibers (asbestos, glass, etc.);
- organic fibers (aramid, carbon, polyester, etc.);
- natural fibers (cotton, sisal, flax, jute); and
- metal fibers.

The used fibers in the composite industry include glass, carbon, Kevlar fibers and organic fibers, such as polyethylene. Figure 8.3 indicates the strain-strain graph for different fibers.

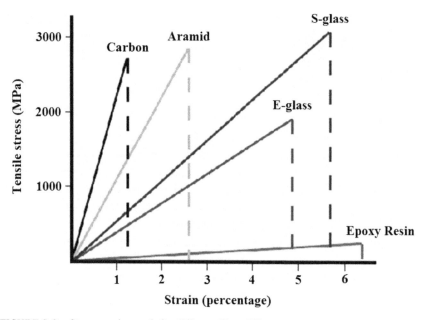

FIGURE 8.3   Stress-strain graph for different fibers [1].

The polymer matrix is the most commonly used matrix in composites. These matrices are used more in the form of various resins in the composite industry. Although different types of resins are available, the most commonly used can be summarized as follows in the construction of composite structures:

- polyester;
- vinylester; and
- epoxy.

Figure 8.4 exhibits the stress-strain behavior of the three listed resins.

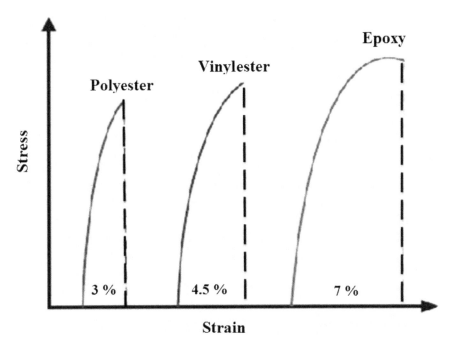

FIGURE 8.4 Stress-strain graph of resins (polyester, vinylester, and epoxy).

8.3 DEPLOYABLE STRUCTURES

Deployable or deformable structures are ones that geometry and their properties can vary according to the environmental loading conditions. Such structures store the elastic strain energy during deformation, which means that these structures into the unconstrained state can be transformed in the free or non-strain state. Since these freely and unconstrained deformations can cause structural failure; hence, the exploitation of such deformations is not common. Therefore, the process of structural deformation is usually controlled using a variety of methods or specific materials.

One of the most widely used deployable structures can be found corrugated structures and multistable structures. Multistable structures have various capabilities, and their applications are developing. Examples of these structures are shown in Figure 8.5.

FIGURE 8.5   A corrugated structure and a multistable structure.
Source: Reprinted with permission from Ref. [2]. © 2013 Elsevier.

8.4   MULTISTABLE STRUCTURES

Multistable structures have several stable states, or more precisely, capable structures with deformation between several stable states. The most important advantage of these structures is that they do not require external energy sources to maintain in a stable state. The simplest multi-stable structures are bistable structures (two stable states). These structures have two non-strained and deformed states that are completely stable. In Figure 8.6, a few numbers of these structures are shown.

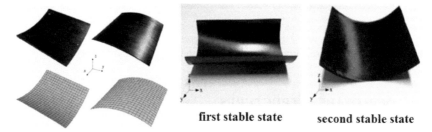

first stable state   **second stable state**

FIGURE 8.6   Different states of a bistable composite plate [4].

Figure 8.6 shows a bistable composite plate with a stacking sequence $[0_2/90_2]$ in two stable states. Other applications of such panels and materials include multi-stable structures, vibrational isolators, energy harvesting, reflectors, receivers, and solar panels and biomechanics.

8.5   LITERATURE REVIEW

In 1981, Hyer [4] performed several thermal tests on composite plates with an asymmetric layer [0/90]. He observed that by applying thermal loads, two cylindrical stable states are observed completely, which is deformed differently from classical linear theory predictions. In addition, the plate snaps through from a stable state to another stable state by apply an external force. In Figure 8.7, the different states of a bistable composite plate are shown.

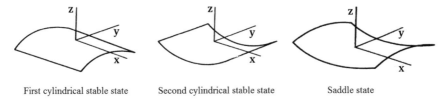

First cylindrical stable state   Second cylindrical stable state   Saddle state

FIGURE 8.7 Different states of a bistable composite plate [4].

In 1981 and 1982 Hyer, including nonlinear strain into classical linear theory, using the principle of minimizing total potential energy and the Rayleigh-Ritz method, by guessing an appropriate displacement field, predicted the deformed shape of a non-symmetric rectangular composite plate with stacking sequence $[0_n/90_n]$ under thermal loads. His method is well-known by the developed classical theory of laminates. Hamamoto [5], like Hyer proposed strain energy based models for bistable plates, assuming flat form surface in the initial state. Galletly et al. [6] abandoned the assumption of constant curvature and used the assumption of the initial Gaussian curvature. They showed that the second equilibrium point, if available, would be non-saddle for split pipes with a symmetric stacking sequence of layers. For layers with isotropic properties, the second equilibrium point is non-saddle and unstable.

Guest and Pellegrino [7], using the assumption of initial Gaussian curvature, firstly considered the bending and membrane components of the total strain energy separately, and then examined the equilibrium and stability of the cylindrical composite shells with the assumption of constant curvature. The advantages of this method are that the bending and membrane components are evaluated separately, which reduces the computations volume. It was also shown that composite shells with isotropic properties are not bistable, while with anisotropic properties, would be bistable.

The Seffen [8] plate using the model provided by Gust and Pellegrino [7] solved membrane problems by combining adaptation conditions and equilibrium equations inside the plane, assuming the uniform curvature for the elliptic plate. It also showed that composite shells with orthogonal properties could be bistable and the range of bistable form decreases with increasing initial twisting. Tawfik et al. [9] studied the effects of the aspect ratio and the length to thickness ratio on the stable states of the rectangular composite plate with non-symmetric using ABAQUS software.

Vidoli and Maurini [10] analyzed shells with tristable states and presented that the geometric parameters and material properties affect the tristable behavior of the orthogonal composite thin shells with uniform curvature. In their study, the range of tristable has highest it values with relation $\beta = v^2$ between Young modulus (β) and Poisson coefficient (v). Diaconu et al. [11] addressed the deformation process between the stable states of a bistable composite plate under concentrated dynamic forces by Hamilton's principle. They also considered the effects of inertia and damping on their modeling. Vogel and Hyer [12] investigated the linear vibration of bistable composite plates using the Hamilton principle and Rayleigh-Ritz methods. Pirrera et al. [13] estimated the transverse displacement of shells using high order polynomials and combining Ritz method and the tracking algorithm of the bistable cylindrical shells, compared the results of MATLAB and ABAQUS softwares. Vidoli [14] solved the equilibrium equations by the Ritz method using the provided model by Guest and Pellegrino for rectangular plates in constant, linear, and second order curvatures.

Eckstein et al. [15] developed the Hyer's research and examined the morphing states of cylindrical shells by applying thermal load conditions for materials under temperature-dependent properties and compared the results with the ABAQUS software. Bowen et al. [16] experimentally investigated the deformation process in bistable composites using various mechanisms such as piezoelectric ceramics, shape memory alloys and carbon nanotubes. Cantera et al. [17] studied the snap through in a square plate under the influence of concentrated force and pinned-fixed point support boundary conditions experimentally and analytically.

In the field of bistable composite plates with initial curvature, few investigations have done as following. Ryu et al. [18] added the initial curvatures in one and two directions by curing a bistable composite plate on a curved surface. They provided both the experimental and new analytical model to determine the stable states of the plate. Lee et al. [19] examined the effect of the initial curvature on the required force to snap through between the stable states of a bistable composite plate.

In the field of hybrid bistable composite plates, for the first time, Daynes, and weaver [20] designed and analyzed hybrid bistable composite plates consisted of the intermediate metal layer. Dai et al. [21] designed a hybrid bistable composite plate consisted of the outer isotropic layer, and analyzed experimentally and analytically its different behaviors. Li et al. [22] developed a new form of hybrid bistable composite plates with symmetrical stacking sequence, and presented a new analytical model for its behavior. Pan et al. [23] investigated the energy harvesting from the symmetric hybrid bistable composite plates designed by Lee [22] experimentally and analytically. In 2000, Iqbal et al., using asymmetric composite plates, designed a band form deployable structure with the capability of compacting. They determined the stable state of this structure using the experimental and analytical methods [24]. Dai et al., in 2012, fabricated a tristable structure consisting of four bistable composite rectangular plates. This structure has three different stable modes. Using ABAQUS software and finite element method, they examined its various behaviors.

8.6   DEVELOPED CLASSICAL LAMINATION THEORY

As mentioned in the first section, as a composite plate with asymmetric stacking sequence is subjected to a temperature variation, due to the differences in mechanical properties and the coefficient of thermal expansion of layers, in addition to the unstable saddle form which is not visible in practice, two stable cylindrical forms occur. Classical lamination theory is not intended to predict cylindrical states due to ignoring the non-linear Green strain terms. In order to eliminate this issue, all the terms of the Green strain are considered first.

Here, according to the assumptions of the classical lamination theory, terms of $\varepsilon = \gamma_{xz}$, and γ_{yz} are neglected. Also, all high-order terms, except $\left(\dfrac{\partial w}{\partial x}\right)^2$, $\left(\dfrac{\partial w}{\partial y}\right)^2$, $\dfrac{\partial w}{\partial x}\dfrac{\partial w}{\partial y}$, are small and can be neglected, which eliminating them, the Von-Karman strain equations are obtained as follows:

$$\varepsilon_{xx} = \frac{\partial u}{\partial x} + \frac{1}{2}\left[\left(\frac{\partial u}{\partial x}\right)^2 + \left(\frac{\partial v}{\partial x}\right)^2 + \left(\frac{\partial w}{\partial x}\right)^2\right]$$

$$\varepsilon_{yy} = \frac{\partial v}{\partial y} + \frac{1}{2}\left[\left(\frac{\partial u}{\partial y}\right)^2 + \left(\frac{\partial v}{\partial y}\right)^2 + \left(\frac{\partial w}{\partial y}\right)^2\right]$$

$$\varepsilon_{zz} = \frac{\partial w}{\partial z} + \frac{1}{2}\left[\left(\frac{\partial u}{\partial z}\right)^2 + \left(\frac{\partial v}{\partial z}\right)^2 + \left(\frac{\partial w}{\partial z}\right)^2\right] \tag{1}$$

$$\gamma_{xy} = \frac{\partial u}{\partial y} + \frac{\partial v}{\partial x} + \frac{\partial u}{\partial x}\frac{\partial u}{\partial y} + \frac{\partial v}{\partial x}\frac{\partial v}{\partial y} + \frac{\partial w}{\partial x}\frac{\partial w}{\partial y}$$

$$\gamma_{xz} = \frac{\partial u}{\partial z} + \frac{\partial w}{\partial x} + \frac{\partial u}{\partial x}\frac{\partial u}{\partial z} + \frac{\partial v}{\partial x}\frac{\partial v}{\partial z} + \frac{\partial w}{\partial x}\frac{\partial w}{\partial z}$$

$$\gamma_{yz} = \frac{\partial v}{\partial z} + \frac{\partial w}{\partial y} + \frac{\partial u}{\partial z}\frac{\partial u}{\partial y} + \frac{\partial v}{\partial z}\frac{\partial v}{\partial y} + \frac{\partial w}{\partial z}\frac{\partial w}{\partial y}$$

$$\varepsilon_{xx} = \frac{\partial u}{\partial x} + \frac{1}{2}\left(\frac{\partial w}{\partial x}\right)^2$$

$$\varepsilon_{yy} = \frac{\partial v}{\partial y} + \frac{1}{2}\left(\frac{\partial w}{\partial y}\right)^2 \tag{2}$$

$$\gamma_{xy} = \frac{\partial u}{\partial y} + \frac{\partial v}{\partial x} + \frac{\partial w}{\partial x}\frac{\partial w}{\partial y}$$

By replacing defined displacements of u, v, and w in accordance with the classical lamination theory in Eq. (2), the Von-Karman equations will be obtained in terms of the displacements of the middle plane as Eq. (3):

$$\varepsilon_{xx} = \frac{\partial u_0}{\partial x} + \frac{1}{2}\left(\frac{\partial w_0}{\partial x}\right)^2 - z\frac{\partial^2 w_0}{\partial x^2}$$

$$\varepsilon_{yy} = \frac{\partial v_0}{\partial y} + \frac{1}{2}\left(\frac{\partial w_0}{\partial y}\right)^2 - z\frac{\partial^2 w_0}{\partial y^2} \tag{3}$$

$$\gamma_{xy} = \frac{1}{2}\left(\frac{\partial u_0}{\partial y} + \frac{\partial v_0}{\partial x} + \frac{\partial w_0}{\partial x}\frac{\partial w_0}{\partial y}\right) - z2\frac{\partial^2 w_0}{\partial x \partial y}$$

Equation (3) can be written in a matrix form (4).

$$\{\varepsilon\} = \{\varepsilon^0\} + z\{\kappa^0\} \tag{4}$$

where, ε^0 and κ^0 are strain and curvature terms in the middle plane, respectively, which are defined as Eqs. (5) and (6):

$$\varepsilon^0 = \begin{bmatrix} \varepsilon^0_{xx} \\ \varepsilon^0_{yy} \\ \varepsilon^0_{xy} \end{bmatrix} = \begin{bmatrix} \dfrac{\partial u_0}{\partial x} + \dfrac{1}{2}\left(\dfrac{\partial w_0}{\partial x}\right)^2 \\[2ex] \dfrac{\partial v_0}{\partial y} + \dfrac{1}{2}\left(\dfrac{\partial w_0}{\partial y}\right)^2 \\[2ex] \dfrac{1}{2}\left(\dfrac{\partial u_0}{\partial y} + \dfrac{\partial v_0}{\partial x} + \dfrac{\partial w_0}{\partial x}\dfrac{\partial w_0}{\partial y}\right) \end{bmatrix} \tag{5}$$

$$\kappa^0 = \begin{bmatrix} \kappa^0_{xx} \\ \kappa^0_{yy} \\ \kappa^0_{xy} \end{bmatrix} = \begin{bmatrix} -\dfrac{\partial^2 w_0}{\partial x^2} \\[2ex] -\dfrac{\partial^2 w_0}{\partial y^2} \\[2ex] -2\dfrac{\partial^2 w_0}{\partial x \partial y} \end{bmatrix} \tag{6}$$

In the Eqs. (5) and (6), terms of u_0, v_0 and w_0 are displacements in the middle plane in X, Y, and Z directions, respectively. Hyer's theory is based on the Rayleigh-Ritz method and total potential energy minimization. In this theory, in order to predict the cylindrical states, Von-Karman strain terms were used, unlike the classical lamination theory, strain nonlinearity effects were considered.

The total stored potential energy in the layers of a composite plate is derived from the Eq. (7).

$$\Pi = \int_{\Omega} \int_{-h/2}^{h/2} \frac{1}{2} \{\sigma\}^T \{\varepsilon\} \, dz d\Omega \tag{7}$$

The relation between stress and strain with regard to thermal or piezo-electric factors for composite layers are following equation (8).

$$\{\acute{o}\} = [\bar{Q}] \left(\{\acute{a}\} - \{\acute{a}^r\} \right) \tag{8}$$

By replacing the Eqs. (4) and (8) in Eq. (7), the total potential energy is obtained as Eq. (9).

$$\Pi = \int_{-L_x/2}^{L_x/2} \int_{-L_y/2}^{L_y/2} \int_{H/2}^{H/2} \left(\begin{array}{l} \frac{1}{2}\bar{Q}_{11}\varepsilon^2_{xx} + \bar{Q}_{12}\varepsilon_{xx}\varepsilon_{yy} + \bar{Q}_{16}\varepsilon_{xx}\gamma_{xy} \\[2mm] +\frac{1}{2}\bar{Q}_{22}\varepsilon^2_{yy} + \bar{Q}_{26}\varepsilon_{yy}\gamma_{xy} + \frac{1}{2}\bar{Q}_{66}\gamma^2_{xy} \\[2mm] -(\bar{Q}_{11}\alpha_{xx} + \bar{Q}_{12}\alpha_{yy} + \bar{Q}_{16}\alpha_{xy})\varepsilon_{xx}\Delta T \\[2mm] -(\bar{Q}_{21}\alpha_{xx} + \bar{Q}_{22}\alpha_{yy} + \bar{Q}_{26}\alpha_{xy})\varepsilon_{yy}\Delta T \\[2mm] -(\bar{Q}_{61}\alpha_{xx} + \bar{Q}_{62}\alpha_{yy} + \bar{Q}_{66}\alpha_{xy})\gamma_{xy}\Delta T \end{array} \right) dz dy dx \tag{9}$$

In Eq. (9), L_x, and L_y are the plate dimensions, h is the total thickness of the plate, Q is the decreasing stiffness matrix, α_{kk} ($k = x, y$) is the thermal expansion coefficient, and ΔT is the difference curing temperature and room temperature.

Selecting suitable field displacements can increase the accuracy of the results. Hence, in order to eliminate the defects of the Hyer shape functions, it is proposed for a new shape function for the displacement field. Using the estimated shape function causes a significant increase in precision, especially in the corners and regions close to edges, in comparison with other shape functions. In this study, the proposed shape functions presented in Eq. (10) were used for displacement fields.

$$W_0(x,y) = \sum_{i=1}^{n}\sum_{j=1}^{m} W_{i,j-i}\, x^i y^{i-j} \quad U_0(x,y) = \sum_{i=1}^{n}\sum_{j=1}^{m} u_{i,j-i}\, x^i y^{i-j}$$

$$V_0(x,y) = \sum_{i=1}^{n}\sum_{j=1}^{m} v_{i,j-i}\, x^i y^{i-j} \tag{10}$$

In Eq. (10), U_0, V_0 and W_0 are displacement fields of the middle plane in the X, Y, Z directions, respectively, $u_{i,j-1}$, $v_{i,j-1}$ and $w_{i,j-1}$ unknown coefficient.

By replacing the defined displacement fields in accordance with Eq. (10) in Eqs. (2) and (3) and including results in Eq. (9) and then integrating, the total potential energy is obtained as a function of the coefficients of displacement fields. According to the principle of potential energy minimization for determining the equilibrium states, is sufficient to obtain the total potential energy changes relative to the unknown parameters and follow equations (Eq. 11) equal to zero.

$$\delta\Pi = \frac{\partial\Pi}{\partial w_{ij}}\delta w_{ij} + \frac{\partial\Pi}{\partial u_{ij}}\delta u_{ij} + \frac{\partial\Pi}{\partial v_{ij}}\delta v_{ij} = 0 \tag{11}$$

The resulting equations form a nonlinear equations system that was solved using the Newton-Raphson method by MATLAB software. To determine stable equilibrium states, the Hessian matrix (Eq. 12) of the total potential energy was calculated, and the equilibrium state with positive Hessian matrix was stable.

$$H = \begin{bmatrix} \dfrac{\partial^2\Pi}{\partial w^2_{ij}} & \dfrac{\partial^2\Pi}{\partial w_{ij}\partial u_{ij}} & \dfrac{\partial^2\Pi}{\partial w_{ij}\partial v_{ij}} \\[3mm] \dfrac{\partial^2\Pi}{\partial u_{ij}\partial w_{ij}} & \dfrac{\partial^2\Pi}{\partial u^2_{ij}} & \dfrac{\partial^2\Pi}{\partial u_{ij}\partial v_{ij}} \\[3mm] \dfrac{\partial^2\Pi}{\partial v_{ij}\partial w_{ij}} & \dfrac{\partial^2\Pi}{\partial v_{ij}\partial u_{ij}} & \dfrac{\partial^2\Pi}{\partial v^2_{ij}} \end{bmatrix} \tag{12}$$

Using the introduced theory, ordinary, and hybrids bistable composite plates with a different stacking sequence of layers are investigated, and the effect of their materials on deformed states has been studied.

8.7 FINITE ELEMENT METHOD FOR BISTABLE COMPOSITE PLATE SIMULATION

In order to model, the bistable composite plate in ABAQUS software, a square plate with a dimensions length of 150×150 mm^2 was modeled. In the next step, the layers and properties of the composite material were introduced. In the step section for static or dynamic analysis, general and static options were selected with nonlinear geometric considerations. In the first step, in order to take into account the geometric imperfections caused by the manufacturing processes and residual stresses, a basic defect was applied into the plate. This defect could also be applied by inserting four equal forces into the four corners of the plate at the initial cooling temperatures [20]. Note that this defect was only applied to a square plate. In the second step, the applied force was removed, and the plate was cooled to the ambient temperature. At the end of this step, the plate was transferred to a stable state. In the first step, if the applied forces was inversed, the plate is transferred to another stable state. The meshing step was conducted using S4RS quadrilateral shell element with reduced integration for the plate with the mentioned dimensions, the number of 10,000 elements and 10201 nodes were obtained. The bistable composite plates have properties according to Table 8.1.

TABLE 8.1   Mechanical and Thermal Properties of Graphite/Epoxy Composite [12]

Properties	Value
Longitudinal elastic modulus (GPa)	181
Transverse Elastic Modulus (GPa)	10.3
Shear Module (GPa)	7.17
Poisson ratio	0.2
Longitudinal coefficient of thermal expansion ($10^{6\circ}$C)	−0.106
Transverse coefficient of thermal expansion ($10^{6\circ}$C)	25.6
Thickness of each layer (mm)	0.127
Density (kg/m^3)	1579

Source: Reprinted with permission from Ref. [12]. © 2011 Elsevier.

8.7.1   SQUARE BISTABLE COMPOSITE PLATE [$0_2/90_2$]

Figure 8.2 shows the equilibrium states which have been obtained from analytical and finite element method for the square composite plate at room temperature. In accordance with the figure below, the bistable square composite plate with [0₂/90₂] arrangement, have two cylindrical stable states and an unstable saddle state. These two stable states are identical, orthogonal, and in the opposite direction.

According to Figures 8.7–8.11, each square bistable composite plate has tree equilibrium states including two stable states and an unstable state. The branch AB represents a saddle state that occurs for length less than a critical length. In this case, the equilibrium of the main curvatures, are the equal transverse curvatures but in the opposite direction. The branches of BD and BE exhibit stable cylindrical equilibrium states. In this case, the plate's equilibrium has two minimum and maximum orthogonal transverse curvatures in the opposite directions, which is orthogonal to the minimum and maximum curvature of the second cylindrical stable state and in the opposite direction. The BC branch represents an unstable equilibrium state, which in practice cannot be detected without applying any external factor.

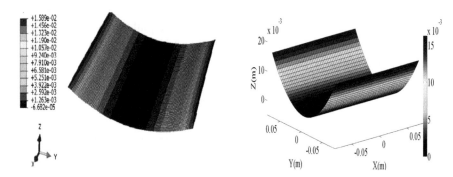

FIGURE 8.8   First stable state of square bistable composite plate [0₂/90₂].

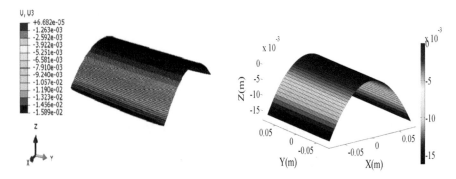

FIGURE 8.9   Second stable state of square bistable composite plate [0₂/90₂].

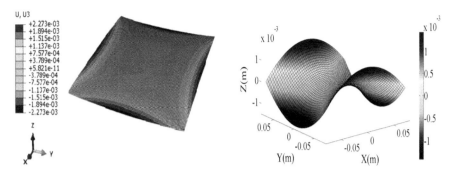

FIGURE 8.10   Saddle state (unstable state) of square bistable composite plate [$0_2/90_2$].

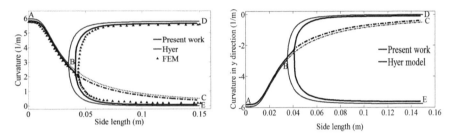

FIGURE 8.11   Curvature variation according to length for square bistable composite plate [$0_2/90_2$].

8.7.2   RECTANGULAR BISTABLE COMPOSITE PLATE [$0_2/90_2$]

The equilibrium modes of a rectangular bistable composite plate with a dimension of 150×300 mm² and stacking sequence [$0_2/90_2$] are shown in the order of the method presented in Figures 8.12–8.14.

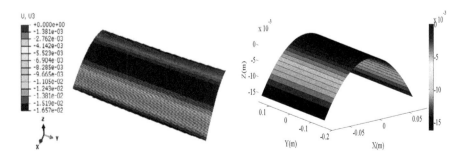

FIGURE 8.12   First stable state of the rectangular bistable composite plate [$0_2/90_2$]

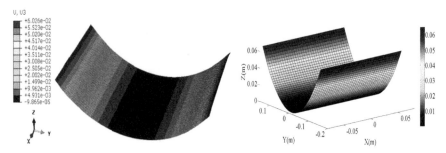

FIGURE 8.13   Second stable state of the rectangular bistable composite plate $[0_2/90_2]$.

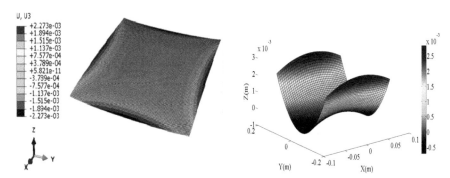

FIGURE 8.14   Saddle state (unstable state) of the rectangular bistable composite plate $[0_2/90_2]$.

As shown in these figures, the rectangular bistable composite plate, like the square plate, have two cylindrical stable equilibrium states. Note that in a bistable composite plate with transverse stacking sequence, in equilibrium states, twisting curvature was very small and close to zero.

8.7.3   RECTANGULAR BISTABLE COMPOSITE PLATE [0/–45/90]

The equilibrium states of the rectangular bistable composite plate with the stacking sequence [0/–45/90] were obtained from the method, presented in Figure 8.15. As seen in this figure, the addition of a layer at an angle between 0 and 90 degrees, resulted in the completely exit of the cylindrical state in stable states. The reason for this state was a deformation of the torsional curvature, which in this stacking sequence has the size between the maximum and minimum transverse curvatures.

The transverse and torsional curvature value for this type of stacking sequence is shown in Table 8.2.

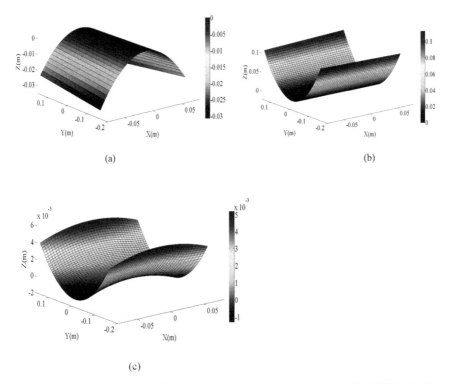

(a)

(b)

(c)

FIGURE 8.15 Obtained equilibrium states for the rectangular plate [0/–45/90], the first stable state (a), the second stable state (b), and the unstable state (c).

TABLE 8.2 Transverse and Torsional Curvatures for Stable States of Rectangular Bistable Composite Plate [0/–45/90]

$\kappa_{xy}(\mathrm{m}^{-1})$	$\kappa_{yy}(\mathrm{m}^{-1})$	$\kappa_{xx}(\mathrm{m}^{-1})$	Equilibrium state
60/0-	008/0-	68/9-	First stable state
60/0	68/9	008/0	Second stable state
0	45/0	45/0-	Unstable state

8.7.4 RECTANGULAR BISTABLE COMPOSITE PLATE [0/90/−45]

The equilibrium states of the rectangular bistable composite plate with a stacking sequence [0/90/–45] obtained from the method presented in Figure 8.16.

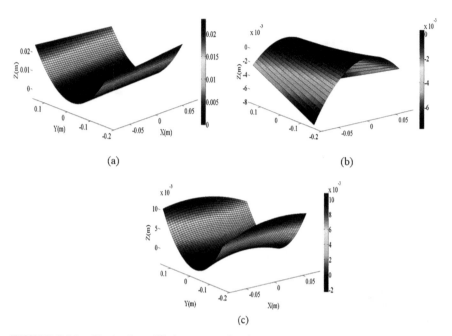

FIGURE 8.16   Obtained equilibrium states for the rectangular plate [0/90/–45], the first stable state (a), the second stable state (b), and the unstable state (c).

8.7.5   RECTANGULAR BISTABLE COMPOSITE PLATE [–30$_2$/60$_2$]

The equilibrium states of the rectangular bistable composite plate with a stacking sequence [–30$_2$/60$_2$] is given in Figure 8.17.

The bistable composite plate with this type of stacking sequence has twisting cylindrical stable modes. This is due to twisting of curvature that is larger than the transverse curvature. The transverse and torsional curvatures values for this type of stacking sequence are shown in Table 8.3.

TABLE 8.3   Transverse and Torsional Curvatures for Stable States of Rectangular Bistable Composite Plate [–30$_2$/60$_2$]

κ_{xy}(m^{-1})	κ_{yy}(m^{-1})	κ_{xx}(m^{-1})	Equilibrium state
01/5	32/4	43/1	First stable state
01/5	32/4-	43/1-	Second stable state
05/1	3/0	3/0-	Unstable state

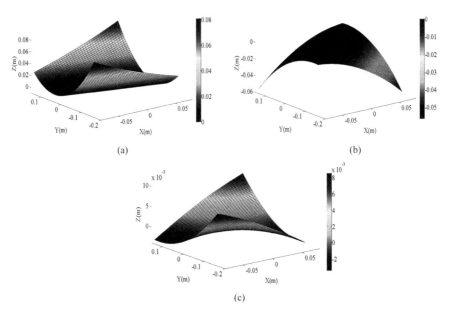

FIGURE 8.17 Obtained equilibrium states for the rectangular plate $[-30_2/60_2]$, the first stable state (a), the second stable state (b), and the unstable state (c).

8.7.6 RECTANGULAR BISTABLE COMPOSITE PLATE [45₂/–45₂]

The equilibrium states of the rectangular bistable composite plate with a stacking sequence $[45_2/–45_2]$ are shown in Figure 8.18.

In the following section, bistable hybrid composite plates with an outer metal layer are addressed.

8.7.7 HYBRID BISTABLE COMPOSITE PLATE [0/90/AL]

Bistable hybrid composite plates require more force value to snap through than ordinary bistable composite plates. They also have a larger out of plane displacement. The metal layer in the bistable hybrid composite plates can be inserted as inner and outer layers. Due to the fact that the bistable hybrid composite plate with an inner metal layer has a similar deformed shape with conventional composite plates, in this case, hybrid composite plates with an outer metal layer [0/90/metal] are investigated. These plates have unique and distinctive features with conventional composite plates (such as CFRP) and bistable hybrid composite plates with an inner metal layer [0/metal/90].

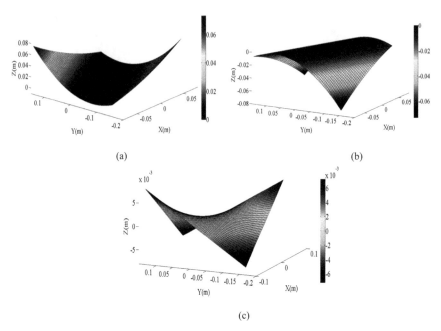

FIGURE 8.18 Obtained equilibrium states for the rectangular plate [45_2/-45_2], the first stable state (a), the second stable state (b), and the unstable state (c).

Bistable conventional and hybrid composite plates with an inner metal layer with transverse stacking sequence in each stable state have a maximum transverse curvature, which is orthogonal to the maximum transverse curvature of the second stable state and is in the opposite direction. While bistable hybrid composite plates with an outer metal layer in each stable cylindrical state have two transverse curves with the same direction and size. Also, due to the twisting, the twisted curve has a considerable amount. In bistable plates with a layer stacking sequence [0/90/metal], unlike the other two types, the main curvatures have an angle with the principle coordinate system. In Figure 8.19, several bistable hybrid plates with outer metal layer are observed.

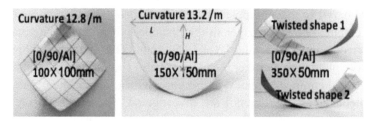

FIGURE 8.19 An example of a bistable plate [0/90/AL], [56]. *Source:* Reprinted with permission from Ref. [56]. © 2013 Elsevier.

Another issue that needs to be considered for addressing bistable hybrid composite plate is the slipping effect. In fact, slipping occurs due to the difference in coefficient of thermal expansion of the metal layer with the coefficients of longitudinal and transverse expansion of the fibers at the contact surface of the metal and the fiber. The slipping coefficient depends on the material and the manufacturing process and is determined experimentally. This coefficient is defined as Eq. (15).

$$\mu = \frac{\varepsilon_f^{cure}}{\varepsilon_m^{cure}}$$

$$\varepsilon_m^{cure} = \frac{E_m \alpha_m \Delta T t}{\frac{\mu}{4} E_f V_f (H - t)(1 - \nu_m) + E_m t}$$

(15)

where, ε_f^{cure}, ε_m^{cure} the strain caused by curing in the fibers and the metal layer, respectively. t is the thickness of the metal layer, H is the total thickness of the plate. Using experimental results for aluminum, μ = 0.5 was selected. For complete connection between metal and fiber μ = 1 and for incomplete connection μ = 0 were considered [55].

Given the slipping coefficient, the new coefficient of thermal expansion is defined as the Eq. (16).

$$\bar{\alpha}_{11} = \frac{\mu \varepsilon_m^{cure}}{2 \Delta T}$$

$$\bar{\alpha}_{22} = \alpha_{22}$$

$$\bar{\alpha}_f = \frac{\varepsilon_m^{cure}}{\Delta T}$$

(16)

Considering the slipping coefficient and the replacing equations, curvature has less difference than neglecting slipping coefficient in experimental results. Square bistable hybrid plates has dimensions of 150×150 mm² and its stacking sequence is [0/90/AL]. Mechanical and thermal properties of CFRP with stacking sequence [0/90] and Aluminum Layer are shown in Table 8.4.

In Figure 8.20, the equilibrium states of the hybrid composite plate with stacking sequence [0/90/AL] are shown. According to Table 8.4, the total thickness of the torsional bistable hybrid composite plate is 0.490 mm. Therefore, for an acceptable comparison, the thickness of the composite plate is considered to be 0.490 mm. According to Figure 8.19, a hybrid composite plate with an outer metal layer has two cylindrical torsional states with equal and same direction transverse curvatures and an unstable equilibrium parabolic stable state for a length greater than the critical length.

The value of the twisted curvature is also equal for both stable states but in the opposite direction. There is also a stable parabolic equilibrium for the lengths less than the critical length, as can be seen from Figure 8.20 (c). The unstable equilibrium parabolic state without applying force cannot be seen.

TABLE 8.4 Mechanical and Thermal Properties of CFRP With Stacking Sequence [0/90] and Aluminum Layer

Properties	Value
Longitudinal elastic modulus of CFRP (GPa)	126
Transverse elastic modulus of CFRP (GPa)	8.8
Shear module of CFRP (GPa)	7.47
Poisson ratio of CFRP	0.3
Longitudinal coefficient of thermal expansion of CFRP (10^{6}°C)	0.25
Transverse coefficient of thermal expansion of CFRP (10^{6}°C)	34
Thickness of each layer of CFRP (mm)	0.125
Volumetric ratio of fiber	0.6
Longitudinal elastic modulus of fiber (GPa)	211
Elastic modules of aluminum (GPa)	70
Coefficient of thermal expansion of aluminum (10^{6}°C)	23.6
Thickness of aluminum layer (mm)	0.240

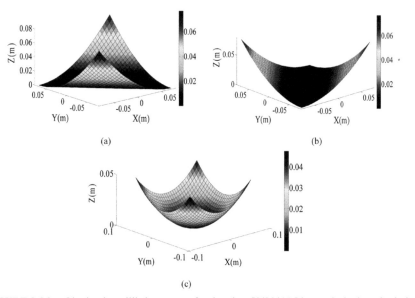

FIGURE 8.20 Obtained equilibrium states for the plate [0/90/AL] by analytical method, the first stable state (a), the second stable state (b), and the unstable state (c).

8.8 CONCLUSIONS

In this research, bistable composite plates with cross-ply, angle ply, and combined layer arrangements were analyzed. After determining the governing equations and solving the nonlinear equations, the equilibrium and stable states for these plates were clarified. The results showed that the bistable composite plate with cross-ply was two completely stable cylindrical states, and adding a layer at an angle between zero and ninety degrees would result in exiting from the complete cylindrical state. In composite plates with angle ply, the stable states are as twisting cylindrical. The curvature was then calculated in stable states for the above-mentioned states. The results indicated that the composite plate with cross-ply arrangement had only transverse curves and would have twisting curvature in addition to the transverse curvatures. In the next step, a hybrid bistable composite plate was examined with an external metal layer, and it was found that this kind of arrangement has of twisting stable cylindrical states.

KEYWORDS

- **bistable**
- **composites**
- **layers arrangement**
- **material effect**
- **plate**
- **stable state**

REFERENCES

1. Kaw, A. K., (2006). *Mechanics of Composite Materials*, Taylor and Francis Press.
2. Fuhong, D., Hao, L., & Shanyi, D., (2013). A multi-stable lattice structure and its snap-through behavior among multiple states, *J. Composite Structures, 97*, 56–63.
3. Tawfik, S., Xinyuan, T., Ozbay, S., & Armanios, E., (2007). Anticlastic stability modeling for cross-ply composites, *J. Composite Materials, 41*(11), 1325–1338.
4. Hyer, M. W., (1981). Some observations on the curved shape of thin unsymmetric laminates, *J. Composite Materials, 15*, 175–194.
5. Hamamoto, A., & Hyer, M., (1987). Non-linear temperature-curvature relationships for unsymmetric graphite-epoxy laminates. *Int. J. Solids Struct., 23*, 919–935.

6. Galletly, D. A., & Guest, S. D., (2004). Bistable composite slit tubes. ii. a shell model, *Int. J. Solids Struct., 462*, 4503–4516.

7. Guest, S. D., & Pellegrino, S., (2006). Analytical models for bistable cylindrical shells, *Proc. R. Soc. A: Math Phys. Eng. Sci., 462*, 839–854.

8. Seffen, K. A., (2007). Morphing bistable orthotropic elliptical shallow shells, *Proc. R. Soc. A: Math Phys. Eng. Sci., 463*, 67–83.

9. Tawfik, S., Xinyuan, T., Ozbay, S., & Armanios, E., (2007). Anticlastic stability modeling for cross-ply composites, *J. of Composite Materials, 41*(11), 1325–1338.

10. Vidoli, S., & Maurini, C., (2008). Tristability of thin orthotropic shells with uniform initial curvature, *Proc. Roy. Soc. A: Math Phys. Eng. Sci., 464*, 2949–2966.

11. Diaconu, C. G., Weaver, P. M., & Arrieta, A. F., (2009). Dynamic analysis of bistable composite plate, *J. of Sound and Vibration, 322*, 987–1004.

12. Vogl, G. A., & Hyer, M. W., (2011). Natural vibration of unsymmetric cross-ply laminates. *J. of Sound and Vibration, 330*, 4764–4779.

13. Pirrera, A., Avitabile, D., & Weaver, P. M., (2012). On the thermally induced bistability of composite cylindrical shells for morphing structures, *Int. J. Solids Struct., 49*, 685–700.

14. Vidoli, S., (2013). Discrete approximation of the Föppl-Von Kármán shell model: From coarse to more refined models. *Int. J. Solids Struct., 50,* 1241–1252.

15. Eckstein, E., Pirrera, A., & Weaver, P. M., (2014). Multi-mode morphing using initially curved composite plates, *Compos. Struct., 109*, 240–245.

16. Bowen, C. R., Kim, H. A., & Salo, A. I. T., (2014). Active composites based on bistable laminates, *Procedia Engineering, 75*, 140–144.

17. Cantera, M. A., Romera, J. M., Adarraga, I., & Mujika, F., (2015). Modeling and testing of the snap-through process of bi-stable cross-ply composites, *Composite Structures, 120*, 41–52.

18. Ryu, J., Kong, J. P., Kim, S. W., Koh, J. S., Cho, K. J., & Cho, M., (2013). Curvature tailoring of unsymmetric laminates with an initial curvature, *Journal of Composite Materials, 47*(25), 3163–3174.

19. Lee, J. G., & Junghyun, R., (2015). Effect of initial tool-plate curvature on a snap-through load of unsymmetric laminated cross-ply bistable composites, *Composite Structures, 122.* 82–91.

20. Daynes, S., & Weaver, P., (2010). Analysis of unsymmetric CFRP-metal hybrid laminates for use in adaptive structures, *Composite Structures, 81*, 1712–1718.

21. Dai, F., Li, H., & Du, S., (2013). Cured shape and snap-through of bistable twisting hybrid [0/90/metal] laminates, *Composites Science and Technology, 86*, 76–81.

22. Li, H., Dai, F., Weaver, P. M., & Du, S., (2014). Bistable hybrid symmetric laminates, *Composites Structures, 116*, 782–792.

23. Pan, D., Fuhong, D., & Hao, L., (2015). Piezoelectric energy harvester based on bi-stable hybrid symmetric laminate, *Composite Structures, 119*, 34–45.

24. Iqbal, K., Pellegrino, S., & Daton-Lovett, A., (2000). Bi-stable composite slit tubes, *IUTAM-IASS Symposium on Deployable Structure, 80*, 153–162.

25. Bowen, C. R., Giddings, P. F., Salo, T, A. I., & Kim, H. A., (2011). Modeling and characterization of piezoelectrically actuated bistable composites, *IEEE Trans. Ultrason., Ferroelectr., Freq. Control, 58*, 1737–1750.

CHAPTER 9

One Stage Production of Superconductive MgB_2 and Hybrid Transmission Lines by the Hot Explosive Consolidation Technology

A. PEIKRISHVILI[1], T. GEGECHKORI[2], B. GODIBADZE[3], G. MAMNIASHVILI[2], and V. PEIKRISHVILI[1]

[1]F. Tavadze Institute of Metallurgy and Materials Science, E. Mindeli str., 10, 0186, Tbilisi, Georgia, E-mail: apeikrishvili@yahoo.com; vaxoo3@gmail.com

[2]Ivane Javakhishvili Tbilisi State University, E. Andronikashvili Institute of Physics, Tamarashvili str., 6, 0177, Tbilisi, Georgia, E-mail: tatagegegechkori@yahoo.com, mgrigor@rocketmail.com

[3]G. Tsulukidze Mining Institute, Mindeli Str., 7, 0186, Tbilisi, Georgia, E-mail: bgodibadze@gmail.com

ABSTRACT

We applied the original hot shock-assisted consolidation method combining a high temperature with the two-stage explosive process without any further sintering which produced superconducting materials with high density and integrity. The consolidation of MgB_2 billets was made at temperatures above the melting point of Mg up to 1000°C in the partially liquid condition of Mg-2B blend powders. The influence of isotope B composition on critical temperature and superconductive properties was evaluated as well as the first successful application of this method for production of hybrid power transmission lines for simultaneous transport of hydrogen and electric energy was demonstrated.

9.1 INTRODUCTION

The superconductive properties of MgB_2 with structure C32 and critical temperature of transformation $T_c = 39K$ was discovered in 2001 [1]. Since that time the intensive investigation toward of development different type of MgB_2 superconductive materials in the forms of films, sheets or bulk rods and increasing their critical temperature of transformation T_c above 39K takes place at different laboratories worldwide [2–5]. The technology of development superconductive materials belongs to traditional powder metallurgy: preparing and densification Mg & B powder blends in static conditions with their further sintering processes [6, 7]. Results described in Ref. [8], where Mg-2B blend powders were first compacted in cylindrical pellets ($\emptyset = 63$ mm, H = 20 mm) and after were secondary loaded in hot conditions (600–1200°C) with 2 GPa pressure, also seem interesting. The observation of a clear correlation between the syntheses condition and crystal structure of formed two phases MgB_2-MgO composites as well as between their superconductive properties allowed to conclude that redistribution of oxygen in the MgB_2 matrix structure and formation of MgO phase may be considered as a positive effect too.

Existing data of the application of shock wave consolidation technology to fabricate high dense MgB_2 billets with higher T_c temperature practically gave the same results and limit of $T_c = 40K$ still is maximal.

Additionally, as the published data show, sintering processes after the shock wave compression are highly recommended providing full transformation of consolidating blend phases into the MgB_2 composites.

The goals of the current investigation are as follows:

- the development of the technology of hot shock wave fabrication of high dense hybrid billets from MgB_2 without any further sintering processes.
- to develop cylindrical combined Cu-MgB_2-Cu composites using copper substrate materials.
- the investigation of the role of temperature on the process of consolidation and sintering MgB_2.
- the consolidation of MgB_2 billets above the melting point Mg up to 1000°C in a partially liquid matrix of Mg-2B blend powders.
- the evaluation and investigation of structure/property relationship.

9.2 EXPERIMENTAL

The novelty of the proposed nonconventional approach relies on the fact that the consolidation of the samples from coarse (under 10–15 µ) Mg–2B blend powders was performed in two stages [9]. The explosive pre-densification of the powders was made at room temperatures. In some cases before dynamic pre-densification, the loading of precursors into the containers were performed by static means or by vibro densification. In all cases, the second stage was done by the hot explosive compaction (HEC) but at temperatures under 1000°C with the intensity of loading around 5 GPa. Cylindrical compaction geometry was used in all of the HEC experiments.

At first stage, the Mg–2B bland powders were placed inside a steel-tube container. The container was sealed at both ends with threaded steel plugs. A concentric cardboard box was filled with the powdered explosive materials from ANFO, AC–4 or ammonium nitrate and was placed around the cylindrical sample container (Figure 9.1).

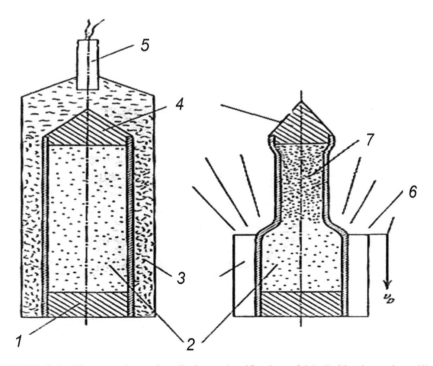

FIGURE 9.1 The procedure of preliminary densification of Mg-B blend powders. (1) bottom plug of steel tube; (2) precursor powders; (3) explosive powder; (4) upper plug of steel tube; (5) electric detonator; (6) products of detonation; (7) consolidated powders.

The key operational component of the planning experiments HEC with the vertical configuration of an explosive charge that allows to consolidate coarse and nanoscale precursors at elevated temperatures is presented in Figure 9.2. Application of vertical configuration of charge allows an increase without limitation the sizes of explosive charges. As a result, the pulse duration during the compression (loading) will be increased resulting of obtaining samples with higher densities. From the other hand increasing of pulse duration will allow to decrease consolidating temperature and to compress samples at 800–1000°C and as a result to reduce the cost of obtained billets too.

The HEC device (Figure 9.2) consists from the main three parts: heating system–cylindrical heating furnace; feeding cylindrical system to forward heated billets; set-up of the explosive charge.

The preliminary pre-densified cylindrical billet (1) were located in the central hole of the heating furnace (4). The heating billet is fixed in the furnace by the opening and closing movement mechanism (6). After heating of billet up to necessary temperature, the opening (6) sheet will opens furnace and billet moves through the cylindrical feeding system (9–11) to the set-up of an explosive charge (17).

After receiving a signal that billet passed feeding system and is located in final position (13) the detonation through the detonators and detonation cords there takes place, and explosive compression of heated billets there takes place.

Determined by the volume, type, and density of the sample composition, the heating lasted about 60 sec. The temperature was measured using a Chromel-Alumel thermocouple whose tip was situated inside the heating furnace. When the set temperature was reached the furnace was switched off by remote control and feeding mechanism was open. Due to the weight, billets were passed through the feeding tube inside of cylindrical charge. As soon as the billet reaches the bottom of the explosive charge and the remote control light fixed the arrival of billets in requested position the detonation circuit was switched on automatically and the explosive was detonated through the detonator (Figure 9.3). The corresponding pressure at the wall of the steel container was around 5 GPa.

Figure 9.3 represents the construction of an explosive charge in detail. The HEC sample is consisting from preliminary compacted powder (1) in a cylindrical container (2) and closed from the bottom by a plug (3) after movement is located in setting up of explosive charge consisting from feeding tube 5, explosive charge 6 and detonator/detonation cord 7. The additional steel tablets at the end of the container were used for avoiding cutting of container end during the HEC experiments and maintaining whole billets without damages.

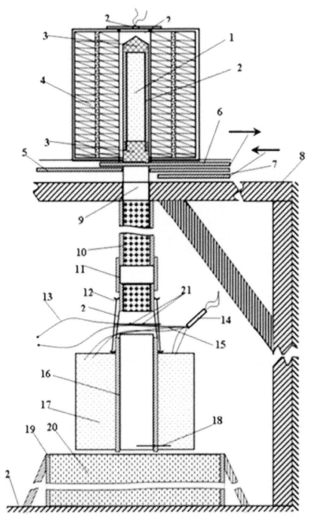

FIGURE 9.2 Set-up of HEC device. (1) Consolidating powder material; (2) cylindrical steel container; (3) plugs of steel container; (4) heating wires of furnace; (5) opening and closing movement of furnace; (6) opening sheet of furnace; (7) closing sheet of furnace; (8) basic construction of HEC device; (9) feeding steel tube for samples; (10) movement tube for heated container; (11) connecting tube from rub; (12) accessory for fixing explosive charge; (13) circle fixing passing of steel container; (14) El-detonator; (15) detonating cord; (16) flying tube for HEC; (17) explosive charge; (18) lowest level of steel container; (19) bottom fixing and stopping steel container; and (20) sand.

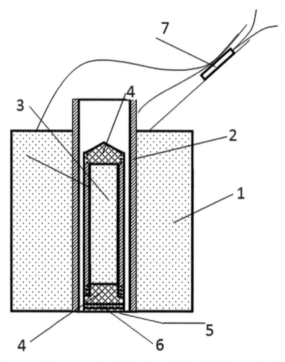

FIGURE 9.3 Experimental set-up for HEC of cylindrical billets explosive charge at the bottom according to the general view of HEC device (Figure 9.1).

Figure 9.4 represents the view of HEC billets after first stage pre-densification and secondary consolidation at 1000°C.

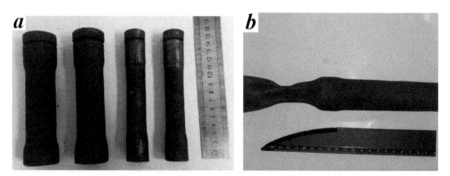

FIGURE 9.4 The view of HEC billets from Mg-B precursors consolidated at 1000°C with the intensity of loading under 10 GPa. (a) First stage pre-densification by shock waves at room temperature. (b) Consolidation of the same billets again at 1000°C with the intensity of loading under 10 GPa.

9.3   RESULTS

In order to consolidate high dense superconductive MgB$_2$ cylindrical billets, the cylindrical tubes from steel and copper were used. As further investigation showed, the high-temperature consolidation of Mg–2B precursors in steel containers have a positive effect and allows one to fabricate two-phase MgB$_2$-MgO near to theoretical density with critical temperatures around T$_c$ = 38K. In contrast to steel containers, the shock wave fabrication of Mg–2B powder blends in Cu containers leads to the formation of undesirable phases such as MgCu$_2$. The mentioned takes place due to the diffusion of copper atoms under the shock wave front towards of center of the container with further chemical reactions with Mg and formation of MgCu$_2$ phases. The observation of HEC billets and outward investigation confirms that there took place too intensive reactions between the Cu and Mg. The reaction was so intensive and exothermic that fully melting of the contact surface between the container's wall and consolidated Mg–2B precursors there took place. The mentioned effect connected with the formation of MgCu$_2$ under the shock wave front is not new and was described in Ref. [3] also.

In order to prevent the movement of Cu atoms on the shock wave front towards of center of the container and further chemical reactions with formation of MgCu$_2$ phases the tantalum foils with a thickness of 100 μ were used as the intermediate layers between the copper container's wall and consolidated Mg–2B precursors. As a result, no MgCu$_2$ phases inside of HEC Mg-B precursors were observed.

Figure 9.5 represents microstructure of HEC Mg–2B precursors obtained at 940°C temperature in the copper container.

As it's seen from microstructures, the central part and edge of HEC precursors differ from each other and traces of high temperature, and melting/crystallization processes may be observed. The existence of microcracks on the edge of HEC precursors (Figure 9.5b) may be explained as a result of thermal stresses during the rapid cooling process.

Figure 9.6 represents of the macrostructure of HEC sample of Mg–2B precursors with the application of the intermediate layer from Ta foil at different magnifications. The application of Ta intermediate layer between the Cu container's wall and consolidated Mg–2B precursors provides full protection of consolidated powders from transportation Cu atoms by shock wave front towards of Centrum and as a result formation of MgCu$_2$ phases. The mentioned was confirmed by X-ray analyzes carried out for HEC Mg–2B

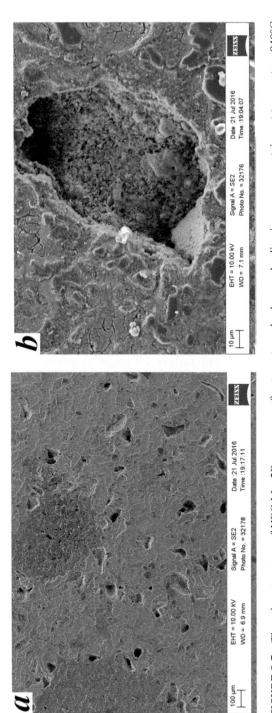

FIGURE 9.5   The microstructures of HEC Mg–2B precursors after two-stage shock wave loading in a copper container at temperature 940°C with the intensity of loading 5GP: (a) central part; (b) edge.

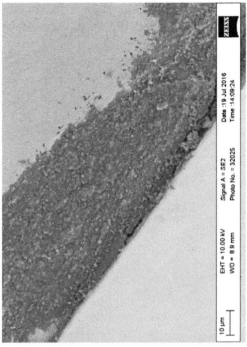
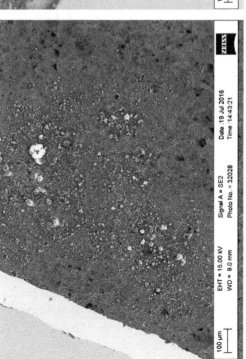

FIGURE 9.6 The macrostructure of HEC MgB$_2$-based superconductive composites in a copper container with application of intermediate layer from tantalum (Ta).

precursors where was demonstrated that application of thin Ta intermediate layer fully prevents the diffusion of Cu atoms toward of consolidated
compositions.

Figures 9.7 and 9.8 represents the diffraction pictures of HEC Mg–2B
precursors demonstrating phase formation in consolidated composites and
efficiency of application tantalum intermediate layer.

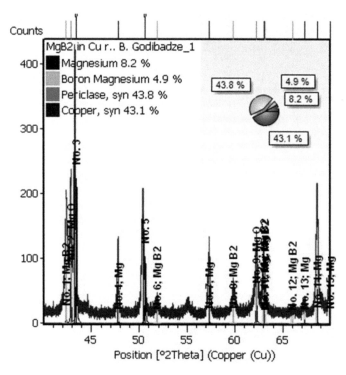

FIGURE 9.7   The diffraction picture of HEC Mg–2B precursors obtained after consolidation
at 1000°C in a copper container.

As it's seen from diffraction pictures (Figures 9.7 and 9.8) the application of intermediate layer from thin tantalum foils with thickness 100 μ fully
prevents the diffusion of copper inside of HEC Mg–2B precursors and no
traces of existing Cu or $MgCu_2$ phases may be observed (Figure 9.7). In
contrast to mentioned, the existing of copper inside of HEC Mg–2B precursors may be easily observed. The identification of phase lines from diffraction
picture (Figure 9.7) shows that after HEC of Mg–2B precursors only two
phase composition consisting from MgB_2 and MgO phases were formed.

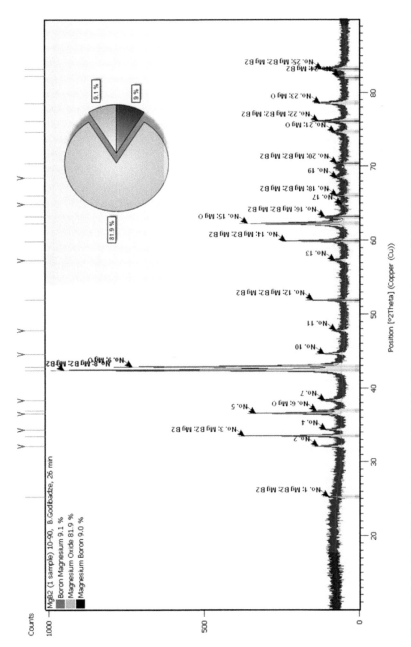

FIGURE 9.8 The diffraction picture of HEC Mg–2B precursors obtained at 1000°C in a copper container using protecting intermediate layer from tantalum.

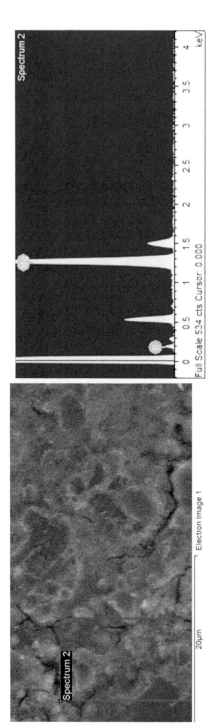

FIGURE 9.9   The macrostructure of HEC Mg−2B precursors obtained at 940° using intermediate Ta layer. As it's seen from spectral analyzes and distribution of atoms, there are fully prevented the diffusion of copper atoms into the HEC Mg−2B precursors.

The mentioned is confirmed by SEM investigations where it was demonstrated advantages of HEC technology towards of fabrication two-phase MgB_2-MgO composites near the theoretical density without porous and any other visible defects of structure (Figure 9.9).

Figure 9.10 represents the microstructures of HEC Mg–2B precursors obtained at 940° with correspondent spectral analyzes with element identification.

As it's seen from spectral analyzes and distribution of atoms, it is fully prevented the diffusion of copper atoms into the HEC Mg–2B precursors. The identification of elements on microstructures shows that we may only consider the existence of two phases. The light (bright) phase of spectrum-1 on the microstructure (Figure 9.10a) belongs to Mg when a dark region on the spectrum-2 shows only both elements Mg and B.

Taking into account the mentioned and results of diffraction analyzes we may be sure that after HEC of Mg–2B precursors based on chemical reactions under shock wave front there takes place the full transformation of starting elements into the two-phase composition from MgB_2 and MgO. The observation of eutectic colonies on the microstructure confirms the fact of melting/crystallization processes behind of shock wave front.

In order to evaluate the superconductive characteristics of obtained billets, there were investigated magnetic moment temperature dependence in zero-field-cooled (ZFC) and field-cooled (FC) modes depending on experimental conditions and type of boron precursors.

As it was reported in earlier investigation [9] application of low temperatures up to 900°C and HEC of Mg–2B precursors in steel containers did not give results. In spite of high density and uniform distribution of phases, there were not obtained superconductive characteristics. The investigation of HEC processes for Mg–2B precursors in copper container gave same results and no superconductive characteristics below 900°C were observed in samples.

Figure 9.11 represents the data of measurements for HEC Mg–2B precursors consolidated above 900°C temperatures in a copper container with the intensity of loading under 5 GPa.

The investigation of influence type of boron isotope onto the final superconductive characteristics of magnesium diborides (MgB_2) after the HEC at 1000°C shows that in contrast to ^{11}B isotope application of ^{10}B isotopes in Mg-B precursors provides increasing of critical temperature on 1K. Such a difference may be explained by lower mass of ^{10}B nucleus in contrast to ^{11}B.

The mentioned confirms the important role of temperature in formation of superconductive MgB_2 phase in the whole volume of the sample and corresponds with literature data, where only after sintering processes above 900°C the formation of MgB_2 phase with $T_c = 40$ K there took place.

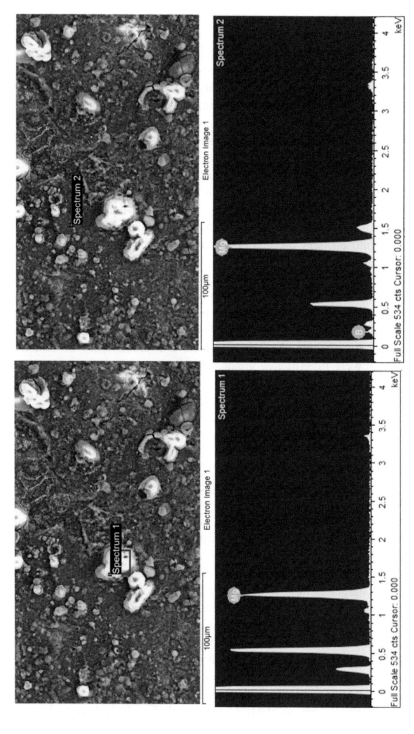

FIGURE 9.10 The microstructures of HEC Mg–2B precursors and their correspondent spectral analyzes with element identification obtained at 940° with the application of intermediate Ta layer. The intensity of loading onto the wall of the container was around 5 GPa.

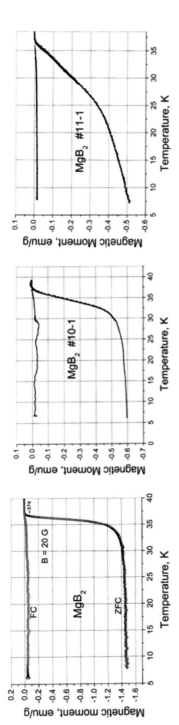

FIGURE 9.11 Magnetic moment temperature dependence measurements in zero-field-cooled (ZFC) and field-cooled (FC) modes, showing the superconducting transition depending on container material and type of boron precursors. a) HEC in a steel container at 1000°C; b) HEC in a copper container with ^{10}B isotope; c) HEC in a copper container with ^{11}B isotope.

The difference of T_c between the HEC and sintered MgB_2 composites may be explained with rest of nonreacted Mg and B phases or existing of some oxides in precursors Figure 9.11a.

The mentioned could be checked by increasing HEC temperature or application of further sintering processes. The careful selection of initial Mg and B phases is important too and in case of consolidation Mg–2B precursors with the above-mentioned corrections the chance to increase T_c in the HEC samples essentially increases.

In Figure 9.4, the views of MgB_2 billets in steel jackets after the previous densification (Figure 9.4a) and after the HEC procedure (Figure 9.4b) are shown.

In further experiments, the application of pure Mg and crystalline and amorphous B powder blend prevented the formation of MgO in HEC billets and increased T_c of the obtained MgB_2 composites up to 38.5 K, Figure 9.12, in case of pure amorphous boron powder without any post-sintering of obtained samples.

FIGURE 9.12 Hybrid Cu-MgB_2–Cu superconductivity tubes.

The experiments for HEC of precursors were performed under and above the melting point of Mg phase. The consolidation was carried out at 500, 700, 950, and 1000°C temperatures with the loading intensity of 10 GPa.

It was experimentally established that the comparatively low-temperature consolidations at 500°C and 700°C give no results and obtained compacts have no superconducting properties.

The HEC technology allows one also to produce multilayer cylindrical tubes (pipes) with the Cu/MgB$_2$/Cu structure which could find important applications for the production of superconducting cables for simultaneous transport of hydrogen and electrical power in hybrid MgB$_2$-based electric power transmission lines filled with liquid hydrogen [11].

An example of a practical realization of hybrid Cu-MgB$_2$–Cu superconductive cylindrical tubes for hybrid power transmission lines are demonstrated in Figure 9.12.

9.4 DISCUSSION

The HEC of Mg-B precursors were performed under and above of melting point Mg phase. The consolidation were carried out at 500, 700, 950 and 1000°C temperatures with the intensity of loading 10 GPa.

As it was established based on investigation the low-temperature consolidation at 500°C and 700°C gives no results and obtained compacts has no superconductive properties.

The application of more high temperatures and consolidation at 1000°C provides formation of MgB$_2$ composition in the whole volume of HEC billets with a maximal value of $T_c = 38.5$K without any post sintering processes of samples. The mentioned confirms the important role of temperature in formation of superconductive MgB$_2$ phase in the whole volume of sample and corresponds with literature data where only after sintering processes above 900°C the formation of MgB$_2$ phase with $T_c = 40$K there took place. The difference of T_c between the HEC and sintered MgB$_2$ composites may be explained with rest unreacted Mg and B phases or existing some oxides in precursors. The mentioned could be checked by increasing HEC temperature or application of further sintering processes. The careful selection of initial Mg and B phases is important too and in case of consolidation Mg–2B precursors with mentioned above corrections the chance to increase T_c of HEC samples essentially increases. The next stage experiments to fabricate MgB$_2$ superconductive materials will be implemented in this direction.

9.5   CONCLUDING REMARKS

The liquid phase HEC of Mg-B precursors under the 1000°C temperature provides formation MgB_2 phase in the whole volume of billets with maximal $T_c = 38.5$ K.

The type of applied B powder has an influence on the final result of superconductive characteristics MgB_2, and in the case of amorphous B precursors, better results are fixed (38.5 K against 37.5).

The purity of precursors is an important factor and existing of oxygen in the form oxidized phases in precursors leads to reducing T_c and nonuniformity of HEC billets.

ACKNOWLEDGMENTS

This research was supported by Shota Rustaveli National Science Foundation (SRNSF) [Grant number: 217004].

KEYWORDS

- **Cu-MgB$_2$-Cu composites**
- **Mg–2B**
- **MgB$_2$**
- **MgO**

REFERENCES

1. Nagamatsu, J., Nakagawa, N., Muranaka, T., Zenitani, Y., & Akimitsu, J., (2001). Superconductivity at 39 K in magnesium diboride. *Nature, 410*(6824), 63–64.
2. Jiang, C. H., Nakane, T., Hatakeyama, H., & Kumakura, H., (2005). Enhanced J$_c$ property in nano-SiC doped thin MgB$_2$/Fe wires by a modified in situ PIT process. *Physica. C., 422*(3–4), 127–131.
3. Mali, V. I., Neronov, V. A., Perminov, V. P., Korchagin, M. A., & Teslenko, T. S., (2005). Explosive incited magnesium diboride synthesis. *Chemistry for Sustainable Development, 13*(3), 449–451.

4. Orlinska, N., Zaleski, A., Wokulski, Z., & Dercz, G., (2008). Characterization of heat treatment MgB$_{c2}$ rods obtained by PIT technique with explosive consolidation method. *Archives of Metallurgy and Materials, 33*(3), 927–932.

5. Holcomb, M. J., (2005). Supercurrents in magnesium diboride/metal composite wire. *Physica C: Superconductivity, 423*(3 & 4), 103–118.

6. Priknha, T. A., Gavwalek, W., Savchuk, Ya. M., Moshchil, V. E., Sergienko, N. V., Surzenko, A. B., et al., (2003). High-pressure synthesis of a bulk superconductive MgB$_2$-based material. *Physica C: Superconductivity, 386*, 565–568.

7. Mamalis, A. G., Vottea, I. N., & Manolakos, D. E., (2004). Explosive compaction/cladding of metal sheathed/superconducting grooved plates: FE modeling and validation. *Physica C: Superconductivity, 408–410*, 881–883.

8. Shapovalov, A. P., (2013). High-pressure syntheses of nanostructured superconducting materials based on magnesium diboride. *High-Pressure Physics and Engineering, 23*(4), 35–45.

9. Mamniashvili, G., Daraselia, D., Japaridze, D., Peikrishvili, A., & Godibadze, B., (2015). Liquid-phase shock-assisted consolidation of superconducting MgB$_2$ composites. *J. Supercond. Nov. Magn., 28*(7), 1926–1929.

10. Daraselia, D., Japaridze, D., Jibuti, A., Shengelaya, A., & Müller, K. A., (2013). Rapid solid-state synthesis of oxides by means of irradiation with light. *J. Supercond. Nov. Magn., 26*(10), 2987–2991.

11. Kostyuk, V. V., Antyukhov, I. V., Blagov, E. V., Vysotsky, V. S., Katorgin, B. I., Nosov, A. A., Fetisov, S. S., & Firsov, V. P., (2012). Experimental hybrid power transmission line with liquid hydrogen and MgB$_2$ based superconducting cable. *Technical Physics Letters, 38*(3), 279–282.

12. Gegechkori, T., Godibadze, B., Peikrishvili, V., Mamniashvili, G., & Peikrishvili, A., (2017). One stage production of superconducting MgB$_2$ and hybrid power transmission lines by the hot shock wave consolidation technology. *International Journal of Applied Engineering Research (IJAIR), 12*(14), 4729–4734.

PART II
Polymer Synthesis and Application

CHAPTER 10

Novel Aliphatic Polyester-Based Macromonomers

E. ÇATIKER[1], M. ATAKAY[2], B. SALIH[2], and O. GÜVEN[2]

[1]Faculty of Art and Science, Department of Chemistry,
Ordu University, 52200, Ordu, Turkey, E-mail: ecatiker@gmail.com

[2]Faculty of Science, Department of Chemistry,
Hacettepe University, 06800 Ankara, Turkey

ABSTRACT

Oligomer of poly(3-hydroxy propionate) (P3HP) was synthesized via base-catalyzed hydrogen transfer polymerization (HTP) of acrylic acid using sodium tert-butoxide as a strong basic catalyst. Structural analyzes based on ^{1}H-NMR and MALDI-MS revealed that the oligomeric products obtained possess olefinic chain-ends that are open to further end-group functionalization. End-group modifications were performed through monobromination by HBr in acetic acid, debromination by bromine and epoxylation by hydrogen peroxide. Extents of conversions of olefinic end-groups to the corresponding functional groups were followed by using ^{1}H-NMR and MALDI mass spectrometry. According to the analyzes, the three modification approaches resulted with end-group modified oligomers with high yield. Monobrominated oligomer was then converted to azide-ended oligomer through treatment with sodium azide. The exchange of bromide atom at the end group to azide group was also studied using by MALDI mass spectrometry. The all end-group modified products may be regarded as novel macromonomers for applications such as synthesis of amphiphilic block copolymers, surface modification and grafting agents.

10.1 INTRODUCTION

Poly(β-propiolactone), also known as poly(3-hydroxypropionate) (P3HP), is a well-known biodegradable [1] thermoplastic polyester with high tensile strength and moisture permeability. There are some reports on its potential utilization as a scaffold material in tissue engineering [2], as a polymeric matrix for drug delivery [3] and dielectrically functional application [4]. Its traditional preparation methods [5–14] are ring opening polymerization of β-propiolactone by ionizing radiation, enzyme-catalyst, and nucleophilic catalysts. However, it is not commercially feasible since β-propiolactone is a human carcinogen and shows high-cost characteristics.

Although hydrogen transfer polymerization (HTP) is a well-known method for synthesis of poly-β-alanine [15], the synthesis of P3HP through HTP of acrylic acid was also reported by Saegusa et al. [16]. However, this method has not been paid attention possibly due to the low degree of polymerization (DP_n was reported as low as 17.4). They have used pyridine and triphenylphosphine (TPP) as the initiators to prepare P3HP from acrylic acid and proved the existence of the initiator at the end-group of the oligomeric P3HP using ^{1}H-NMR spectroscopy. Yamada et al. [17] reported that DP of P3HP might be elevated when HTP was carried out in the presence of crown ether as a co-catalyst. Moreover, Yamada et al. [17] revealed that HTP of acrylic acid yielded P3HP oligomers with olefinic end-group.

Well-defined oligomers with specific functional groups have many applications in synthetic chemistry and material science such as graft onto approaches [18–20], surface modification [21, 22], synthesis of block or diblock copolymers via click chemistry [23, 24], making telechelic polymers [25, 26] and bioconjugation [27]. This paper describes in detail the structural characterization of the oligomeric product obtained from HTP of acrylic acid and its end-group functionalization via monobromination, debromination, epoxylation, and azidation.

10.2 EXPERIMENTAL

10.2.1 MATERIALS

Acrylic acid (99%, Aldrich), sodium tert-butoxide (97%, Aldrich), diethyl ether (99.9%, Acros), chloroform (99.8%, Sigma-Aldrich), hydrogen bromide in acetic acid (33%, Sigma-Aldrich), sodium azide (99.5%,

Sigma-Aldrich), bromine (reagent grade, Sigma-Aldrich) and hydrogen peroxide (30%, Sigma-Aldrich) were obtained commercially and used without purification.

10.2.2 POLYMERIZATION

Oligomer of P3HP was obtained via HTP of acrylic acid using tNaBuO ([I] = [M]/30) as the basic catalyst. The reaction was carried out under nitrogen flux at 90–100°C for a week. Then, the reaction mixture was poured into an excess amount of dry diethyl ether to precipitate out P3HP crystals. P3HP crystals were separated by a filtration kit connected to a vacuum pump, dried in vacuum overnight oven at ambient temperature.

10.2.3 END-GROUP MODIFICATION

10% (w/v) solutions of P3HP oligomers were prepared using chloroform as a solvent. Monobromination (by HBr in acetic acid), debromination (by bromine) and epoxylation (by hydrogen peroxide, 30%) of P3HP oligomers with olefinic chain-ends were carried out in 2mL-vials with continuous stirring at ambient temperature for 24 hours. Excess bromine and HBr was removed by using a water jet pump, and the remaining solutions were analyzed without precipitating the modified oligomers. The monobrominated oligomer was treated with excess sodium azide at 60°C for two days to convert the alkyl bromide to azide group.

10.2.4 MEASUREMENTS

^{1}H-NMR spectra of P3HP oligomers were recorded using Bruker 400 MHz NMR Spectrometer and trifluoroacetic acid (TFA) as a solvent. MALDI mass spectra were acquired on a Voyager-DE™ PRO MALDI-TOF mass spectrometer (Applied Biosystems, USA) equipped with a nitrogen UV-Laser operating at 337 nm. Spectra were recorded in positive ion and linear mode with an average of 500 shots. MALDI matrix, 2,5-dihydroxybenzoic acid (DHB) was prepared in THF:ACN mixture (1:1, v/v) at a concentration of 20 mg/mL. The solution of cationization reagent, LiTFA (10 mg/mL in THF), was added to the matrix solution (1.0 % (v/v) of the total solution). All samples were dissolved in THF and initially spotted (1.0 μL) on MALDI

target. After air-drying, the matrix solution (1.0 µL) was deposited on each sample spot. Sample spots were finally allowed to air-dry prior to the MALDI-MS analyzes.

10.3   RESULTS AND DISCUSSION

10.3.1   *STRUCTURAL CHARACTERIZATION OF PRODUCT OF HTP OF ACRYLIC ACID*

10.3.1.1   *NMR SPECTROSCOPY*

^{1}H-NMR spectrum of P3HP given in Figure 10.1 was recorded to elicit chemical structure in detail. The peaks at about 2.90 and 4.50 ppm with equal intensity belong to methylene protons next to the carbonyl ($COCH_2$) and the oxygen ($O-CH_2$), respectively. The peaks between 6.0–7.0 ppm are attributed to olefinic protons ($CH_2 = CH$) at the chain-ends. The singlet peak at 11.50 ppm belongs to the proton from the solvent (CF_3COOH). The absence of a peak for t-BuO, used as an initiator for HTP, indicates that the polymerization is initiated by hydrogen abstraction from acrylic acid. Considering the relative intensities of olefinic protons and protons from repeating units, average DP of P3HP was estimated to be 10–11. Hence, the chemical structure of P3HP oligomer may be given as in Scheme 10.1.

10.3.1.2   *MALDI MASS SPECTROMETRY*

MALDI mass spectrum of P3HP oligomer was obtained using positive ion and linear mode in DHB matrices. MALDI-MS spectrum of P3HP has single distribution as shown in Figure 10.2. The mass difference in the repeating units of P3HP was found to be 72 Da, which corresponds to the repeating unit weight of P3HP.

The randomly selected m/z 583.9, 727.8, 871.6, 1015.4, 1087.4, 1231.2, 1303.1, and 1518.8 Da can be assigned to $[M_n\text{-Li}]^+$ ions for $n = 8$, 10, 12, 14, 15, 17, 18, and 21, respectively. In other words, the m/z values obey the equation (72* n + 7 Da). Here, 72 Da is mass of the repeating unit of P3HP, n is the number of repeating units at the corresponding peak, and 7 is the atomic mass of Li.

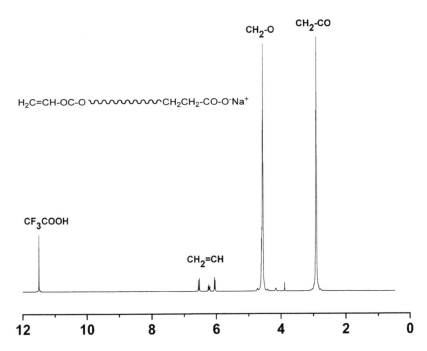

FIGURE 10.1 ¹H-NMR spectrum of P3HP oligomers.

Similarly, randomly selected m/z 599.9, 743.8, 887.6, 1031.4, 1103.4, 1247.2, 1319.1, and 1534.8 Da can be assigned to $[M_n\text{-Na}]^+$ ions for $n = 8, 10, 12, 14, 15, 17, 18,$ and 21, respectively. In other words, the m/z values obey the equation ($72* n + 23$ Da) where 23 is the atomic mass of Na. The result proves not only the initiation mechanism proposed by Yamada et al. [17] (hydrogen abstraction by the initiator) but also sodium carboxylate ended chains (the mobile sodium ion of carboxylate at chain-ends) as proposed in Scheme 10.1.

$$CH_2=CH-CO-O\left[CH_2-CH_2-CO-O\right]_{9-10}CH_2-CH_2-COO^-Na^+$$

SCHEME 10.1 Chemical structure of living P3HP oligomer.

10.3.2   END-GROUP FUNCTIONALIZATION

NMR and MALDI analyzes outlined above proved that product of HTP of acrylic acid was olefin-ended oligomer with approximately 10–11 repeating

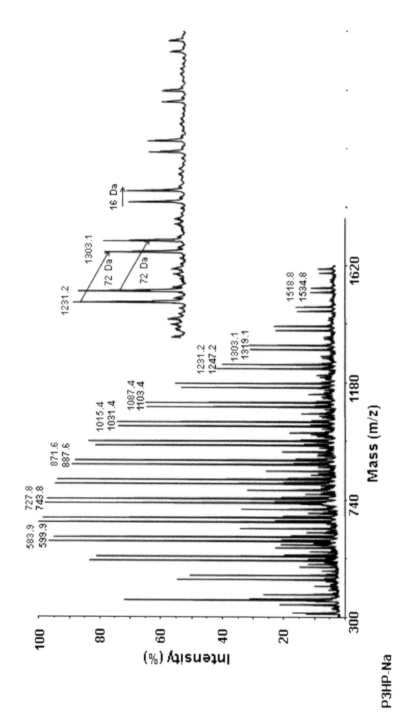

FIGURE 10.2 MALDI mass spectrum of P3HP oligomers.

units. Since olefinic end-groups have relatively low reactivity, end-group functionalization of the oligomer was aimed to obtain novel hydrophobic macromonomers with enhanced reactivity. Monobromination [28], dibromination [29, 30] and epoxylation [31, 32] of the P3HP oligomer were carried out according to the literature [33] and outlined as in Scheme 10.2.

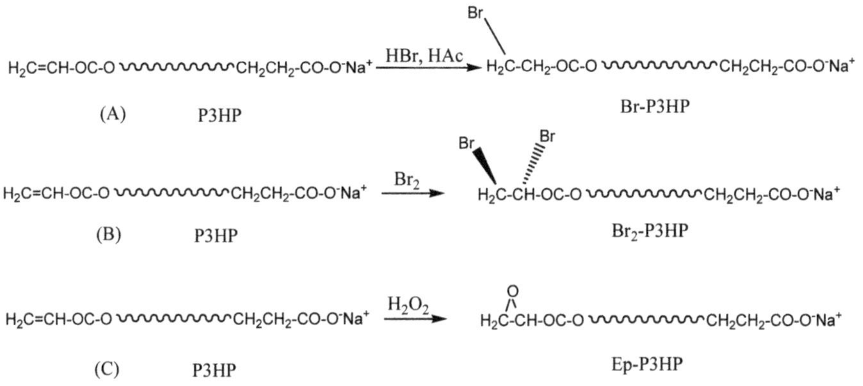

SCHEME 10.2   End-group functionalization of P3HP oligomers through mono-bromination, debromination, and epoxylation, respectively.

10.3.2.1   NMR SPECTROSCOPY

^{1}H-NMR spectra of pristine and modified P3HP oligomers (Figure 10.3) were given together to compare changes after modification. As mentioned before, the trace peaks between 6.0 and 7.0 ppm (Figure 10.3A) belong to the vinyl protons at chain-ends. Almost complete disappearance of these peaks on the spectra (Figure 10.3B–10.3D) of P3HP-Br, P3HP-Br$_2$, and P3HP-Ep, proves that end-groups were functionalized successfully.

10.3.2.2   MALDI MASS SPECTROMETRY

MALDI mass spectra of the end-group functionalized oligomers were recorded not only to reveal the extents of functionalization but also to check stabilities of the new reactive sites. Figure 10.4 shows a MALDI mass spectrum of brominated P3HP. The signals (labeled blue) at 735.5–737.6 and 807.6–809.6 Da belong to monobrominated P3HP oligomers (together with

Li⁺ adduct) with 9 and 10 repeating units, respectively. The 2Da between the neighbor peaks are characteristics of the isotopic distribution of bromine atom. Similarly, the signals (labeled blue) at 751.6–753.6 and 823.5–825.5 belong to monobrominated P3HP oligomers (together with Na⁺ adduct) 9 and 10 repeating units, respectively. Structures of monobrominated P3HP oligomers with adduct ions discussed were given in Scheme 10.3A and B.

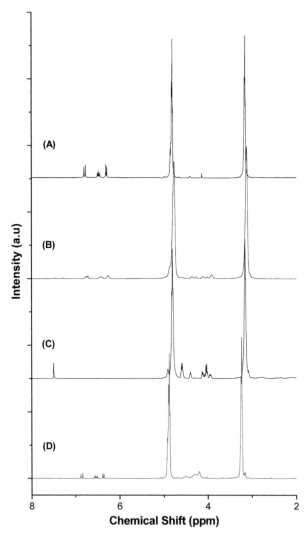

FIGURE 10.3 ¹H-NMR spectra of (A) pristine, (B) dibrominated, (C) monobrominated, (D) epoxylated P3HP oligomers.

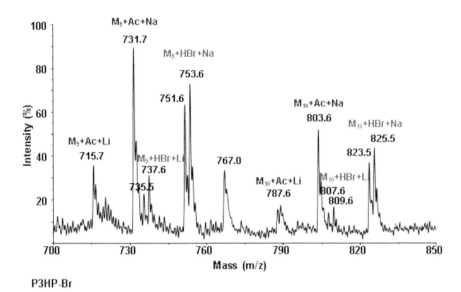

P3HP-Br

FIGURE 10.4   MALDI mass spectrum of monobrominated P3HP oligomer.

As mentioned above the monobromination reactions were conducted by HBr in acetic acid. We think that immediately after bromination of olefinic end-groups substitution reaction between bromides and acetate ions occurs since direct addition of acetic acid to C-C double bonds is not favorable. Considering this assumption, the signals (labeled red) at 715.7 and 787.6 Da were attributed to P3HP oligomers with acetate end-group (together with Li^+ adduct) with 9 and 10 repeating units, respectively. This approach is reasonable mathematically because the signals (labeled red) at 715.7 and 787.6 Da obey to the equation (M_n+59+23). Here, 59 Da is the mass of acetate ion, and n is 9 and 10). Similarly, the signals (labeled red) at 731.7 and 803.6 Da were attributed to P3HP oligomers with acetate end-group (together with Na^+ adduct) with 9 and 10 repeating units, respectively. Structures of P3HP oligomers with acetate end-group (with adduct ions) were given in Scheme 10.3C and 10.3D.

Since the excess amount of strong acid (HBr) was used for monobromination, Na ions at chain-ends exchange (displacement) with H^+ as shown in Scheme 10.3.

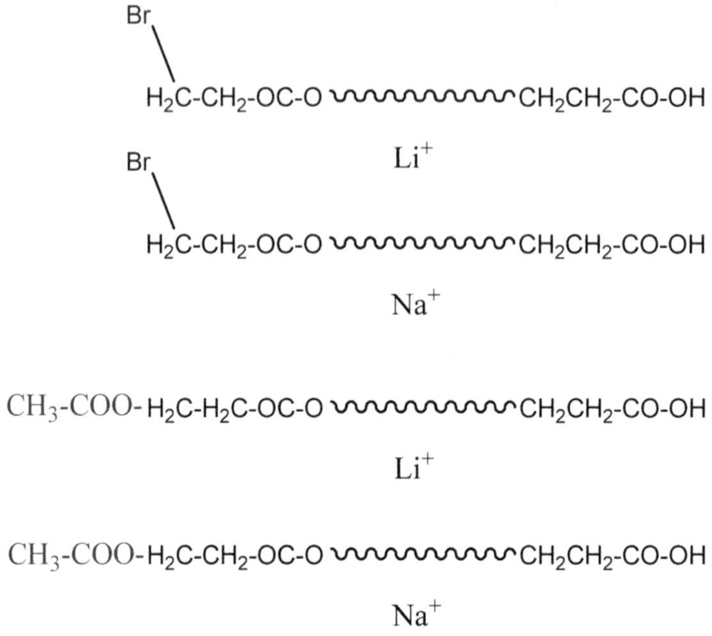

$$
\underset{\text{(A)}}{\text{H}_2\text{C-CH}_2\text{-OC-O} \sim\sim\sim\sim\sim\sim \text{CH}_2\text{CH}_2\text{-CO-OH}}
$$

Br
\
H₂C-CH₂-OC-O ∿∿∿∿∿∿CH₂CH₂-CO-OH   (A)

Li⁺

Br
\
H₂C-CH₂-OC-O ∿∿∿∿∿∿CH₂CH₂-CO-OH   (B)

Na⁺

CH₃-COO-H₂C-H₂C-OC-O ∿∿∿∿∿∿CH₂CH₂-CO-OH   (C)

Li⁺

CH₃-COO-H₂C-CH₂-OC-O ∿∿∿∿∿∿CH₂CH₂-CO-OH   (D)

Na⁺

SCHEME 10.3 Structures proposed for four types of signals in MALDI analysis of monobrominated P3HP oligomers with Li⁺ and Na⁺ adducts.

Figure 10.5 shows a MALDI mass spectrum of dibrominated P3HP. The signals labeled with blue color correspond to pristine oligomers with Li⁺ adduct. The signals labeled with red and green colors correspond to dibrominated P3HP oligomers with Li⁺ and Na⁺ as shown in Scheme 10.4, respectively. The signals at about 1031.3 and 1047.2 Da may be attributed to dibrominated P3HP oligomers with 12 repeating units since a bromine molecule has a molecular mass of about 160 Da. The signals corresponding to dibrominated oligomers of P3HP exhibit a bundle of signal distribution due to the isotopic composition of bromine atoms. Considering the intensities of both pristine and dibrominated oligomers of P3HP, it may be claimed that majority of the double bonds in both oligomers were modified as intended.

Figure 10.6 shows the MALDI mass spectrum of epoxylated P3HP oligomer. As can be seen clearly, the spectrum consists of two different groups of signals with a mass difference of 22 Da. There are no signals belonging to unreacted P3HP molecules, obeying to M_n+Li or M_n+Na formula. The signals labeled with blue color (671.4, 743.3 and 815.2) obey to M_n+16+Li equation. Here, n is a number of repeating units, and 16 comes from the mass of epoxy oxygen. Similarly, the signals labeled with red color (693.3, 765.2 and 837.1 Da) obey to $M_n+16+Na+Li$ equation. The mass differences between the signals labeled with blue and red colors are 22 Da; this may be attributed to the structures given in Scheme 10.5.

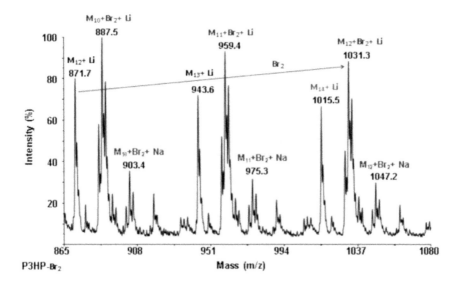

FIGURE 10.5 MALDI mass spectrum of dibrominated P3HP oligomers.

SCHEME 10.4 Structures responsible from the signals labeled with red and green colors in Figure 10.5.

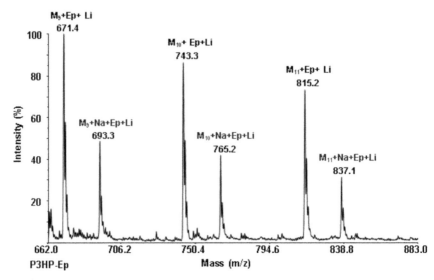

FIGURE 10.6   MALDI mass spectrum of epoxylated P3HP oligomers.

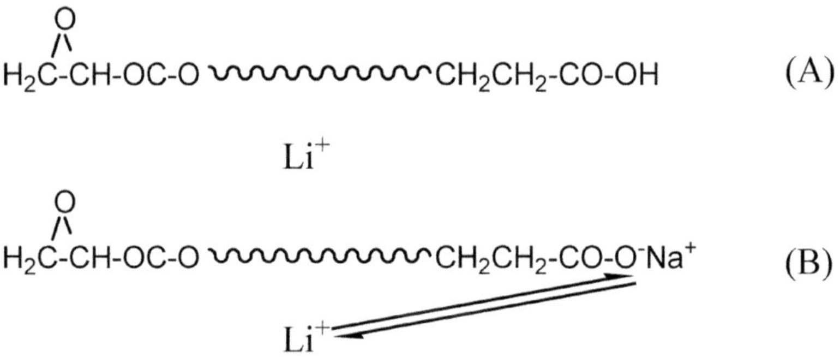

SCHEME 10.5   Structures of P3HP oligomers responsible from the signals in Figure 10.6. (A) carboxylic acid ended; (B) sodium carboxylate ended.

Since the well-known Click reactions [34, 35] between the azide and acetylene groups are of high yield, additional end-group modification study was performed on the monobrominated P3HP macromonomers to convert the bromide atoms into azide groups. Figure 10.7 gives MALDI mass spectrum of azide-ended P3HP macromonomers. The main signals (698.9, 771.1, 843.2, etc.) obey to M_n+azide+Li equation. Here, n is a number of repeating units. As clearly seen, azidation of monobrominated P3HP macromonomers is a successful process in the aspect of high yield.

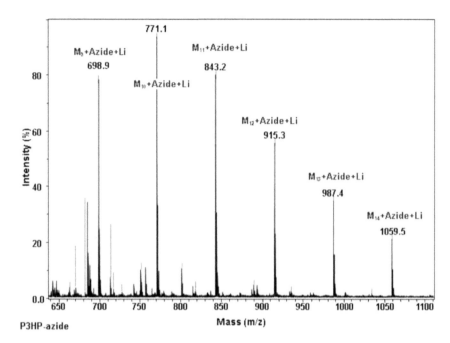

FIGURE 10.7   MALDI mass spectrum of azide-ended P3HP macromonomer.

10.4   CONCLUSIONS

When evaluated collectively, MALDI MS and ^{1}H-NMR analyzes showed that HTP of acrylic acid resulted in P3HP oligomer with olefinic and carboxylate end-groups as given in Scheme 10.1. Tertiary base as initiator behaves as strong base and responsible from the chains with olefinic end-group. This approach can be applied on various potential monomers to obtain oligomers with olefinic end-groups for various applications like "grafting-through" [36].

According to the ^{1}H-NMR and MALDI mass spectra of the modified oligomers, the olefinic end-groups were modified through successful monobromination, debromination, epoxylation, and azidation processes yielding more activated macromonomers. The novel macromonomers with reactive end-groups may find applications as grafting agents, a component of a block copolymer and surface modification reagent in synthetic chemistry and material science. We are currently working on "grafting-onto" application of these novel macromonomers to enhance the elastic properties of some common polymers.

ACKNOWLEDGMENTS

The study has been supported by Ordu University Scientific Research Project Coordination (ODU BAP HD–1606).

KEYWORDS

- 3-hydroxypropionate
- end-group functionalization
- hydrogen transfer polymerization
- hydrophobic macromonomer

REFERENCES

1. Furuhashi, Y., Iwata, T., Kimura, Y., & Doi, Y., (2003). *Macromol. Biosci., 3*(9), 462.
2. Cortizo, M. S., Molinuevo, M. S., & Cortizo, A. M., (2008). *J. Tissue Eng. Regen. Med., 2*, 33.
3. Cortizo, M. S., Alessandrini, J. L., Etcheverr, S. B., & Cortizo, A. M., (2001). *J. Biomater. Sci. Polym. Ed., 12*(9), 945.
4. Tasaka, S., Kawaguchi, M., & Inagaki, N., (1998). *Eur. Polym. J., 34*(12), 1743.
5. Zhang, D. Hillmyer, M. A., & Tolman, W. B., (2004). *Macromolecules, 37*, 8198.
6. Andreessen, B., Lange, A. B., Robenek, H., & Steinbuchel, A., (2010). *Appl. Environ. Microbiol., 76*(2), 622.
7. Suehiro, K., Chatani, Y., & Tadokoro, H., (1975). *Polym. J., 7*(3), 352.
8. Kricheldorf, H. R., & Scharnagl, N., (1996). *Polymer., 37*(8), 1405.
9. Ohnishi, S., Sugimoto, S., Hayashi, K., & Nitta, I., (1964). *Bull. Chem. Soc. Jpn., 37*, 524.
10. Watanabe, M., Togo, M., Sanui, K., Ogata, N., Kobayashi, T., & Ohtaki, Z., (1984). *Macromolecules, 17*(12), 2908.
11. Evstropov, A. A., Lebedev, B. V., Kulagina, T. G., & Lyudvig, E. B., (1980). *Polym. Sci. USSR., 21*, 2240.
12. Rosenoasser, D., Casas, A. S., & Figini, R. V., (1982). *Makromol. Chem., 183*, 3067.
13. Suzuki, Y., Taguchi, S., Hisano, T., Toshima, K., Matsumura, S., & Doi, Y., (2003). *Biomacromolecules, 4*, 537.
14. Nobes, G. A. R., Kazlauskas, R. J., & Marchessault, R. H., (1996). *Macromolecules, 29*, 4829.
15. Breslow, D. S., Hulse, G. E., & Matlack, A. S., (1957). *J. Am. Chem. Soc., 79*, 3760.
16. Saegusa, T., Kobayashi, S., & Kimura, Y., (1973). *Macromolecules, 7*(2), 256.
17. Yamada, B., Yasuda, Y., Matsushita, T., & Otsu, V. J., (1976). *Polym. Sci.: Polym. Lett. Ed., 14*(5), 277.

18. Tan, I., Zarafshani, Z., Lutz, J. F., & Titirici, M. M., (2009). *ACS Appl. Mater. Interfaces, 1*(9), 1869.
19. Hawker, C. J., Mecerreyes, D., Elce, E., Dao, J., Hedrick, J. L., Barakat, I., et al., **(1997).** *Macromol. Chem. Phys., 198,* 155.
20. Sumerlin, B. S., Tsarevsky, N. V., Louche, G., Lee, R. Y., & Matyjaszewski, K., **(2005).** *Macromolecules, 38,* 7540.
21. Goddard, J. M., & Hotchkiss, J. H., (2007). *Prog. Polym. Sci., 32,* 698.
22. Chaimberg, M., Parnas, R., & Cohen, Y. J., (1989). *Appl. Polym. Sci., 37*(10), 2921.
23. Zarafshani, Z., Akdemir, Ö., & Lutz, J. F., (2007). *Macromol. Rapid Commun., 29,* 1161.
24. Lemechko, P., Renard, E., Volet, G., Colin, C. S., Guezennec, J., & Langlois, V., (2012). *React. Funct. Polym., 72,* 160.
25. Tasdelen, M. A., Kahveci, M. U., & Yagci, Y., (2011). *Prog. Polym. Sci., 36*(4), 455.
26. Fray, M. E., Skrobot, J., Bolikal, D., & Kohn, J., (2012). *React. Funct. Polym., 72,* 781.
27. Thompson, M. S., Vadala, T. P., Vadala, M. L., Lin, Y., & Riffle, J. S., (2008). *Polymer, 49,* 345.
28. Hmamouchi, M., & Prudhomme, R. E., (1988). *J. Polym. Sci.: Part A: Polym. Chem., 26,* 1593.
29. Haldar, U., Ramakrishnan, L., Sivaprakasam, K., & De, P., (2014). *Polymer, 55,* 5656.
30. Ballard, N., Salsamendi, M., Santos, J. I., Ruipérez, F., Leiza, J. R., & Asua, J. M., (2014). *Macromolecules, 47,* 964.
31. Tong, K. H., Wong, K. Y., & Chan, T. H., (2003). *Org. Lett., 5,* 3423.
32. Parent, E. E., Dence, C. S., Jenks, C., Sharp, T. L., Welch, M. J., & Katzenellenbogen, J. A., (2007). *J. Med. Chem., 50,* 1028.
33. Çatıker, E., Güven, O., & Salih, B., (2017). *Polym. Bull.,* https://doi.org/10.1007/s00289–017–2014–2.
34. Presolski, S. I., Hong, V. P., & Finn, M. G., (2011). *Curr. Protoc. Chem. Biol., 3*(4), 153. doi: 10.1002/9780470559277.ch110148.
35. Shengtong, S., & Peiyi, W., (2010). *J. Phys. Chem. A., 114*(32), 8331.
36. Bhattacharya, A., & Misra, B. N., (2004). *Prog. Polym. Sci., 29,* 767.

CHAPTER 11

Photochemical Conversion of Polycyclopropanes

K. G. GULIYEV, A. E. RZAYEVA, A. M. ALIYEVA, G. Z. PONOMAREVA, R. D. DZHAFAROV, and A. M. GULIYEV

Institute of Polymer Materials of Azerbaijan National Academy of Sciences, S. Vurgun Str.124, Sumgayit AZ5004, Azerbaijan, E-mail: ipoma@science.az

ABSTRACT

The esterification of polyvinyl alcohol (PVA) with chloranhydride of phenylcyclopropane acid with the aim of preparation of a different-link polymer containing fragments of the vinyl alcohol and group with cyclopropane ring in the side links has been studied. As a result of the investigation the new polymer has been synthesized, and its composition, structure, and properties on the basis of data of the IR- and NMR-spectroscopy have been determined. The photochemical structuring of the synthesized copolymer has been studied, and it has been established that the modified polymer possesses photosensitivity of the order 48–51 cm^2/J, which allows its use as a negative photoresist.

11.1 INTRODUCTION

Now, the ability of polymers to lose solubility after irradiation with ultraviolet light is widely used in photochemical processes both in polygraphy and for the preparation of semiconductor devices, printed circuits [1, 2].

The mechanism of light action on photosensitive polymers containing cyclopropane fragments in macromolecule is still very little studied. The functionally substituted cyclopropyl styrenes are very perspective in terms of creation of the new photosensitive materials [3–5].

11.2   EXPERIMENTAL

11.2.1   MATERIALS

An interaction of polyvinyl alcohol (PVA) with chloranhydride of phenylcyclopropane carboxylic acid is accompanied by the formation of a different-link polymer containing fragments of vinyl alcohol and cyclopropane in combination with the carbonyl group in the side chains.

In this work, PVA was used, prepared by alkaline saponification with residual molar content of the acetate groups 0.72% and molecular weight 56000.

The chloranhydride of phenylcyclopropane acid was prepared by the interaction of the corresponding acid with an excess of thionyl (sulfinyl) chloride.

11.2.2   METHODS

The photochemical structuring of the polymer was carried out as follows: at first there have been made 2–10% solutions of the polymers. Then by a method of centrifugation at 2500 rev/min, it was carried out applying the films on a glass substrate (K–8) by size 60 x 90 mm. The thickness of prepared film-resists was measured by micro interferometer "LINNIKAMII–4." The layer thickness of resist after its drying for 10 min at room temperature and for 20 min at 40–45°C/10 mm was 0.15–0.20 mcm.

As a source of UV-irradiation it was used the mercury lamp DPT – 220 (current intensity – 2.2 A, distance from radiation source – 15 cm, mobile shutter rate of exponometer – 720 mm/h) exposition time – 5–20 sec. The content of the insoluble polymer was calculated on residue weight.

The IR-spectra of the copolymer were registered on spectrometer "Specord" M–80, NMR-spectra-on spectrometer BS–487B Tesla (80MHz) in the solution of deuterated chloroform.

11.3   RESULTS AND DISCUSSION

In this work, we have studied the action of UV-light on polymers containing cyclopropane fragment in the macromolecule. This task included a systematic study of the influence of irradiation on the photosensitivity of the polymer containing various substituents at cyclopropane ring. With this aim, the polyvinyl cyclopropane has been synthesized, and its photosensitivity has been investigated.

The total scheme of synthesis of the polymer forms can be presented as follows:

$$\left(CH_2\text{-}CH\right) + C_6H_5\text{-}CH\text{-}CH\text{-}C \overset{O}{\underset{Cl}{\diagdown}} \xrightarrow{\text{Pyridine}} \left(CH_2\text{-}CH\right)\left(CH_2\text{-}CH\right)$$

In the IR-spectrum of the prepared modified PVA, the absorption bands characteristic for cyclopropane group (1035–1040 cm^{-1}) have been detected. An availability of the absorption band in the spectrum in the field of 1700 cm^{-1} characterizes the valence vibrations of carbonyl groups present in the synthesized polymer and also the absorption bands in the field of 3500–3600 cm^{-1}, referring to hydroxyl groups.

The data of the IR-spectroscopy confirm the structure of the modified polymer. A formation of such structure is also confirmed by data of NMR-spectroscopy. It has been established that the NMR-spectra contain the resonance signals for protons of benzene ring (δ = 6.85–7.25 ppm.) and cyclopropane group (δ = 0.75–1.75 ppm.).

It is assumed considering the acylation of PVA with chloranhydrides in a medium of the solvents of amide type that the solvent not only blocks an intensive intermolecular interaction of PVA, favoring its solubility, but also favors nucleophilic substitution reaction at the carbon atom of the chloranhydride group.

As a solvent it was used the dimethylacetamide allowing to prepare sufficiently concentrated solutions of PVA at 30–40°C. It has been revealed that a degree of substitution depends on reaction conditions and has been determined on data of the elemental analysis. With the aim of establishment of structure and composition of the synthesized polymer prepared by modification of polyvinyl ester, the elemental and spectral analyzes (IR- and NMR-spectroscopy) have been carried out.

Attempts to use dimethylformamide as a solvent showed that it has the worst dissolving capacity in relation to PVA and allows to prepare the concentrated solutions of the latter one only at higher temperatures.

It has been found during the investigation of reaction in the solution that the process is practically finished via 1 h, and a degree of substitution essentially depends on the reaction temperature and also concentration and ratio of reagents.

As is seen from Figure 11.1, a dependence of the degree of substitution on temperature during the reaction of PVA and chloranhydride of cyclopropane carboxylic acid in a medium of dimethylacetamide has an extreme character with the maximum at 70°C.

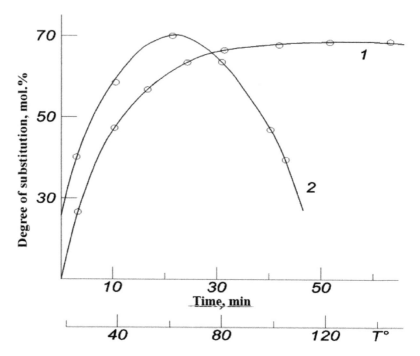

FIGURE 11.1 Dependence of degree of substitution on time (1) and temperature of reaction (2); concentration 0.7 M and 70° (1), time – 1 h (2).

At higher temperatures, a degree of substitution was gradually decreased due to side interaction of solvent and chloranhydride [6].

The dependence of a degree of substitution on the concentration of the initial reagents also has extreme character (Figure 11.2). And in this case, the growth of a degree of substitution with an increase in concentration is determined by a rate increase of the main reaction. The decrease of a degree of substitution with a further increase in concentration can be explained by the difficulty of transition of a larger quantity of PVA to a solution. In Figure 11.2, a dependence on the concentration of yield of forming polymer is also presented. Its character corresponds to the character of the dependence of a degree of substitution on the concentration of reagents.

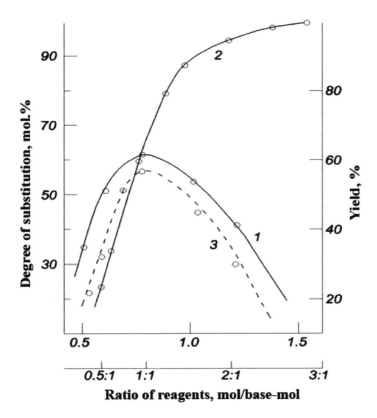

FIGURE 11.2 Dependence of degree of substitution on concentration (1), the ratio of the initial reagents (2) and dependence of yield of end polymer on the concentration of reagents (3); time – 1 h, 60°; concentration – 0.7 M (2).

It was known that the polyvinyl esters of some acids, chloranhydrides of which possess higher stability to hydrolysis, can be prepared in the interaction of latter ones with PVA in the conditions of Schotten-Bauman reaction [7].

In this work, we have also used this reaction. The data are presented in Figure 11.3.

As is seen from the figure, an interaction of chloranhydride of phenyl cyclopropane acid and PVA in these conditions proceeds sufficiently rapidly and is finished for 40 min.

The reaction temperature rise to 40° leads to a noticeable growth of degree of substitution. Apparently, in these conditions, an increase of the rate of the main reaction affected to a greater extent than an increase in the rate of hydrolysis of the chloranhydride.

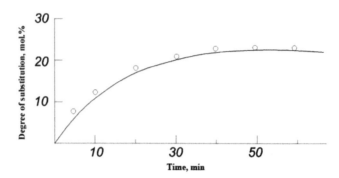

FIGURE 11.3 Dependence of degree of substitution on esterification at the border of phase division. Organic phase-benzene, chloranhydride: PVC = 1 mol: 1 base-mol, concentration–0.4 M, 30°, quantity NaOH–1 mol/mol of chloranhydride.

An availability of fragments of the vinyl alcohol and cyclopropane ring in combination with the carbonyl group in the side chain of the polymer should favor its structuring in the influence of UV-irradiation.

We have investigated the photochemical conversion of the modified polymer under the influence of UV-irradiation. It has been shown the availability of two processes at the photochemical influence on polymers: formation of more complex cyclic and spatial chain structures.

It has been revealed that the modified polymer is very high photosensitive material (48–51 cm²/J), possesses good adhesion to substrates and low deficiency of the films.

It has been established that the prepared polymer is easily subjected to the photochemical conversions with the formation of cross-linked net structures and allows its use in the lithographical processes as a negative photoresist material.

11.4 CONCLUSIONS

The polyfunctional polymer containing hydroxyl, carbonyl, cyclopropane groups in macrochain has been synthesized by esterification reaction of PVA by chloranhydride of phenylcyclopropane acid, and its composition and structure have been established.

It has been revealed that the synthesized polymer has photosensitivity of the order 48–51 cm²/J and can be used as the photosensitive base of resists of negative type.

ACKNOWLEDGMENTS

We express our gratitude to Dilbar Nurullayeva for her help.

KEYWORDS

- **chloranhydride of phenylcyclopropane acid**
- **esterification**
- **photosensitivity**
- **polyvinyl alcohol**

REFERENCES

1. Glazstein, L. Ya., & Korablin, A. S., (1972). Polymer resists for electrolithography. *M.: TsNII Electronics, 12*, 155.
2. Bokov, Y. U. S., (1982). Photo-, electrono- and roentgeno-resists. *M: Radio and Communication, 136.*
3. Guliyev, K. G., Ponomareva, G. Z., & Guliyev, A. M., (2007). Vysokomolek. *Soed., Ser. B., 49*(8), 1577.
4. Guliyev, K. G., Garayeva, A. A., Ponomareva, G. Z., Aliyeva, A. M., & Guliev, A. M., (2015). *Russian, J. Applied Chem., 88*(6), 1047.
5. Guliyev, K. G., Aliyeva, A. M., Mamedli, S. B., Nurullayeva, D. R., & Guliyev, A. M., (2016). *Eur. Chem. Bull., 5*, 108.
6. Savinov, V. M., & Sokolov, L. B., (1965). Vysokomolek. *Soed., 7*, 772.
7. Korshak, V. V., Shitman, M. I., & Yaroshenko, I. V., (1977). Vysokomolek. *Soed., (B), 19*, 234.

CHAPTER 12

Photosensitive Copolymers on the Basis of Gem-Disubstituted Vinyloxycyclopropanes

RITA SHAHNAZARLI, SHABNAM GARAYEVA, and ABASGULU GULIYEV

Institute of Polymer Materials of Azerbaijan National Academy of Sciences, S. Vurgun str.124, Sumgayit AZ5004, Azerbaijan, E-mail: abasgulu@yandex.ru

ABSTRACT

The synthesis of cyclopropane-containing vinyl ethers with geminal substituents at cycle has been carried out, and their homo-and copolymerization with maleic anhydride (MA) in the radical conditions have been carried out. The interrelation between the process of the copolymerization and formation of donor-acceptor type between comonomers in the system of complexes has been investigated. It has been established that in the conditions of copolymerization there are formed the donor-acceptor complexes between comonomers, an availability of which has been revealed by a method of NMR-spectroscopy. It has been considered the scheme of copolymerization due to the stage of addition to growing chain end both of separate comonomers and complex monomers as a whole. The photosensitivity of films of the synthesized copolymers has been investigated, and it has been shown that it depends on MW, the solubility of the polymer, nature of the functional substituent and a number of other factors.

12.1 INTRODUCTION

It was known that the polymers, macromolecules of which contain unsaturated or cyclic groups both in basic and in the side chains easily pass to three-dimensional net structures at their thermal treatment or in the influence of radiation, electron beam or UV-irradiation on them [1]. During their reaction with various vinyl monomers in the presence of initiators, they can also be structured. Therefore, the polymers containing chemically active groups such as vinyl, epoxide, allyl in the side chains of macromolecules is suggested to use in making the photoresists of negative type.

The vinyl ethers (in particular, vinyl glycidyl ethers of glycols) have proven to be active monomers for the preparation of technically valuable materials with a given complex of properties (strength, thermal stability, plasticity, adhesion, etc.) from them [2]. The copolymers of the vinyl monomers and maleic anhydride (MA) also possess a wide spectrum of important characteristic and have a large scientific-practical value. Consequently, such copolymers are very interesting and attractive macromolecular compounds for their use as a base in the creation of photosensitive materials.

12.2   EXPERIMENTAL

The IR-spectra of the synthesized monomers and homo- and copolymers prepared from them were taken on apparatus "Cary 630 FTIR" of firm Agilent Technologies (crystal ZnSe). The NMR-spectra were taken on spectrometer "Fourier" (frequency 300MHz) of firm "Bruker" in the various solvents, internal standard – hexamethyldisiloxane. The chemical shifts of signals are presented in a scale δ (ppm.).

The purity of the synthesized compounds was determined by a method of gas-liquid chromatography on the chromatogram of mark LKhM–8 MD.

The characteristic viscosities of the polymer products were determined in Ostwald viscometer in the various solvents.

The parameters of MWD were determined on high-performance liquid chromatography of firm "Kovo" (Czech Republic) with the refractometric detector. There were used two columns by size 3.3 x 150 mm, "Separon-SGX" with the size of the particles 7 mcm and porosity 100Å served as an immobile phase. The calibration dependence lg M on V_R was measured in the range of M = (3–100) x 10^2 with use of polyethylene glycol standards, temperature –20–25°C, 1 account –0.13 ml.

The photostructuring process of the polymers was studied by irradiation of UV-light of the films applied by centrifugation on glass substrates of the polymer solution in MEK or acetone (film thickness 0.4–0.5 mcm) for 15–180 min at room temperature.

The synthesis of monomers 1–6 has been carried out by the interaction reactions of diethoxycarbonyl- and dichlorocarbenes with one of two double bonds of divinyl ether with subsequent conversion of the prepared adducts into corresponding compounds [3, 4].

The polymerization of the compounds 1–6 were carried out in single-chamber ampoule in the presence of 2,2′-azo-bis-isobutyronitrile (AIBN) and ditertiary butyl peroxide (DTBP) at various temperatures (60–120°C), depending on type of used initiator both in solution (benzene or toluene) and in mass. Polymerization duration is 30 h.

12.3 RESULTS AND DISCUSSION

With the aim of preparation of the linear polymers with reactive side groups we have synthesized the vinyl ethers with various functional groups in geminal position at cyclopropane ring.

$$Gly = -CH_2-CH-CH_2$$

It should be taken into account that the cyclopropane group due to its specific electron structure gives the electronic influence of the substituent [5]. The synthesized vinylcyclopropyl ethers contain the various substituents at the three-membered cycle, which possess negative (−J) or positive (+J) inductive effect:

$x = H$ (1); CH_2OMe (2); CH_2OGly (3); CO_2Et (4); CO_2Gly (5); Cl (6)

It follows from synthesized compounds 1–6 that depending on nature of substituent, a nucleophilicity of double bond either is decreased (when substituent X possesses positive (+J) effect) or is increased (when X possesses negative (–J) effect).

In the IR-spectra of the synthesized compounds, there are the absorption bands characteristic for corresponding groups and the NMR-spectra (Table 12.1) characterize the suggested structures.

TABLE 12.1   Spectral Characteristics of *Gem*-Disubstituted Vinyloxycyclopropanes

Code of monomer	Chemical shifts of protons and proton-containing groups, δ, ppm							
	H^1	H^2	H^3	H^4	H^5	H^6	X	
							Geminal substituent	Chemical shifts
1	4.63	4.15	6.45	2.22	0.43	0.68	H	0.43
2	4.64	4.13	6.45	2.20	0.44	0.66	CH$_2$OCH$_3$	3.33(CH$_2$O)
								3.30(OCH$_3$)
3	4.26	4.34	6.42	2.26	0.66	0.67	CH$_2$OGly	3.37(CH$_2$-O)
								3.38;3.63(CH$_2$)
								2.86(CH)
								2.63;2.38(CH$_2$)
4	4.28	4.35	6.45	3.30	1.40	1.40	CO$_2$CH$_2$CH$_3$	4.12 (qt.CH$_2$),
								1.29(tp.CH$_3$)
5	4.24	4.36	6.42	2.28	0.68	0.68	CO$_2$Gly	4.34;4.00(CH$_2$)
								3.13(CH)
								2.63;2.38(CH$_2$)
6	4.21	4.46	6.43	2.0	1.40	1.40	Cl	–

The monomers 1–6, owing to a combination of the vinyl group with ether oxygen and cyclopropane ring possess specific properties. A substitution of alkyl group in the vinyl ether for cyclopropane one having *p*-character favors a weakening of the positive influence of ether group [6], as a result of which an inclination of these monomers to polymerization in the conditions of radical initiation is rather improved.

The results of the model reactions on the addition of thiols to vinyloxycyclopropanes in the conditions of radical initiation [7] allow to estimate the possibility of the behavior of the radical polymerization of the synthesized compounds 1–6 on scheme excluding the opening of three-membered cyclopropane ring preliminarily. The polymerization proceeds purely only on double bond:

It has been experimentally established that the radical polymerization of the synthesized monomers, independently of reaction conditions, leads to the formation of oligomers with low MW (degree of polymerization 15–20). This is confirmed by viscous characteristics of the prepared polymers and determination of their MW by a method high-performance liquid chromatography (Figure 12.1 and Table 12.2).

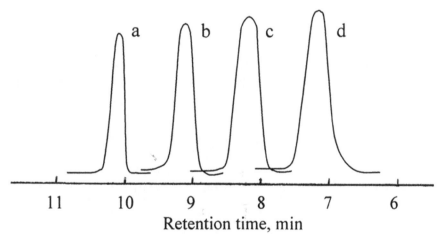

Retention time, min

FIGURE 12.1 Curves of MWD of a homopolymer of the compound 3, prepared in the presence of AIBN (a), DTBP at 100°C (b) and at 120°C (c), and $BF_3 \cdot O(C_2H_5)_2$ (d).

A comparison of the IR-spectra of the synthesized monomers 1–6 and polymers prepared on their basis showed that the absorption bands at 1640–1645 cm^{-1}, available in the initial monomers, characteristic for vinyl group disappear in the polymerization process. The absorption bands characterizing the availability of other functional groups and skeleton vibrations of cyclopropane ring, in this case, are kept in the spectrum of the polymer. The analogous results are obtained in consideration of NMR-spectra of the initial monomers and polymers prepared on their basis. These data confirm that the polymerization of the synthesized monomers proceeds due to opening only double C = C-bond, in this case, the cyclopropane group is kept.

TABLE 12.2   Conditions and Results of Homopolymerization of Vinyloxycyclopropanes

Code of monomer	T, °C	Time, h	Initiator		Solvent	Yield of polymer, %		MW
			Type	Quantity, mol.%		Soluble fraction	[η], dl/g	
1	70	30	AIBN	0.5	Benzene	100	0.13	1200–1700
2	60	30	AIBN	0.6	Benzene	100	0.15	2600–3400
3	60	30	AIBN	0.8	Benzene	96.2	0.23	1000–1500
4	100	30	DTBP	0.8	Toluene	100	0.13	3400–4500
5	120	30	DTBP	0.8	Toluene	97.5	0.26	1100–1700
6	70	30	AIBN	0.8	Benzene	100	0.16	2300–3000

It should be noted that during polymerization in the presence of radical initiators with temperature rise a yield of the polymer is increased. In particular, at temperature rise from 100 to 120°C a yield of the polymer for 6 h is increased approximately in 3 times. A calculation of activation energies on the Arrhenius equation showed that they correspond to the values characteristic for the polymerization of the vinyl monomers (104 J/mol).

An increase of the concentration of initiator doesn't lead to the increase of MW of the prepared polymers (as well as their viscosities), these indices remain almost on the same level, and some cases are even decreased. An increase of the polymerization duration doesn't influence on MW of the prepared polymers; it influences only on the yield of polymer. The greatest MW are observed in a case of polymerization of monomers having substituents with –J-effect.

Thus, in a case of the radical polymerization of monomers 1–6 the process proceeds selectively on the vinyl group with the preparation of oligomers of the linear structure with side functionally substituted cyclopropane groups able to further conversions in more mild conditions.

For revealing the dependence between the structure of *gem*-disubstituted compounds and their polymerization activity we have carried out the polymerization of these compounds in the same conditions both in mass and in benzene solution in the presence of AIBN. It has been found in this case that the monomer 4 shows the greatest activity, and the monomer 1 shows the least activity. According to the results of the obtained data a number of activities of monomers (4 > 5 > 6 > 3 > 2 > 1) has been constructed. It means that an introduction of the polar ethoxycarbonyl group in cyclopropane ring due to -J effect of substituent an activity of monomer in the radical reactions is increased. The least activity of the monomer 1 has been connected with the absence of the functional group at cyclopropane ring.

In the NMR-spectra of the polymers, two partially signals overlap with the same intensity. These signals are referred to protons of CH_2-group of cyclopropane ring ($\delta = 1.54$ ppm) and protons of CH_2-groups of the polymer chain ($\delta = 1.9$ ppm). The analogous signals of equal intensity are appeared for methane protons of the three-membered cycle ($\delta = 3.60$ ppm.) and protons of CH-groups of the polymer chain ($\delta = 4.12$ ppm).

The polymerization of the compounds 3 and 5 was carried out in the solution of benzene (toluene) and in mass at various temperatures depending on the type of used initiator. As a result of polymerization the soluble oligomer products (degree of polymerization of the compounds 3 and 5 was 4–6) of the linear structure with epoxide groups and cyclopropane fragments in the side, chain have been prepared. The prepared oligomers are the viscous resins with molecular weight $1000 \div 1700$ and relative viscosity $63 \div 77$ sec (depending on the type of monomer), soluble in acetone, benzene, dioxane, diethyl ether and insoluble in water (Table 12.3).

TABLE 12.3 Some Physical-Chemical Properties of Epoxy-Oligomers on the Basis of the Compounds 3 and 5

Indices	Epoxy-oligomer on the basis	
	Compound 3	**Compound 5**
Viscosity	From low viscosity to viscous	
Color	From light yellow to yellow	
Content of epoxide groups, %	32.7–33.2 (33.6)	28.3–28.8 (30.3)
Relative viscosity*, sec	63–66	72–77
Characteristic viscosity, cm³/g	0.05–0.09	0.07–0.11
Molecular weight**	1000–1500	1100–1700

* Determined on ball viscometer (d = 2.37 mm at 30°C).
** Determined by high-performance liquid chromatography.

The synthesized oligomer products containing epoxide groups besides cyclopropane groups can be used as the reactive oligomers. For example, they easily participate in the hardening reaction with amine and anhydride hardeners forming cross-linked compounds. In this way, one can prepare the various glues, coatings, and other products.

In our further investigations, the synthesized *gem*-disubstituted vinyloxycyclopropanes have been included in radical copolymerization with MA. Since in the radical conditions of the copolymerization each of comonomers of this system is not able to homopolymerization, the chain growth is possible only due to intercoupling of comonomers, in which can be participated both free monomers and such ones connected with complexes.

The calculated MW values (accepting $K = 4.11 \cdot 10^{-4}$ and $\alpha = 0.89$ [8]) of the copolymers prepared in the various conditions indicate to their relatively low MW; their characteristic viscosity changed in the ranges from 0.27 to 0.60 dl/g.

The spectral analysis showed that the copolymers, prepared both in mass and in solution, are identical on composition, intramolecular distribution of links and chemical structure. It has been established on data of IR-spectroscopy and elemental analysis that the composition of the prepared copolymers practically doesn't depend on the ratio of the initial monomers and close to equimolar one (Table 12.4).

As a result of the copolymerization of *gem*-disubstituted vinyloxycyclopropanes with MA, there have been prepared the copolymers insoluble in aromatic hydrocarbons being of light-yellow crystalline substances. However, the copolymers were well dissolved in polar organic solvents -in acetone, ethers, DMF, owing to which one can cast the thin films from them. The maximum yield of the copolymer corresponded to a ratio of the monomers in the initial mixture 1:1. The method of carrying out of copolymerization (in solution or in mass) shows a slight influence on the yield of copolymers.

The IR-spectra of the prepared copolymers showed an availability of the absorption bands in the field of 1765–1845 cm^{-1}, characteristic for carbonyl group of anhydride fragment. An availability of the absorption bands at 1020–1050 cm^{-1} characterizes the skeleton vibrations of the three-membered carbonic cycle. In a case of the copolymerization with the participation of the compounds 3 and 5 there have been prepared the copolymers, in the IR-spectrum of which along with other groups, the availability of the absorption bands of epoxide groups at 830–840 cm^{-1} and 1260–1265 cm^{-1} was observed. At the same time in the IR-spectra of the copolymers the absorption bands in the field of 1640–1645 cm^{-1}, corresponding to valence vibrations of double C = C-bond, are absent. Our experiments showed that there are not the essential changes in the IR-spectrum of copolymers prepared at various ratios of the initial comonomers. All these data allow to conclude that: firstly, in the copolymerization process, only double bond of vinyloxycyclopropane (in this, the three-membered cycle is not touched) takes part. Secondly, the copolymerization proceeds with the participation of both comonomers with the formation of the copolymers of equimolar composition and alternating structure and thirdly, a ratio of comonomers in the initial mixture doesn't influence on the composition of the prepared copolymers.

TABLE 12.4 Copolymerization of *Gem*-Disubstituted Vinyloxycyclopropanes with Maleic Anhydride ([AIBN] = 0.02 mmol/l; T = 70°C)

Code of monomers	Composition of the initial mixture, mol.% of monomers	Copolymerization time, min	Conversion, %	Composition of copolymer, mol.% of monomers
1	90.0	20	8.6	50.1
	70.0	20	10.8	50.0
	50.0	20	15.2	50.0
	30.0	20	12.1	49.9
	10.0	20	9.4	49.8
3	90.0	20	8.0	49.8
	75.0	20	9.92	49.95
	50.0	20	13.6	50.0
	25.5	30	11.1	49.9
	10.0	60	8.7	50.2
4	90.0	20	7.8	50.7
	75.0	20	9.3	50.0
	50.0	20	14.1	49.5
	25.5	30	12.4	48.5
	10.0	60	11.0	48.0

It was known that in the copolymerization of the vinyl ethers with electron-acceptor monomers, for ex. MA, there is the donor-acceptor interaction between the comonomers, which extremely increases the polar effect of the reaction. In this case, the complex-formation between comonomers in the initial mixture is reflected on characteristic peculiarities of proceeding of the copolymerization process [9]. For the study of the complex-formation process there were taken the NMR-spectra both of pure MA and MA in the presence of comonomers 1–6 in the solution $CDCl_3$ at various temperatures (Table 12.5).

TABLE 12.5 Complex-Formation Constants of the Vinylethers 1, 2 and 4 with MA (Solvent -$CDCl_3$, [MA] –0.03 mol/l, [Vinyloxycyclopropane]–0.65–2.20 mol/l)

Complex	T, °C	K, l/mol
Monomer **1** - MA	25	0.25
	45	0.20
	65	0.16
Monomer **2** - MA	25	0.45
	45	0.41
	65	0.38
Monomer **4** - MA	25	0.37
	45	0.32
	65	0.27

As shown in the data of NMR-spectroscopy, the chemical shift of protons at double bond of MA (δ = 7.25 ppm) in the presence of comonomer of vinyloxycyclopropane undergo the displacement to a weaker field (δ = 7.13 ppm.). Such change of the chemical shift of protons of MA in the presence of comonomers listed above has been connected with the formation of the donor-acceptor complexes in the system. In other words, in mixing of comonomers in the initial mixture it takes place a charge transfer from donor (vinyloxycyclopropane) to acceptor (MA), which evidences about formation of complex between them and as a result a displacement of the chemical shift of protons in double bond in a molecule of MA occurs [10].

The equilibrium constants (K) of the complexes of the vinyl ethers with MA were determined analogously to methodology presented in work [11].

The investigation of the temperature dependence of value K allowed to estimate the thermodynamical parameters of complex-formation on Vant-Hoff equation [12] (Table 12.6).

TABLE 12.6 Values of Enthalpy and Entropy at Complex-Formation of Monomers 1,2 and 4 with MA

Complex	-ΔH, kcal /mol	-ΔS, e.un.
Monomer 1 - MA	2.03	9.74
Monomer 2 - MA	1.23	5.28
Monomer 4 - MA	1.39	6.76

The value ΔH evidences of very weak donor-acceptor interactions in the studied complexes.

For investigation of the photosensitive properties of the prepared copolymers, we have carried out the photochemical structuring. The films of the synthesized copolymers poured off from solutions and dried at 60°C, were irradiated by xenon lamp by capacity 2 kWt for 30 min. In this case, we have detected that during irradiation the films partially lost their solubility. This was clearly observed in the comparison of the IR-spectra of these copolymers' films before and after irradiation. The spectral investigations revealed the decrease of intensity of the absorption bands of cyclopropane and epoxide groups at 1040 cm^{-1} and 836 cm^{-1}, respectively. It was simultaneously observed the displacement of the absorption bands of carbonyl group from 1720 cm^{-1} to 1730 cm^{-1} without change of its intensity. The change of intensities of the absorption bands after UV-irradiation of the films has been probably connected with the formation of "bridge bonds" between macromolecules due to the opening of cyclopropane and epoxide groups.

The cross-linking process under the action of photoirradiation was investigated by immersion of irradiated films to solvent (acetone) for 3 min. After drying and weighting of the films a quantity of insoluble part of the film has been determined (Figure 12.2). Since the mass losses of the film depend on the degree of its cross-linking, with the increase in the degree of cross-linking the mass losses fall.

The unirradiated areas of the films were completely dissolved in acetone. On the increase of irradiation duration (in this the degree of cross-linking increased), the solubility of the films fell, and after 30 min. of irradiation, the mass of the soluble part was minimal and constant. It has also been established that the addition of approximately 1.0% of benzoin as sensitizer or cobalt-naphthenate to the system led to the sharp fall of the mass loss. In this case, the degree of cross-linking of the copolymer films reached 80–90% already for 2–4 min. This value depended on nature of the functional substituent at cyclopropane fragment, MW, solubility of polymer and a number of other factors.

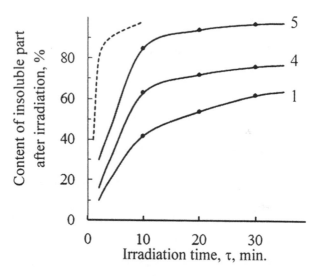

FIGURE 12.2 Dependence of mass loss of the copolymer films 1, 4 and 5 with MA after irradiation on exposition time (for copolymer 5 with MA with the addition of sensitizer).

12.4 CONCLUSIONS

1. By interaction reaction of diethoxycarbonyl- and dichlorocarbenes with divinyl ether there have been synthesized the *gem*-disubstituted vinylcyclopropanes, on the basis of which a number of their derivatives has been prepared. The homopolymerization of the synthesized substituted vinyloxycyclopropanes in the presence of radical initiators both in mass and in solution has been carried out, and it has been shown that in the polymerization only vinyl group participates. The composition and structure of the synthesized monomers and polymers prepared on their basis have been established by data of chemical and spectral analyzes.

2. The joint polymerization of the synthesized monomers with MA has been carried out, and the interaction between comonomers has been investigated. It has been shown that the copolymerization proceeds with formation of complexes of donor-acceptor type. It has been established that the composition of the prepared copolymers doesn't depend on composition of the initial mixture and in all cases corresponds to equimolar composition of comonomers. There have been determined the complex-formation constants, the values of which evidence about the appearance of the weak complexes. The temperature dependence

of complex-formation constants has been established, and thermodynamic parameters of the process have been calculated.

3.   Some properties, including photosensitivity of the prepared homo- and copolymers have been investigated. It has been established that the content of insoluble part of the copolymer films after irradiation depends on irradiation time, structure of polymer, nature of functional substituent in cyclopropane ring, MW, and its solubility.

KEYWORDS

- **copolymers with maleic anhydride**
- **cyclopropane-containing epoxy oligomers**
- **donor-acceptor complexes**
- **negative photoresist**
- **photosensitivity**
- **substituted vinyloxycyclopropanes**

REFERENCES

1.   Winterling, H., Haberkerna, H., & Swiderek, P., (2001). Electron-induced reactions in thin solid films of cyclopropane. *Phys. Chem. Chem. Phys., V.3, N., 20,* 4592–4599.

2.   Raskulova, T. V., Sviridov, D. P., Tupota, K. V., & Antokhina, E. Y., (2008b). Investigation of strength properties of coatings on the basis of copolymers of vinyl chloride and vinyl ethers. *Vestnik Angarskogo GTU., 2*(1), 16–19.

3.   Shahnazarli, R. Z., & Guliyev, A. M., (2015). Cationic polymerization of vinylcyclopropyl ethers and photo-structurization of the prepared polymers. *SBU, J. of Sci., 11*(3), 349–352.

4.   Shahnazarli, R. Z., & Guliyev, A. M., (2017). Synthesis and radical polymerization of glycidyloxycarbonyl- and glycidyloxymethyl-substituted vinyloxycyclopropanes. *Collection of Papers of the Winners of the VI International Scientific-Practical International Innovation Research," Penza,* p 26–32.

5.   Yanovskaya, L. A., Dombrovskiy, V. A., & Khusid, A. K., (1980). *Cyclopropanes With Functional Groups.* M.: Nauka, p. 223.

6.   Plemenkov, V. V., (1997). Electronic and spatial structure of multicomponent cyclopropanes. *J. Org. Chem., 33*(6), 849–859.

7.   Shahnazarli, R. Z., (2015). Adducts of thiols with allylcyclopropoylmethyl ethers– biocide additions for polyvinyl chloride. *4ᵗʰ International Polymeric Composites Symposium, Exhibition & Brokerage Event "IPC–2015,"* Izmir, Turkey, (PK–025).

8. Antonovich, F. F., Kruglova, V. A., Skobeeva, N. I., et al., (1980). Copolymerization and complex-formation in the system of 2-trichloromethyl–4-methylene–1,3-dioxolane–maleic anhydride. *Vysokomolek. Soyed., 12-A*(10), 2273–2278.

9. Gülden, G., & Rzaev, Z. M. O., (2008). Synthesis and characterization of copolymers of N-vinyl–2-pyrrolidone with isotructural analogs of maleic anhydride. *Polymer Bulletin, 60,* 741–752.

10. Hanna, M. W., & Ashbaugh, A. L., (1964). Nuclear magnetic resonance study of molecular complexes of 7,7,8,8-tetracyanoquinodimethane and aromatic donors. *J. Phys. Chem., 68*(4), 811–816.

11. Kalinina, F. E., Mognonov, D. M., Radnaeva, L. D., & Vasnev, V. A., (2002). Alternating copolymers of vinylglysidyl ether of ethylene glycol and imides. *Vysokomolek. Soyed. Ser. A., 44*(3), 401–406.

12. Guryanova, E. I., Goldstein, N. G., & Romm, I. P., (1973). Donor-acceptor bond. *M.: Khimiya*, p. 100.

CHAPTER 13

Reaction Hydrosilylation of Allyl-2,3,5,6-Di-O-Isopropylidene-D-Mannofuranose with Methyl- and Phenylcyclodisilazanes

NELI SIDAMONIDZE, RUSUDAN VARDIASHVILI, and
MAIA NUTSUBIDZE

*Iv. Javakhishvili Tbilisi State University, Department of Chemistry,
Ilia Chavchavadze Ave., 0128 Tbilisi, Georgia*

ABSTRACT

By hydrosilylation of 1-O-allyl-2,3;5,6-di-0-isopropylidene-D-mannofuranose
(1) with 1,3-bis(dimethylsilyl)–2,2,4,4-tetra-methylcyclodisilazane (2) and
1,3-bis(diphenylsilyl)–2,2,4,4-tetra-phenylcyclodisilazane (3) in the presence
of the catalyst $Co_2(CO)_8$, we obtained 1,3-di[3-(2,3;5,6-di-0-isopropylidene-
D-mannofuranosyloxy)propyldimethylsilyl]–2,2,4,4-tetramethylcyclodisila-
zane (4) and 1,3-di[3-(2,3;5,6-di-0-isopropylidene-D-mannofuranosyloxy)
propyldiphenylsilyl]–2,2,4,4-tetra-phenylcyclodisilazane (5). After removal
of isopropylidene protections, we obtained 1,3-di[3-(β-D-mannofuranosyloxy)
propyldimethylsilyl]–2,2,4,4-tetramethylcyclodisilazane (6) and 1,3-di[3-(β-
D-mannofuranosyloxy)propyldiphenylsilyl]–2,2,4,4-tetraphenylcyclodisila-
zane (7). The reaction mainly occurs according to Farmer's rule, although a
small amount of Markovnikov addition product is also formed.

13.1   INTRODUCTION

For some time past wide opportunities have arisen for the use of organic
compounds containing silicon, sulfur, arsenic, and other heteroatoms in agriculture, medicine, and industry. It is known that organosilicon compounds
serve for the prophylaxis of the formation and growth of tumors (a various

form of leukemia, Ehrlich's tumor, sarcoma–180, etc.) [1]. At the present time, various organic and inorganic silicon-containing compounds have been approved. These possess bactericidal and fungicidal properties and are successfully applied against a series of pathogenic microorganisms participating in biodegradation.

Synthesis of the low-toxicity compound has become important in biological and pharmacological studies, and so there is interest in using carbohydrates to modify linear and cyclolinear siloxanes, which may lead to a substantial change in the nature of the drug action. The effect of glycosides on the organism is mainly conditioned by aglycones (non-sugar part of glycosides). Presence of traces of sugar contributes to the improvement of solubility, a decrease of toxicity, conductivity in biological membranes, which contributes to the creation of favorable conditions for a decrease of active concentration of some hard pharmacologic preparations and increase of the range of therapeutic effect [2–5].

13.2   GENERAL RESULTS

We have studied the reaction of hydrosilylation of 1-O-allyl–2,3;5,6-di–0-isopropylidene-D-mannofuranose (1) with 1,3-bis(dimethylsilyl)–2,2,4,4-tetramethylcyclodisilazane (2) and 1,3-bis(diphenylsilyl)–2,2,4,4-tetraphenylcyclodisilazane (3). The corresponding 1,3-di[3-(2,3;5,6-di–0-isopropylidene-D-mannofuranosyl-oxy)propyldimethylsilyl]–2,2,4,4-tetramethylcyclodisila-zane (4) and 1,3-di[3-(2,3;5,6-di–0-isopro-pylidene-D-mannofuranosyloxy)-propyldiphenylsilyl]–2,2,4,4-tetra-phenylcyclodisilazane (5) have been synthesized. The reaction was carried out in dry chloroform with a mole ratio of the reacting components equal to 2.5:1 at a temperature of 60–65⁰ in the presence of the catalyst $Co_2(CO)_8$ according to Scheme 13.1.

The reaction mainly occurs according to Farmer's rule, although a small amount of Markovnikov addition product is also formed.

The course of the reaction was monitored from the decrease in active hydrogen on the silicon over time. We have established that in 1.5 h, the hydrogen in τηε ≡Si-H group is completely removed, which is supported by the IR spectrum.

After removal of isopropylidene protections of compounds 4 and 5 by heating with cation exchanger KY–2 (H), we obtained 1,3-di[3-(β-D-mannofuranosyloxy)propyldimethylsilyl]–2,2,4,4-tetramethylcyclodisilazane (6) and 1,3-di[3-(β-D-mannofuranosyloxy)propyldiphenyl-silyl]–2,2,4,4-tetraphenylcyclodisilazane (7) according to Scheme 13.2.

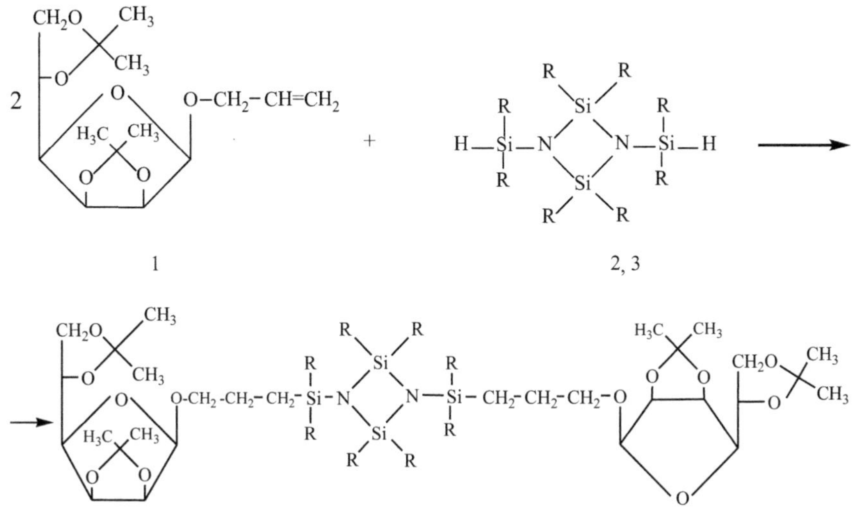

4, 5

R = CH_3 (1, 2, 4);

R = C_6H_5 (1, 3, 5).

SCHEME 13.1 The reaction of hydrosilylation of 1-O-allyl–2,3;5,6-di–0-isopropylidene-D-mannofuranose (1) with cyclodisilazanes 4 and 5.

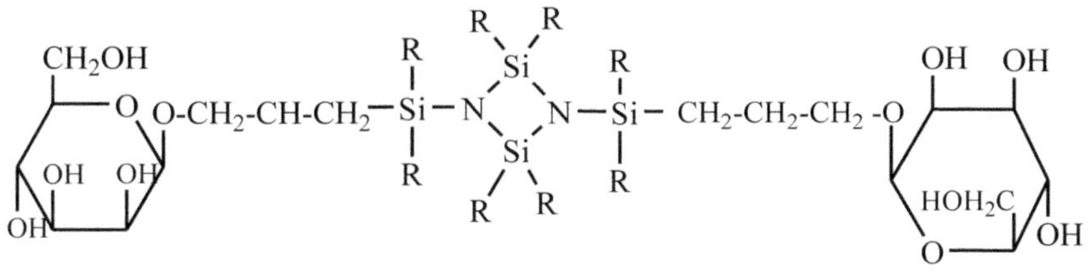

6, 7

R = CH_3 (6);

R = Ph (7).

SCHEME 13.2 Removal of isopropylidene protections of compounds 4 and 5.

The structures of obtained compounds were established by physical-chemical methods of analysis. Physical-chemical characteristics of synthesized compounds are presented in Table 13.1.

TABLE 13.1 Physical-Chemical Characteristics of Synthesized Compounds

Compound	Reaction duration hr	m.p. °C	$[\alpha]_D$, CHCl$_3$	R_f	Elemental analysis found, % Calculated, %				Yield, %
					C	H	N	Si	
4	2.5	122–123.5C	+66.5 (c 0.5; t = 17°)	0.53*	53.45	7.22	2.96	12.78	58.8
					53.0	8.58	3.24	13.0	
5	2.5	144–144.5C	+91.5° (c 0.62; t = 17°)	0.44*	68.02	7.12	2.63	8.78	52.3
					67.56	6.5	2.02	8.08	
6	0.5	139–140°C	+47° (c 0.54; t = 17°)	0.7**	44.95	8.71	3.44	15.25	61.2
					44.44	8.26	4.0	15.95	
7	1.0	165–166°C	+36 (c 1.2; t = 17°)	0.85**	65.21	6.32	1.74	8.77	54.7
					65.88	6.48	2.31	9.27	

*Benzene-chloroform 2:1.
**Chloroform-methanol 5:3.

Addition of 1,3-bis-(dimethylsilyl)–2,2,4,4-tetramethylcyclodisilazane (2) to 1-O-allyl–2,3;5,6-di–0-isopropylidene-D-mannofuranose (1) was selected to study the pathway and mechanism of the model reaction.

The reactions in the presence of $Co_2(CO)_8$ probably occurred according to the following mechanism (Scheme 13.3). The $Co_2(CO)_8$ itself cannot catalyze a hydrosilylation process of alkyl glucosides. The first step in the formation of the active catalyst species, $HCo(CO)_4$, is important and results from the reactions of $Co_2(CO)_8$ with compound 2. It is known that $HCo(CO)_4$ adds readily to olefins to form intermediate 8, that is unstable because of the excess of electron density on the metal and tends to convert into the saturated Co complex 8a by the loss of the most labile ligand, in this instance hydride. The last step of the mechanism is the formation of the final product and regeneration of the catalysts:

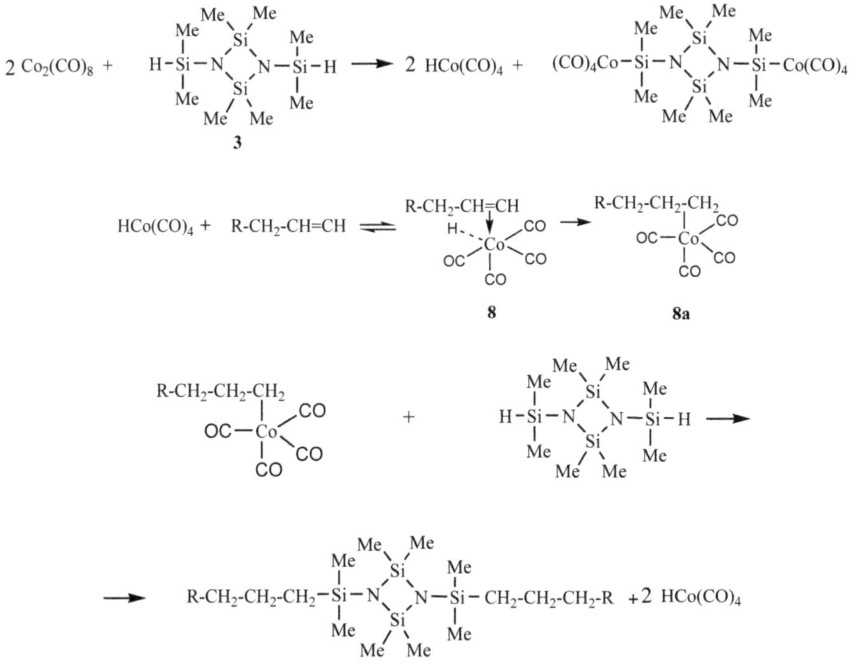

SCHEME 13.3 The mechanism of cyclodisilazane addition to allyl glycosides.

The proposed mechanism of cyclodisilazane addition to allyl glycosides agrees well with the experimental results.

13.3 EXPERIMENTAL

The IR spectra were obtained on a UR–20 in KBr disks. The ^{1}H NMR spectrum was taken on a Bruker WM–250 spectrometer (250 MHz); the ^{13}C NMR spectrum was taken on Bruker AM–300 spectrometer (75 MHz) in CDCl$_3$. The purity of the compounds obtained and the R_f values were determined on Silufol UV–254. The optical rotation was measured on an SU–3 general-purpose saccharimeter at 20±2°C.

1,3-di[3-(2,3;5,6-di–0-isopropylidene-D-mannofuranosyl-oxy) propyldimethylsilyl]–2,2,4,4-tetramethylcyclodisilazane (4). 1.31 g (5 mmol) 1,3-bis(dimethylsilyl)–2,2,4,4-tetramethylcyclodisilazane (2) in dry chloroform (35 ml) and Co$_2$(CO)$_8$ (0.15 g) were added drop wise to a solution of 7.5 g (12.5 mmol) compound (1) in dry chloroform (45 ml). The reaction was carried out under a nitrogen atmosphere with constant stirring for 2.5 h (60–65°C). After cooling and separating on a column (2:1 benzene-chloroform system, silicagel L (50/100), a chromatographically pure product was obtained in a yield of 6.29 g (58.8%); m.p. 122–123.5°C. R_f 0.53 (2:1 benzene-chloroform system). $[\alpha]_D^{18}$ +66.5° (c 0.5, chloroform). IR spectrum, ν, cm^{-1}: 690 (Si-C); 1020, 1070 (C-O-C); 920 (Si-N); 1460 (CH$_3$);1370–1380 (–C(CH$_3$)$_2$). ^{1}H NMR spectrum, δ, ppm (*J*, Hz): 4.35 (1H, d, $J_{1,2}$ = 8, H–1); 5.55 (1H, d, $J_{1,2}$ = 4, H–1'); 4.38 (1H, dd, $J_{2,1}$ = 8.1; $J_{2,3}$ = 9.4, H–2); 4.78 (1H, dd, $J_{2,1}$ = 4; $J_{2,3}$ = 10.6, H–2'); 5.67 (1H, dd, $J_{3,2}$ = 9.4; $J_{3,4}$ = 10, H–3); 5.60 (1H, dd, $J_{3,2}$ = 10.6; $J_{3,4}$ = 9.8, H–3'); 4.22–4.28 (dd, $J_{4,3}$ = 10; $J_{4,5}$ = 12.3, H–4); 5.29–5.30 (dd, $J_{4,3}$ = 9.8; $J_{4,5}$ = 9.9, H–4'); 3.62–3.68 (1H, m, H–5); 3.90–4.00 (1H, m, H–5'); 4.80–4.10 and 4.11–4.14 (2H, d, H–6 and H–6' CH$_2$O); 3.55–3.60 and 3.78–3.84 (2H, 2m, RO-CH$_2$–CH$_2$–CH$_2$ –SiR$_3$); 1.85–1.90 and 1.92–1.98 (2H, 2m, RO-CH$_2$–CH$_2$–CH$_2$ –SiR$_3$); 1.62–1.70 and 1.73–1.82 (2H, 2m, RO-CH$_2$–CH$_2$–CH$_2$ –SiR$_3$); 1.05–1.10 (24H, m, 8 Si- CH$_3$); Found, %: C 53.45; H 7.22; N 2.96; Si 12.78. C$_{38}$H$_{74}$N$_2$O$_{12}$Si$_4$. Calculated, %: C 53.0; H 8.58; N 3.24; Si 13.

1,3-di[3-(2,3;5,6-di-0-isopropylidene-D-mannofuranosyloxy) propyldiphenylsilyl]–2,2,4,4-tetraphenylcyclodisilazane (5) were obtained similarly from compound (1) (7.5 g, 12.5 m mol) and 1,3-bis (diphenylsilyl)–2,2,4,4-tetraphenylcyclodisilazane(3) (3.79 g, 5 m mol) with a yield of 8.83 g (52.3%); m.p. 144–144.5°C. R_f 0.44 (2:1 benzene-chloroform system). $[\alpha]_D^{17}$ +91.5° (c 0.62, chloroform). IR spectrum, ν, cm^{-1}: 710 (Si-C); 1020, 1050, 1110 (C-O-C); 899 (Si-N); 1445 (CH$_3$); 1460, 1743 (C = C$_{arom}$); 730, 840 (C-H$_{arom}$); 1390(–C(CH$_3$)$_2$). ^{13}C NMR spectrum, δ, ppm. CDCl$_3$: 92.212 and 98.543 (C–1 and C–1'); 38. 756 (-C(CH$_3$); 20.62 (C(CH$_3$); 70.82–71.05 (RO-CH$_2$-CH$_2$-CH$_2$-SiR$_3$); 29.38–29.73 (RO-CH$_2$-CH$_2$-CH$_2$-SiR$_3$);

20.78–22.72 (RO-CH$_2$-CH$_2$-CH$_2$-SiR$_3$); 8.05–13.40 (Si-CH$_3$); 91.90 and 100.88 (C$_{(1)}$ and C$_{(1')}$); 60.98 and 61.80 (C$_{(6)}$ and C$_{(6')}$); 66.80–77.51 (C$_{(2-5)}$ and C$_{(2'-5')}$). Found, %: C 68.02; H 7.12; N 2.63; Si 8.78. C$_{26}$H$_{58}$N$_2$O$_{12}$Si$_4$. Calculated, %: C 67.56; H 6.5; N 2.02; Si 8.08.

1,3-di[3-(β-D-mannofuranosyloxy)propyldimethylsilyl]–2,2,4,4-tetramethylcyclodisilazane (6). 2 g of the substance (4) in 20 mL of chloroform was heated in a boiling water bath with cation exchanger KY–2 (H). The reaction was carried out under a nitrogen atmosphere with constant stirring for 0.5 h (70–80°C). After filtration, the filtrate was concentrated in vacuum until the crystal emerged with a yield of 0.99 g (61.2%).%); m.p. 139–140°C. R$_f$ 0.7 (chloroform-methanol 5: 3); [α]$_D$^{17} +47° (c 0.54, chloroform). IR spectrum, ν, cm^{-1}: 1220–1248 (Si-CH$_3$); 1080, 100, 1150 (C-O-C); 915 (Si-N); 3580–3650 (OH) 1390–1410 (-CH$_2$-); Found, %: C 44.95 H 8.71; N 3.44; Si 15.25. C$_{42}$H$_{74}$N$_2$O$_{20}$Si$_4$. Calculated, %: C 44.44; H 8.26; N 4.0; Si 15.95.

1,3-di[3-(β-D-mannofuranosyloxy)propyldiphenylsilyl]–2,2,4,4-tetraphenylcyclodisilazane (7). 2 g of the substance (5) in 20 mL of chloroform was heated in a boiling water bath with cation exchanger KY–2 (H). The reaction was carried out under a nitrogen atmosphere with constant stirring for 1.0 h (70–80°C). After filtration, the filtrate was concentrated in vacuum until the crystal emerged with a yield of 0.9 g (54.7%). m.p. 165–166°C. R$_f$ 0.85 (chloroform-methanol 5: 3); [α]$_D$^{17} +36° (c 1.2, chloroform). IR spectrum, ν, cm^{-1}: IR spectrum, ν, cm^{-1}: 702 (Si-C); 1100, 1040, 1030 (C-O-C); 940 (Si-N); 1460, 1743 (C = C$_{arom}$) 732, 839 (C-H$_{arom}$) 3430–3380 (OH); 50 (OH); 1350–1430 (-CH$_2$-). Found, %: C 44.95; H 8.71; N 3.44; Si 15.25. C$_{42}$H$_{74}$N$_2$O$_{20}$Si$_4$. Calculated, %: C 44.44; H 8.26; N 4.0; Si 15.95.

KEYWORDS

- **1-O-allyl–2,3;5,6-di–0-isopropylidene-D-mannofuranose**
- **1,3-bis(dimethylsilyl)–2,2,4,4-tetra-methylcyclodisilazane**
- **1,3-bis(diphenylsilyl)–2,2,4,4-tetra-phenylcyclodisilazane**
- **1,3-di[3-(2,3;5,6-di–0-isopropylidene-D-mannofuranosyloxy) propyldimethylsilyl]–2,2,4,4-tetra-methylcyclodisila-zane**
- **1,3-di[3-(2,3;5,6-di-0-isopropylidene-D-mannofuranosyloxy) propyldiphenylsilyl]–2,2,4,4-tetra-phenylcyclodisilazane**

REFERENCES

1. Ignatenko, M. A., (1984). *Pharmaceutical Chem. Journal, 28*, 401.
2. Kochetkov, N. K., Bochkov, A. F., Dmitriev, B. A., Usov, A. I., Chizhov, O. S., & Shibaev, V. N., (1967). *Carbohydrate Chemistry.* Publisher: Moscow "Chemistry," p. 674.
3. Lonas, G., & Stadler, R., (1994). *J. Acta Polymer, 45*, 14.
4. Sidamonidze, N. N., & Gakhokidze, R. A., (1987). *Chan Van Tan, and Khidzsheli, Z. G. J. Zashchita Rastenii, N., 7*, 41.
5. Sidamonidze, N. N., Vardiashvili, R. O., & Isakadze, J. M. O. L. K., (2007). *Pharmaceutical Chem. Journal, 41*(3), 131–134.

CHAPTER 14

Peculiarities of Polymerization of Vinyl Acetate and Styrene in Static Heterogeneous Monomer: Water System in the Absence of Emulsifier

A. A. HOVHANNISYAN[1], M. KHADDAJ[2,3], I. A. GRITSKOVA[3],
G. K. GRIGORYAN[1], N. G. GRIGORYAN[1], and A. G. NADARYAN[1]

[1]*Scientific and Technological Center for Organic and Pharmaceutical Chemistry of the National Academy of Sciences of Armenia, 0014, Yerevan, Azatutyan Ave., 26, Russia, E-mail: hovarnos@gmail.com*

[2]*Peoples' Friendship University of Russia (RUDN), 117209, Moscow, Mikloukho-Maklaya Str., 6, Russia*

[3]*Moscow Technological University of Fine Chemical Technologies, 117571 Moscow, Vernadsky Ave., 86, Russia*

ABSTRACT

The topology of formation of polymer-monomer particles, polymerization of vinyl acetate (VA) and styrene in a heterogeneous static monomer-water system, in the presence of hydroquinone and azobisisobutyronitrile (AIBN) has been studied by photo and spectrophotometric observation methods. It was established that in a static heterogeneous monomer-water system, regardless of the nature of the monomer and initiator, the dispersion of the monomer and the generation of latex particles in the form of microdroplets of the monomer occur at the interface. It is suggested that the stability of the resulting colloidal system is due to the terminal hydrophilic groups of the polymer and oligomeric molecules formed as a result of radical reactions taking place in the aqueous phase. The concluded assumption is that the stability of the obtained colloidal system is due to the polymer and oligomer end hydrophilic groups of in the aqueous phase.

14.1   INTRODUCTION

Polymerization is a static heterogeneous styrene-water system that leads to the formation of latex particles [1–3]. A hypothesis was set forth that the work of dispersing a monomer, and thus the generation of latex particles, is due to the heat of the polymerization reactions proceeding at the interface [2, 3]. It remains unclear how realistic this mechanism is, since in the aqueous phase it is possible to accumulate polymer molecules and their homogeneous nucleation [4]. This issue is especially relevant for the polymerization of monomers, the solubility of which in water may lead to the formation of large polymer molecules. The reason of stability of latex particles to coagulation is still unclear.

To clarify these circumstances, the topology of dispersed particle formation in polymerization processes of vinyl acetate (VA) in emulsifier-free statistic monomer-potassium persulfate (PP) aqueous solution was investigated. VA-water and styrene-water systems were also investigated, in the presence of an inhibitor and an oil-soluble initiator. Under these conditions, the polymerization processes are localized in separate zones of the heterogeneous system, and such an approach helps in obtaining a more advanced topological picture of the latex particles formation.

14.2   EXPERIMENTAL PART AND DISCUSSION OF THE RESULTS

Polymerization in a static monomer-water system was carried out in thermostated U-shaped tubes, which allowed sampling for analyzes from the aqueous phase without perturbing the phase boundary. The temperature of the experiments is 60°C.

The processes of polymerization and the formation of dispersed particles in the static monomer-water system are accompanied by turbidity of the aqueous phase, which makes it possible to follow the dynamics of this process by photographing the tubes and measuring the optical density of the aqueous phase.

The topological picture of the dynamics of turbidity of the aqueous phase in the VA system is an aqueous solution of PP is shown in Figure 14.1. The dynamics of the development of the process is clearly visible on these pictures: the dispersed particles that originate in the polymerization process are first localized in a narrow zone of the water phase near the interface, and then gradually spread into the interior of the water phase. Earlier, an analogous picture was obtained in the polymerization of styrene [2, 3] and

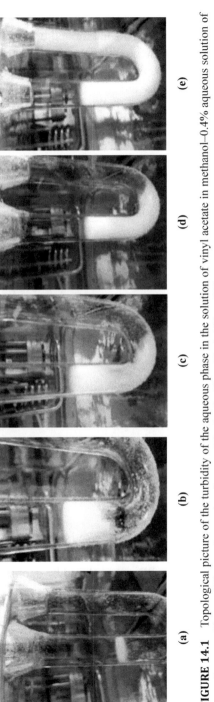

(a)   (b)   (c)   (d)   (e)

FIGURE 14.1 Topological picture of the turbidity of the aqueous phase in the solution of vinyl acetate in methanol–0.4% aqueous solution of potassium persulfate at different observation times: a–30 min., b–50, c–70, d–100, e–130.

the results of Figure 14.1 suggest that the process of nucleation of dispersed particles at the phase interface does not depend on the nature of the monomer but the result of polymerization reactions in the boundary layer monomer.

To locate possible polymerization reactions in the aqueous phase, an inhibitor (hydroquinone) was introduced into the system. Hydroquinone was initially dissolved in the monomeric phase (0.4% by weight of the monomer). Systems such as styrene-water and VA-water were studied.

In the styrene-water system, the change in the optical density of the aqueous phase and the formation of the emulsion began after approximately 60 minutes of observation, and in the VA-water system, the turbidity of the aqueous phase was not observed at all. In Figure 14.2 photo of test tubes of the two systems after 10 days of temperature control is illustrated.

FIGURE 14.2 VA–water and styrene–water systems containing hydroquinone in the monomeric phase and 0.4% potassium persulfate $K_2S_2O_8$ in the aqueous phase, after 10 days of temperature control at T 60°C.

The absence of the process of the formation of dispersed particles in the VA-water system is possible due to the formation of water-soluble oligomers with sulfate end groups. To clarify this issue, PP was replaced by azobisiso-butyronitrile (AIBN).

AIBN was introduced into the aqueous phase as a methanol solution. The studies were carried out in U-shaped test tubes. On the one side, VA was layered on the water, and on the other, a saturated solution of AIBN in methanol. Thus, polymerization was initiated by the diffusion of AIBN into the aqueous phase. The phase compositions are VA – 5 ml, water phase: water – 20 ml, saturated solution of AIBN in methanol – 5 ml.

In these series of experiments, samples were taken from the test tubes, and the viscosity of the upper zone of the aqueous phase from the VA-water side was measured. The results of the measurement are shown in Figure 14.3. These results show that the viscosity of the upper zone of the water phase increases with time. The increase in viscosity was not accompanied by the turbidity of the medium, which indicates the accumulation of water-soluble oligomers in the water phase. A change in the viscosity of the monomeric phase, during the time of the experiments, was not observed.

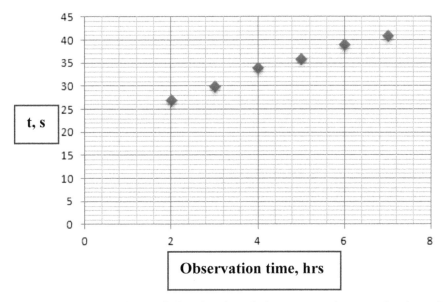

FIGURE 14.3 Dependence of the viscosity of the aqueous phase on the time of thermostating (observation) of the system, t is the time of the outflow of aqueous phase samples in a viscometer.

The inhibition of the formation of polyvinyl acetate latex particles with hydroquinone indicates that the length of polymer molecules produced in the aqueous phase and the nature of their terminal groups play a decisive role in the process of formation and stabilization of latex particles. From this result, it follows that the chemistry of stability of emulsifier-free latex should be directly dependent on both the nature of the monomer and the initiator.

According to Ref. [5], the thermal decomposition of AIBN into dimethyl-cyanomethyl radicals proceeds according to the monomolecular mechanism and does not depend on the nature of the solvent:

$$(CN)(CH_3)_2CN = NC(CH_3)_2(CN) \Rightarrow 2(CN)(CH_3)_2C^* + N_2 \qquad (1)$$

In the static system, with the diffusion of AIBN and VA molecules into the aqueous phase, the dimethylcyanomethyl radicals interact with both the monomer molecules and the water molecules:

$$(CN)(CH_3)_2C^* + M \Rightarrow CN)(CH_3)_2CM^* \qquad (2)$$

$$(CN)(CH_3)_2C^* + HOH (CN)(CH_3)_2CH + HO^* \qquad (3)$$

where, M is the monomer molecule, * is the sign of the radical. As a result of reaction 3, polymerization of BA in the aqueous phase can be initiated by hydroxyl radicals, as a result of which surface-active oligomers with OH end groups can be formed:

$$HO^* + M \Rightarrow HOM^* \qquad (4)$$

The flow of reaction 3 was confirmed by the presence of an absorption band characteristic of hydroxyl groups in the obtained IR spectra of the polymers.

The possibility of obtaining stable polymer dispersions in the polymerization of VA will largely depend on the hydrophilic-lipophilic balance (HLB) of the oligomers formed in the aqueous phase. Usually, the calculation of HLB of diphilic molecules is carried out according to Davis formula [6]:

$$HLB = 7 + aA - bB$$

where A and B are numbers characterizing the hydrophilicity of the polar and the hydrophobicity of alkyl groups in the molecule. a and b are equal to the numbers of the polar and nonpolar groups of the molecule.

The values of A and B for different groups of organic diphilic molecules are given in [6]:

A for $(-O-)$, $(-OH)$ and $(-C = O)$ groups are 1.3, 1.9 and 0.3, respectively, and B for CH_3-, $-CH_2-$ and $-C =$ groups is equal to -0.475.

Substituting these values of A and B in the Davis formula for the HLB of VA oligomers with one and two hydroxyl end groups, we obtain:

$$HLB = 7 + 1.9 + n(1.6) + n(-1.9) = 8.9 + n(-0.3)$$

$$HLB = 7 + 3.8 + n(1.6) + n(-1.9) = 10.8 + n(-0.3)$$

where n number of monomer units in oligomers.

The calculated values of the HLB are given in Table 14.1.

TABLE 14.1 Dependence of HLB Oligomers of VA with Hydroxyl-End Groups on the Number of Monomer Units (n) of Oligomers

n	HLB oligomers of VA with one terminal OH group	HLB oligomers of VA with two OH groups
1	8.6	10.5
4	7.7	9.6
10	5.9	7.8
15	4.4	6.3
20	2.9	4.8

According to literature data, substances whose HLB values are in the range of 5 to 20 can be used as stabilizers. The calculated HLB values of the oligomers shown in Table 14.1 show that in the static VA-water system, when polymerization is initiated by an oil-soluble initiator, stable latexes can be obtained under the condition of initiation of polymerization by hydroxyl radicals. To create such conditions, systems containing the minimum and maximum amount of VA in the aqueous phase were investigated. Polymerization was carried out in a two-phase system of VA–(water + a saturated solution of AIBN in methanol).

Two series of experiments were carried out. In the first series, the monomer concentration in the aqueous phase, at the beginning of the polymerization, was zero. In the second series, the water was initially saturated with a monomer. Measurements of induction periods of turbidity of aqueous phases for both systems gave the same result, which indicated the limiting role of diffusion of oligomeric radicals during the formation of the dispersed phase. Figure 14.4 shows photographs of the resulting dispersions, after 10 days of storage, at room temperature. As can be seen from the figure, the dispersion obtained without preliminary dissolution of the VA in the aqueous phase (tube 2) was fairly stable, whereas in the second series (tube 1) in 10 days, the polymer completely separated from the aqueous phase. In both systems, the dry residue of the aqueous phase was 2%.

The obtained experimental results allow us to conclude that when using oil-soluble initiators (as well as inhibitors and other chain transmitters) stable polyvinyl acetate latexes can be synthesized. The optical density of the solutions was measured on a SF–26 spectrophotometer, the dry residue was determined by the gravimetric method, the IR spectra were examined with a SPEKORD UR 75 instrument, the viscosity of the solutions was measured using Ubbelohde viscometer, electron micrographs were obtained with the Tescan Vega electron microscope.

FIGURE 14.4 Latexes obtained by the polymerization of VA in the monomer-AIBN solution in alcohol-water, after 10 days of storage. (1) The aqueous phase at the beginning of the polymerization was saturated with the monomer. (2) No monomer was in the aqueous phase at the beginning of the polymerization.

14.3 CONCLUSION

The polymerization of VA and styrene in a heterogeneous static monomer-water system shows that regardless of the nature of the monomer and initiator, polymerization processes proceeding at the interface disperse the system and produce a latex whose stability to coagulation is due to the hydrophilic balance of the oligomeric molecules synthesized in the aqueous phase.

KEYWORDS

- **heterogeneous styrene**
- **latex particles**
- **oligomer**
- **vinyl acetate**

REFERENCES

1. Oganesyan, A. A., Khaddazh, M., Gritskova, I. A., & Gubin, S. P., (2013). *Theoretical Foundations of Chemical Engineering, 47*(5), 600–603.
2. Hovhannisyan, A. A., Grigoryan, G. K., Khaddazh, M., & Grigoryan, N. G., (2015). *J. Chem. Chem. Eng., USA, 9*(5), 363–368.
3. Hovhannisyan, A. A., & Khaddazh, M., (2016). On the theory of emulsion polymerization lambert, Academic Publishing, *Germ*, p. 68.
4. Fitch, R. M., Ross, B., & Tsai, C. H., (1970). *Amer. Chem. Soc. Polym. Prepr., II*(2), 807–810.
5. Lewis, F. M., & Matheson, M. S., (1949). *J. Am. Chem. Soc., 71*, 747.
6. Adamson, A. A., (1990). *Physical Chemistry of Surfaces* (p. 777). Weily, USA.

CHAPTER 15

Sol-Gel Processing of Precursor for Synthesis of Mercury-Based Superconductors

I. R. METSKHVARISHVILI[1], T. E. LOBZHANIDZE[2], G. N. DGEBUADZE[1], M. R. METSKHVARISHVILI[3], B. G. BENDELIANI[1], V. M. GABUNIA[1,4], and L. T. GUGULASHVILI[1]

[1]*Department of Cryogenic Technique and Technologies, Ilia Vekua Sukhumi Institute of Physics and Technology 0186 Tbilisi, Georgia, E-mail: metskhv@yahoo.com*

[2]*Department of Chemistry, Faculty of Exact and Natural Sciences, Ivane Javakhishvili Tbilisi State University, 0179 Tbilisi, Georgia*

[3]*Department of Engineering Physics, Georgian Technical University, 0175 Tbilisi, Georgia*

[4]*Petre Melikishvili Institute of Physical and Organic Chemistry of the Iv. Javakhishvili Tbilisi State University, Jikia str 5, 0186, Tbilisi, Georgia*

ABSTRACT

Formation of the $HgBa_2Ca_2Cu_3O_{8+\delta}$ superconducting materials critically depends on the used $Ba_2Ca_2Cu_3O_x$ precursor. On this basis, the influence of precursor on the synthesis and properties of Hg-based superconductors has been examined. For comparisons two methods we synthesized precursors separately by sol-gel and as well as ordinary solid-state reaction methods (SSRs). The results showed that the superconductors have been prepared by sol-gel route with an excellent homogeneity and high T_c.

15.1 INTRODUCTION

$HgBa_2Ca_2Cu_3O_{8+\delta}$ superconductors exhibit the highest critical temperature among of the oxide superconductors. If these samples are prepared at ambient conditions, their superconducting transition temperatures are $T_c \approx 135$ K [1] and $T_c \approx 165$ K [2] under extremely high pressures. This peculiar property makes the Hg-1223 phase as a winsome material for its practical utilization. The volatility of the mercury oxide at high temperature is one of the problems for the practical application, respectively, do not appear possible for applications of this superconductive materials. In spite of this, an intense research activity until the present days doesn't decrease [3–11].

The conventional synthesis of Hg-1223 phase consists two-step process, first, a mercury-free precursor $Ba_2Ca_2Cu_3O_x$ is formed and, secondly, the reaction with HgO is promoted. This system is very toxic and what is more important for this family, to achieve high purity superconductivity phase, critically depends on the used precursor and synthesis conditions. In this connection, for the synthesis of precursors have been reported different methods in various works [6, 12–14].

Loureiro et al. [12] investigated the importance of average copper valence (n) in the precursor $Ba_2Ca_{n-1}Cu_n$ for synthesized homologous series of Hg–1223 superconductors under high pressure. Have to note that the precursor was prepared by solid-state reaction methods (SSRs) and in order to avoid problems elimination residual barium carbonate in the precursor, therefore in work, carbonate-free oxides and nitrates was used as starting materials. They show that average copper valence is a crucial dependence on the oxygen intake of non-stoichiometric phases during the precursor synthesis at a specific temperature, time, and cooling process treatment. In the results was an establishment that, lower precursor copper valences produce higher members of the Hg–1223 phase while higher precursor copper valence has the opposite effect. Lee and coauthors [13] proposed method of freeze-dried for synthesized highly homogeneous precursors. In this paper were compared two different drying methods of solution: freeze and thermal drying. For both methods $Ba(NO_3)_2$, $Ca(NO_3).4H_2O$ and $Cu(NO_3).3H_2O$ reagents dissolved and mixed in distilled water with a ratio of Ba:Ca:Cu = 2:2:3. In the freeze-dried method, the solution was sprayed through a nozzle into stirred LN_2. The frozen droplets were placed in a freeze drier under vacuum 10^{-2} Torr at –40°C for 24 h and slowly warmed to room temperature. In the second method, a hot plate was used to dry the solution. From this point, the processes of both methods were the same, and both precursors were sintered in flowing O_2 at 910°C for 24 h. As a result, for

Hg–1223 samples prepared from different precursors was obtained: freeze-dried precursor–75% of Hg–1223 superconducting fazes, 25% of $BaCuO_2$ and Tc (zero) ~133 K; thermal-dried precursor 50% of Hg–1223 supercon-ducting fazes, 10% of Hg–1212, 40% of $BaCuO_2$ and Tc (zero) ~129 K. Sin et al. [14] have studied the influence of the rhenium (ReO_6) addition on the synthesis of precursors; using a sol-gel method involving the gelification of a solution by an in-situ polymerization of acrylamide monomers. XRD anal-ysis showed at 700°C temperature $BaCO_3$ is removed from the Re-addition precursors. Authors conclude that the rhenium stabilizes the barium against its carbonation. Therefore, Re-based precursor is much more stable against moisture and carbonation than $Ba_2Ca_2Cu_3O_x$. The urea combustion method as an alternative sol-gel route for synthesizing precursors are presented by T.M. Mendonca and coauthors [6]. In this paper, precursors were prepared separately by urea combustion and sol-gel methods, respectively. The initial reactants ($BaCO_3$, $CaCO_3$, CuO) was dissolved individually in nitric acid, for first methods were added urea (H_2NCONH_2) and in second methods as a complexing agent was added acrylamide. Thermogravimetric measure-ments show the urea powder loss total weight of 8% regarding the 5% of the acrylamide powder. The authors think this will be the reason for better efficiency of the urea in formation of the precursor powders allowing a complete decarbonation at lower temperatures regarding acrylamide. The results showed that the in urea samples observed the single phase 99 % vol., whereas for acrylamide observed 80% of the Hg-1223 phase.

In this paper, the influence of $Ba_2Ca_2Cu_3O_x$ precursor on the synthesis and properties of Hg-based superconductors has been examined. $HgBa_2Ca_2Cu_3O_y$ superconducting samples were prepared by a two-step method. In the first step, was prepared multiphase ceramic precursors, for comparisons two methods we synthesized precursors separately by sol-gel method (SG) and as well as ordinary SSR. In the second step on both samples, HgO was added, and finale synthesis of $HgBa_2Ca_2Cu_3O_y$ was carried out in a sealed quartz tube.

15.2 EXPERIMENTAL

For synthesized $HgBa_2Ca_2Cu_3O_{8+\delta}$ samples, we used the two-step method. In the first stage, the Hg-free precursor was prepared before proceeding to the second stage where HgO was added to the precursor before the final sintering. We note that for both methods, starting materials were used powders materials $BaCO_3$ (99.0% Oxford Chem Serve), $CaCO_3$ (99.98% Oxford Chem Serve), and CuO (99.999 % Sigma-Aldrich).

The preparation of Ba:Ca:Cu = 2:2:3 multiphase ceramic precursors are:

- **Sol-gel method (SG)**: The initial reactants were dissolving $BaCO_3$ and $CaCO_3$ in acetic acid and CuO in nitric acid separately. When the complete dissolution was achieved, all the solutions were mixed, and the ε-Caprolactam ($C_6H_{11}NO$, 99% Sigma-Aldrich) was added. The solution dried at 100°C temperature for 12 h and after the obtained green gel was dreaded at 300°C. The xerogels are then homogenized in an agate mortar and calcined at 900°C, during 4 h by using of the temperature programmed muffle furnace (KSL–1100X-S, MTI-Corporation, temperature accuracy ±1°C). Resulting powders were ground and pressed into the pellets by a hydraulic press (Holzmann-Maschinen, Type: WP10H) with about 500 MPa. For eliminating the CO_2 from the precursor, the pellet is annealed home made a programmable tube type furnace at (temperature accuracy ±1°C) at 900°C in flowing oxygen partial pressure of 0.3 bar for 8 h.

- **Solid-state reaction method (SSR)**: The materials $BaCO_3$, $CaCO_3$ and CuO were mixed in the stoichiometric ratio Ba:Ca:Cu = 2:2:3 and then they were mixed and ground carefully in an agate mortar. The resulting powder mixture was calcined in an alumina crucible in the air in a muffled furnace, with four intermediate grindings at 900°C for 60 h. The resulting powders were ground and pressed into the pellets by a hydraulic press with about 500 MPa. The obtained pellet is annealed in tube type furnace at 900°C and 950°C in flowing oxygen partial pressure of 0.3 bar on 8 h.

In the second step both $Ba_2Ca_2Cu_3O_x$ precursors prepared by SG and SSR methods separately was mixed with HgO according to the composition $HgBa_2Ca_2Cu_3O_{8+\delta}$ and After final grinding the powder was pressed into a disc-shaped pellet 6 mm in diameter, and 1 mm thick, by using a hydraulic press under a pressure of 400 MPa. The pellets were put in a gold capsule then put into a quartz tube and from quartz tube was evacuated up to 10^{-3} Torr and sealed. Thereafter, a quartz tube was inserted into a programmed muffle furnace. The temperature of the furnace was raised at a rate of 300°C/h up to 700°C and thereafter at a rate of 120°C/h up to 900°C and held at this temperature for 1 h. The furnace was cooled at the rate of 60°C/h to room temperature. Finally, the samples were oxygenated in tube type furnace at 300°C in flowing oxygen for 15 h.

The prepared patterns were characterized by X-ray diffraction (Dron–3M) with CuKa radiation.

The Fourier transformed IR of the samples was taken in the region between 400–4000 cm^{-1} on a Cary 600 series FTIR Spectrometer using the KBr disc technique, with scanning resolution 0.5 cm^{-1}. The samples were pulverized into a fine powder and then mixed with potassium bromide powder using a weight ratio of 1:100. The IR absorption spectra were measured immediately after preparing the discs.

The phase method was used to study the real parts of the linear suscep-tibility [15]. The errors in the determination of χ' at higher frequencies than 1 kHz does not exceed 1%. For the measurements of intergranular critical current densities, we used the method of high harmonics [16].

15.3   RESULTS AND DISCUSSION

An important step of the preparation of highly pure phases $HgBa_2Ca_2Cu_3O_y$ superconductors is the sintering of the $Ba_2Ca_2Cu_3O_y$ multiphase precursor. To obtain the superconducting phase with optimal properties, there are two, critical precursor parameters: the cation homogeneity and the oxygen content. The thermal annealing at the oxygen partial pressure provides homogeneity and stipulates an elimination of the carbonates, remained in the multiphase precursor sample.

The sensitivity of X-ray diffraction to carbonate might be not sufficient; therefore we used IR spectroscopy for fixed anything remain CO_3^{2-} impurity species. Absorption spectra of the precursors, with annealing at a different temperature, are shown in Figure 15.1 for the range 400–4000 cm^{-1}.

The signature of carbonate is clear for SSR-precursor annealed at 900°C. The CO_3^{2-} ion which occurs at 692 cm^{-1}, at 855 cm^{-1}, at 1,058 cm^{-1}, and at 1,440 cm^{-1} [17]. By contrast, for SG-precursor annealed at 900°C remain CO_3^{2-} impurity species are not observed and also for SSR-precursor annealed at 950°C. Barium crystalline oxide BaO presents an intense band at ~483 cm^{-1} and a low one at ~503 cm^{-1} [18–20]. The appearance of a strong IR absorption band at 424 cm^{-1} may be attributed to the lattice vibrations of CaO [20]. According to McDevitt and Baun [21], there are two IR absorptions characterizing Ca–O lattice vibrations of pure CaO, a broad, strong absorption centered around 400 cm^{-1} and a medium strong band at 290 cm^{-1}, which, if present, are involved in the blackout absorption displayed in the present spectrum at <600 cm^{-1}. For CuO quantum dots, three main vibrational modes are observed at 468, 529, and 590 cm^{-1} [22].

The X-ray diffraction pattern of $Ba_2Ca_2Cu_3O_y$ precursor powder is plotted in Figure 15.2. As one can see, the sample consists of only two phases of $BaCuO_2$ and Ca_2CuO_3. The existence of $BaCuO_2$ and Ca_2CuO_3 phases only is a good indicator for qualitative preparation precursor powder with Ba:Ca:Cu = 2:2:3 cation ration [23].

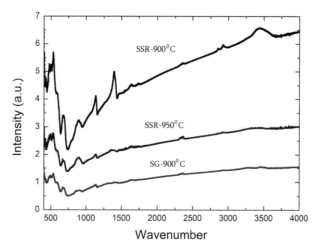

FIGURE 15.1 IR analysis of $Ba_2Ca_2Cu_3O_y$ after annealing in flowing oxygen partial pressure of 0.3 bar: For SG-precursor at 900°C and for SSR-precursor at 900°C and at 950°C.

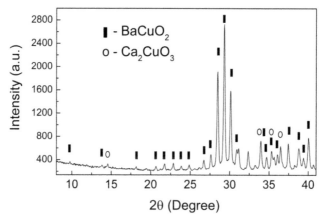

FIGURE 15.2 The x-ray diffraction pattern of $Ba_2Ca_2Cu_3O_y$ precursor powder after calcinations at 900°C at flowing oxygen partial pressure of 0.3 bar for 8 h.

Figure 15.3 shows the temperature dependences of the real $(-4\pi\chi')$ part of *ac* susceptibility for SG-sample and SSR-samples, measured in zero magnetic fields $(H = 0)$ at $h = 1$ Oe, $f = 20$ kHz. The diamagnetic onset temperature of the superconducting transition for SG-sample is about $T_c \approx$ 134 K, and a full screening of applied *ac* magnetic is observed at $T \sim 110$ K. As we see for SSR-samples synthesized at 900°C and 950°C temperature diamagnetic onset are at $T_c \approx 124$ K and $T_c \approx 130$ K, respectively.

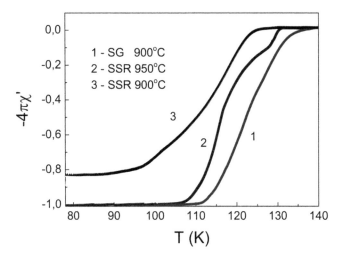

FIGURE 15.3   Temperature dependences of the real χ' parts of *ac* susceptibility for SG- and SSR-samples.

Figure 15.4 presents the dependence of the transport critical current densities on temperature j_c value measured by high harmonics method [16], at $h = 1$ Oe and $H = 0$. As we see, the largest value of critical current densities observed in SG-sample (175 A/cm²), whereas for SSR-samples synthesized at 900°C and 950°C temperature are at 93 A/cm² and 78 A/cm², respectively.

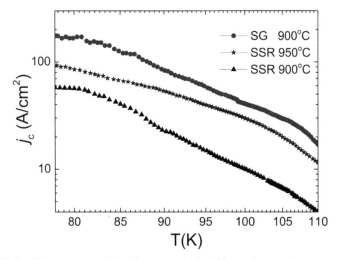

FIGURE 15.4   Dependence of the critical current densities on temperature.

15.4   CONCLUSIONS

The effect of the precursor on Hg-1223 system has been examined. Better chemical homogeneity and higher reactivity of the precursor powder were successfully obtained using the modified sol-gel method. A polycrystalline sample of the $HgBa_2Ca_2Cu_3O_{8+\delta}$ superconductor was obtained via the thermal treatment of Hg-free precursor powders in the mercury vapor environment using the sealed quartz tube technique. The diamagnetic onset temperature of the superconducting transition for sol-gel-sample was observed about $T_c \approx 134$ K and for solid-state reaction-samples $T_c \approx 130$ K. The critical current density for the SG-sample at liquid nitrogen temperature was obtained approximately 175 A/cm^2, whereas for SSR-samples is 93 A/cm^2. As a result, we could conclude that, the superconductors prepared by sol-gel route with excellent physicochemical properties and high T_c.

ACKNOWLEDGMENTS

This work was supported by Shota Rustaveli National Science Foundation (SRNSF), Grant number: 217524, Project title: Influence of the polymerization and various dopants on the Hg-1223 superconductive properties.

KEYWORDS

- **FT-infrared**
- **Hg–1223**
- **physical properties**
- **sol-gel route**
- **solid-state reaction**
- **XRD**

REFERENCES

1. Isawa, K., Tokiwa-Yamamoto, A., Itoh, M., Adachi, S., & Yamauchi, H., (1994). *Physica C., 222*, 33.

2. Gao, L., Xue, Y. Y., Chen, F., Xiong, Q., Meng, R. L., Ramirez, D., Chu, C. W., Eggert, J. H., & Mao, H. K., (1994). *Physical Review B., 50*, 4260.

3. Brylewski, T., Przybylski, K., Morawski, A., Gajda, D., Cetner, T., & Chmis, J., (2016). *Journal of Advanced Ceramics, 5* 185.

4. Jasim, K. A., & Supercond, J.., (2012). *Nov. Magn., 25,* 1713.

5. Jasim, K. A., Alwan, T. J., Al-Lamy, H. K., & Mansour, H. L., (2011). *J. Supercond Nov. Magn., 24,* 1963.

6. Mendonca, T. M., Tavares, P. B., Correia, J. G., Lopes, A. M. L., Darie, C., & Araujo, J. P., (2011). *Physica C., 471* 1643.

7. Hamdan, N. M., & Hussain, Z., (2009). *Supercond. Sci. Technol., 22* 034007.

8. Sakamoto, N., Akune, T., & Ruppert, U., (2008). *Journal of Physics: Conference Series, 97,* 012067.

9. Martinez, L. G., Rossi, J. L., Corrêa, H. P. S., Passos, C. A. C., & Orlando, M. T. D., (2008). *Powder Diffr. Suppl., 23,* S23.

10. Chen, X. J., Struzhkin, V. V., Wu, Z., Hemley, R. J., & Mao, H., (2007). *Phys. Rev. B., 75* 134504.

11. Giri, R., Tiwari, R. S., & Srivastava, O. N., (2007). *Physica C., 451,* 1.

12. Loureiro, S. M., Stott, C., Philip, L., Gorius, M. F., Perroux, M., Le Floch, S., Capponi, J. J., Xenikos, D., Toulemonde, P., & Tholence, J. L., (1996). *Physica C., 272,* 94.

13. Lee, S., Shlyakhtin, O. A., Mun, M. O., Bae, M. K., & Lee, S. L., (1995). *Supercond. Sci. Technol., 8,* 60.

14. Sin, A., Odier, P., & Nuez-Regueiro, M., (2000). *Physica C, 330,* 9.

15. Metskhvarishvili, I. R., Dgebuadze, G. N., Bendeliani, B. G., Metskhvarishvili, M. R., Lobzhanidze, T. E., & Mumladze, G. N., (2013). *J. Low Temp. Phys., 170,* 68.

16. Metskhvarishvili, I. R., Dgebuadze, G. N., Bendeliani, B. G., Metskhvarishvili, M. R., Lobzhanidze, T. E., & Gugulashvili, L. T., (2015). *J. Supercond Nov. Magn., 28,* 1491.

17. Chaney, J., Santillán, J. D., Knittle, E., & Williams, Q., (2015). *Phys. Chem. Minerals, 42,* 83.

18. Bentley, F. F., Smithson, L. D., & Rozek, A. L., (1968). *Infrared Spectra and Characteristic Frequencies 700–300 cm⁻¹*, Interscience, New York, 779 p.

19. Toderas, M., Filip, S., & Adrelean, I., (2006). *J. Optoelectron. Adv. M., 8,* 1121.

20. Zaki, M. I., KnËozinger, H., Tesche, B., & Mekhemer, G. A. H., (2006). *J. Colloid Interface Sci., 303,* 9.

21. McDevitt, N. T., & Baun, W. L., (1964). *Spectrochim. Acta, 20,* 799.

22. Borgohain, K., Singh, J. B., Rama, R. M. V., Shripathi, T., & Mahamuni, S., (2000). *Phys. Rev. B., 61,* 11093.

23. Whiter, J. D., & Roth, R. S., (1991). *Phase Diagrams for High-T_c Superconductors.* Westerville, Ohio, USA: Amer Ceramic Society, 175 p.

CHAPTER 16

Intergel Systems: Universal Sorbents for Different Nature Ions Sorption

T. K. JUMADILOV and R. G. KONDAUROV

JSC Institute of Chemical Sciences After A.B. Bekturov, Almaty, Republic of Kazakhstan, E-mail: jumadilov@mail.ru

ABSTRACT

The study represents a short review devoted to comparison of extraction degree of intergel system based on rare-crosslinked polymer hydrogels of polyacrylic acid (hPAA) and poly-4-vinylpyridine (hP4VP) in relation to La^{3+}, Ce^{3+}, Nd^{3+}, Mg^{2+}, Pb^{2+}, Cu^{2+}, Au^{3+} ions. Individual PAA and P4VP hydrogels have not sufficiently high sorption degree in relation to these ions (sorption degree is over 70%). Intergel system hPAA-hP4VP has higher values of extraction degree in relation to the mentioned ions. Maximum values of sorption degree relatively to La^{3+}, Ce^{3+}, Nd^{3+}, Mg^{2+}, Pb^{2+}, Cu^{2+}, Au^{3+} ions are observed at certain molar ratios (hPAA:hP4VP, mol: mol): hPAA:hP4VP = 2:4–94.0%; hPAA:hP4VP = 1:5–92.3%; hPAA:hP4VP = 3:3–91.7%; hPAA:hP4VP = 5:1–96.2%; hPAA:hP4VP = 3:3–94.9%; hPAA:hP4VP = 1:5–95.2%; hPAA:hP4VP = 1:5–99.2%, respectively. Such strong difference of extraction degree (over 25–30%) of the intergel system and individual polymer hydrogels is due to transition of polymers into highly ionized state in intergel pairs during their mutual activation at remote interaction.

16.1 INTRODUCTION

Previous investigations [1–10] showed that remote interaction of rare-crosslinked polymer hydrogels provides significant changes of electrochemical, volume-gravimetric, and sorption properties in intergel systems. As it was mentioned earlier [11–13], intergel systems can be determined as multi-component systems, which consist of 2 or more hydro- or organogels in common solvent. It should be noted, that intergel systems and interpolymer complexes are not the same polymer systems. Interpolymer complexes are products of interaction of chemically and/or dimensionally complementary macromolecules [14–18]. Intergel systems have not chemical bonding between hydrogels, due to this fact they cannot be considered as inter-polymer complexes (or systems). Intergel system based on rare-crosslinked polymer hydrogels of polyacrylic acid and poly-4-vinylpyridine is presented in Figure 16.1.

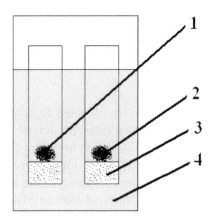

FIGURE 16.1 Intergel system hPAA-hP4VP (1 – hPAA; 2 – hP4VP; 3 – glass filter; 4 – solution).

Presence of the intergel system hPAA-hP4VP in an aqueous medium provides a variety of processes impacting on electrochemical equilibrium in solution.

For hPAA remote interaction with hP4VP occurs in three stages:

1. Ionization, ionic pairs formation, dissociation. At the end of the process there are three forms of functional groups in solution: -COOH, -COO⁻…H, -COO⁻;
2. After binding of a released into solution proton by nitrogen atoms equilibrium in solution in reaction ~COOH↔~COO⁻+ H⁺ shifts to the right according to the principle of "Le Chatelier" (concentration of H^+ decreases due to Eq. 2).

$$\sim COOH \leftrightarrow \sim COO^- + H^+ \tag{1}$$

$$\equiv N + H^+ \rightarrow \equiv NH^+ \tag{2}$$

–COOH groups dissociation takes place in accordance with the traditional representation [19]: firstly, there is the formation of ionic pairs, after that ionic pairs partially dissociates to individual ions. Thus, –COOH groups undergo additional ionization and dissociation.

3. Inter-node links are in the maximally unfolded state are forced to fold as a result of intramolecular and intermolecular interactions of the type $\sim COO^- \ldots H^+ \ldots {}^- OOC\sim$ for maintaining of stable condition.

For hP4VP remote interaction with hPAA occurs in two stages:

1. Association of protons released at –COOH groups association:

$$\equiv N + H^+ \leftrightarrow \equiv NH^+ \tag{3}$$

2. Interaction of $\equiv NH^+$ with water molecules.
    If water medium becomes acidic, interaction occurs in accordance with reaction:

$$\equiv NH^+ + H.OH \leftrightarrow \equiv NH^+.OH^- + H^+ \tag{4}$$

If the interaction of vinylpyridine links with water molecules leads to alkaline medium then the reaction will be:

$$\equiv N + H^+.OH^- \leftrightarrow \equiv NH^+ + OH^- \tag{5}$$

The result of these interactions is stated in which some part of charged functional groups of hydrogels have no counterions. The concentration of ionized groups without counter ions is impacted by initial molar ratios of polymer hydrogels in intergel pairs and other factors.

As known, modern sorption technologies are represented by ion-exchangers [20–24]. The main drawback of ion-exchange resins is selectivity only for one ion. There are no possibilities of the restructuring of sorption ability to other ions. In this regard, the goal of this investigation is to study the possibility of selectivity control in relation to certain ion by change initial molar ratios in the intergel system hPAA-hP4VP.

16.2   EXPERIMENTAL PART

16.2.1   EQUIPMENT

La^{3+}, Ce^{3+}, Nd^{3+} ions concentration in solutions was determined on spectrophotometer Jenway–6305 (UK) and Perkin Elmer Lambda 35 (USA). The concentration of Mg^{2+}, Pb^{2+}, and Cu^{2+} ions was determined on ion-meter I–160 MI (Russian Federation). The concentration of Au^{3+} ions was determined on iCE 3400 atomic absorption spectrometer (USA).

16.2.2   MATERIALS

Studies were carried out in 0.005 M salts solutions ($La(NO_3)_3*6H_2O$; $Ce(NO_3)_3*6H_2O$; $Nd(NO_3)_3*6H_2O$; $MgCl_2$; $Pb(NO_3)_2$; $MgCl_2$; $HAuCl_4$). Polyacrylic acid hydrogel (hPAA) was synthesized in the presence of crosslinking agent *N,N*-methylene-bis-acrylamide and redox system $K_2S_2O_8$-$Na_2S_2O_3$ in a water medium. Synthesized hydrogels were crushed into small dispersions and purified by washing with distilled water. Poly-4-vinylpyridine (hP4VP) hydrogel of Sigma-Aldrich Company (linear polymer crosslinked by divinylbenzene) was used as polybasis. Swelling degree of the hydrogels is: $\alpha_{(hPAA)} = 27.93$ g/g; $\alpha_{(hP4VP)} = 3.27$ g/g.

For the study of sorption properties intergel system hPAA-hP4VP was created. In the system hPAA concentration decreased from 6.2 m mol/L to 0.95 m mol/L (hPAA:hP4VP ratios interval 6:0–1:5, mol: mol). The concentration of hP4VP with a decrease of hPAA share increased from 1.1 m mol/L to 6.8 m mol/L (hPAA: hP4VP ratios interval 1:5–0:6).

Experiments were conducted at room temperature. Studies of the intergel system were made in the following order: the calculated amount of each hydrogel (hPAA and h4VP) of each intergel pair was put in separate glass filters, pores of which are permeable for low molecular ions and molecules, but impermeable for hydrogels dispersion. After that, filters with hydrogels were put in glasses with salts solutions. Measurement of concentration and aliquots sampling was made in the presence of glass filters with hydrogels.

Extraction (sorption) degree was calculated by the equation:

$$\eta = \frac{C_{initial} - C_{residual}}{C_{initial}} * 100\%$$

where $C_{initial}$ is the initial concentration of lanthanum in solution, g/L; $C_{residue}$ is the residual concentration of lanthanum in solution, g/L.

16.3 RESULTS AND DISCUSSIONS

The salts mentioned above $(La(NO_3)_3*6H_2O$; $Ce(NO_3)_3*6H_2O$; $Nd(NO_3)_3*6H_2O$; $MgCl_2$; $Pb(NO_3)_2$; $MgCl_2$; $HAuCl_4)$ present in solution in a dissociated state. Dissociation of each of these salts occurs in 2 or 3 stages. As known, each dissociation stage of salt has a certain value of dissociation rate constant. Dissociation rate constant of the 1st stage is much higher comparatively with 2nd and 3rd stages, due to this fact there is an occurrence of different sorption mechanisms of the ions. Products of the 1st dissociation stage are bind by ionic mechanism; and products of 2nd and 3rd stages by coordination.

The main chemical reactions, which occur in salts solutions, are:

1. Dissociation of carboxyl groups:

$$\sim COOH \leftrightarrow \sim COO^- + H^+ \tag{6}$$

2. Sorption of the metals ions by polyacids:

$$n \sim COO^- + Me^{n+} \rightarrow \sim COO_n Me \tag{7}$$

3. Sorption of the metals ions by polybases:

$$n \equiv N + Me^{n+} \rightarrow \equiv N^+_n Me \tag{8}$$

16.3.1 SORPTION OF RARE-EARTH METALS

Figure 16.2 shows extraction degree of lanthanum ions of the intergel system hPAA-hP4VP. As seen from the figure, mutual activation provides higher ionization of the initial polymers and, as a result, higher extraction degree of La^{3+} ions in comparison with individual hPAA and hP4VP. Sorption of lanthanum does not occur very intensive in the presence of individual hPAA and hP4VP, extraction degree of La^{3+} ions at 48 hours is 67.7% and 66.1%, respectively. Obtained data shows that a high level of mutual activation occurs at hPAA: hP4VP = 3:3, 2:4, and 1:5 ratios. Consequently, extraction degree at these ratios is above 90%. Maximum sorption of lanthanum is observed at hPAA:hP4VP = 2:4, 94.0% of the metal is extracted by the intergel pair at 48 hours of interaction.

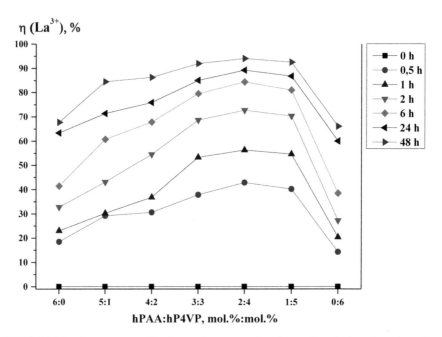

FIGURE 16.2   Dependence of lanthanum ions extraction degree from hydrogels molar ratio in time.

Curves, which describe the sorption of cerium by the intergel system, are presented in Figure 16.3. Low values of extraction degree of Ce^{3+} ions are observed in the presence of only polyacid or polybasis (hPAA: hP4VP = 6:0 and 0:6 ratios), at 48 hours 63.3% and 56.7% of cerium is sorbed. The increase of polybasis' share provides an increase of sorption ability in case of cerium sorption by intergel system hPAA-hP4VP. The highest value of extraction degree is observed at hPAA: hP4VP = 1:5 at 48 hours, at this ratio 92.3% of cerium ions are extracted.

Sorption of neodymium ions by the intergel system based on polyacrylic acid and poly-4-vinylpyridine hydrogels is shown in Figure 16.4. Not very high sorption degree of individual PAA and P4VP hydrogels (61.6% and 54.6% respectively) is due to the absence of mutual activation phenomenon. High values of extraction degree are seen at hPAA:hP4VP = 4:2 and 3:3 ratios, wherein maximum sorption occurs at hPAA:hP4VP = 3:3 ratio. At this ratio, 91.7% of lanthanum is extracted at 48 hours of interaction between the intergel system and salt solution.

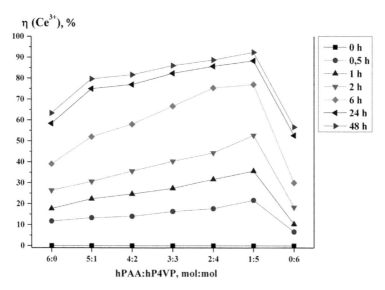

FIGURE 16.3   Dependence of cerium ions extraction degree from hydrogels molar ratio in time.

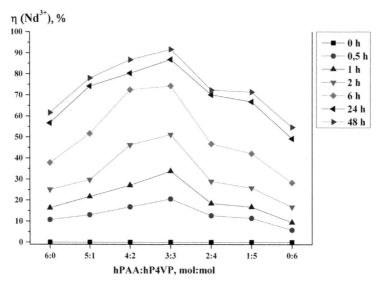

FIGURE 16.4   Dependence of neodymium ions extraction degree from hydrogels molar ratio in time.

16.3.2　SORPTION OF ALKALINE-EARTH METALS

Sorption of alkaline-earth metals is represented by extraction of magnesium. As seen from Figure 16.5, the equilibrium between individual hPAA and hP4VP is reached rather fast, as a result, not high values of extraction degree–70.4% for hPAA and 67.3% for hP4VP at 48 hours. Maximum sorption of magnesium occurs at hPAA: hP4VP = 5:1 ratio (extraction degree is 96.2% at 48 hours).

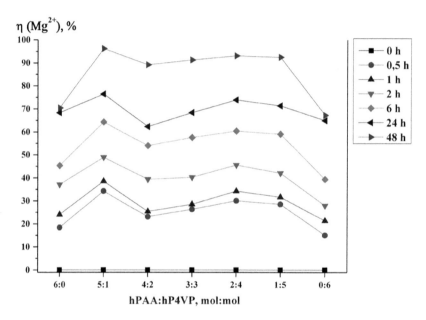

FIGURE 16.5 Dependence of magnesium ions extraction degree from hydrogels molar ratio in time.

16.3.3　SORPTION OF NON-FERROUS METALS

Figure 16.6 shows extraction degree of intergel system hPAA-hP4VP in relation to lead ions. Obtained data shows that individual acid and basic polymer hydrogels do not have high sorption degree (68.7% for hPAA, 66.9% for hP4VP). Mutual activation provides high ionization degree at hPAA:hP4VP = 4:2, 3:3 and 2:4 ratios. At these ratios, the extraction degree is higher than 92%. Maximum extraction degree is observed at hPAA:hP4VP = 3:3 at 48 hours, it is 94.9%.

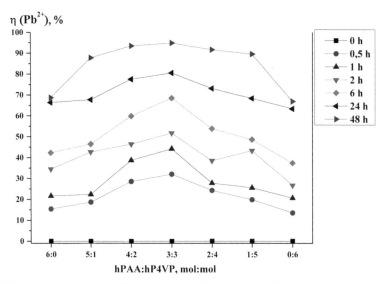

FIGURE 16.6 Dependence of lead ions extraction degree from hydrogels molar ratio in time.

16.3.4 SORPTION OF TRANSITION METALS

Dependence of copper ions extraction degree from the hydrogels molar ratio in time is presented in Figure 16.7. Individual polymers extract no more than 70% of copper (69.1% of copper ions is extracted by hPAA; 67.3% is extracted by hP4VP). Low (not sufficiently high) sorption of copper ions occurs at hPAA:hP4VP = 4:2 ratio. This phenomenon is a result of conformational changes in the structure of both hydrogels during their remote interaction. High sorption degree is observed at hPAA: hP4VP = 5:1 and 2:4 values. The maximum amount of copper is sorbed at 48 hours at hPAA: hP4VP = 5:1, sorption degree is 95.2%.

16.3.5 SORPTION OF NOBEL METALS

Figure 16.8 presents extraction of gold ions by the intergel system hPAA:hP4VP. As seen from the obtained data, individual PAA and P4VP hydrogels do not have high sorption degree (sorption degree is 69.9% and 67.5%, respectively). Exact maximums are seen at hPAA: hP4VP = 5:1 and 4:2, wherein more than 95% of gold is extracted by the intergel system. Almost all Au^{3+} ions are sorbed by the intergel system at hPAA:hP4VP = 4:2 ratio (extraction degree is 99.2%).

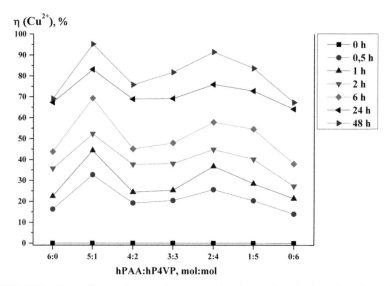

FIGURE 16.7   Dependence of copper ions extraction degree from hydrogels molar ratio in time.

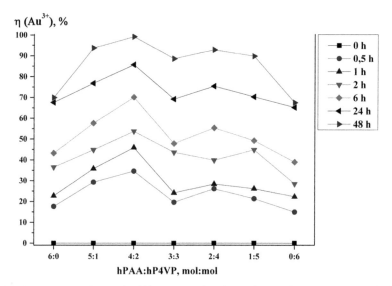

FIGURE 16.8   Dependence of gold ions extraction degree from hydrogels molar ratio in time.

Comparative analysis of extraction degree of the mentioned above intergel pairs (ratios) with individual hydrogels is presented in Table 16.1.

TABLE 16.1 Extraction Degree of hPAA-hP4VP Intergel System and Individual PAA and P4VP Hydrogels in Relation to La^{3+}, Ce^{3+}, Nd^{3+}, Mg^{2+}, Pb^{2+}, Cu^{2+}, and Au^{3+} Ions

	Molar ratio, $\eta(La^{3+})$	Molar ratio, $\eta(Ce^{3+})$	Molar ratio, $\eta(Nd^{3+})$	Molar ratio, $\eta(Mg^{2+})$	Molar ratio, $\eta(Pb^{2+})$	Molar ratio, $\eta(Cu^{2+})$	Molar ratio, $\eta(Au^{3+})$
hPAA:hP4VP	2:4, 94.0%	1:5, 92.3%	3:3, 91.7%	5:1, 96.2%	3:3, 94.9%	5:1, 95.2%	4:2, 99.2%
hPAA	67.7%	63.3%	61.6%	70.4%	68.7%	69.1%	69.9%
hP4VP	66.1%	56.7%	54.6%	67.3%	66.9%	67.3%	67.5%

As seen from the table, there is a significant increase (over 25–30%) of extraction degree at certain hPAA:hP4VP molar ratios in the intergel system comparatively with individual hydrogels. Such an increase in sorption properties of the rare-crosslinked polymer hydrogels is due to different conformational changes in their macromolecular structure during their mutual activation. It is seen, that some of the considered metals are maximally extracted at same molar ratios (ex. neodymium and lead at hPAA:hP4VP = 3:3 ratio; magnesium and copper at hPAA:hP4VP = 5:1 ratio). As known, all ions differ from each other by atomic radius, charge density, and polarizability. Also, it should be noted, that not only change of molar ratios change the selectivity of the intergel system towards the aimed ion, selectivity is also impacted by the external conditions.

16.4 CONCLUSION

Based on the obtained data the following can be concluded:

1. The result of rare-crosslinked polymer hydrogels of polyacrylic acid and poly-4-vinylpyridine is their transition into the highly ionized state with a consequent significant increase of sorption properties.
2. Individual hydrogels of PAA and P4VP do not have high sorption degree in relation to La^{3+}, Ce^{3+}, Nd^{3+}, Mg^{2+}, Pb^{2+}, Cu^{2+}, Au^{3+} ions. Extraction degree does not exceed 70%.
3. Intergel system hPAA-hP4VP have high sorption degree in relation to La^{3+}, Ce^{3+}, Nd^{3+}, Mg^{2+}, Pb^{2+}, Cu^{2+}, Au^{3+} ions. Extraction degree is in the interval from 91.7% to 99.2% in dependence of ion.
4. As seen from obtained results, varying of molar ratios of the hydrogels in the intergel system allows to control selectivity in relation to certain ion.

ACKNOWLEDGMENTS

The work was financially supported by the Committee of Science of Ministry of Education and Science of the Republic of Kazakhstan.

KEYWORDS

- ions
- poly-4-vinylpyridine
- polyacrylic acid
- remote interaction
- sorption

REFERENCES

1. Jumadilov, T. K., (2011). *Industry of Kazakhstan, 2*, 70.
2. Jumadilov, T. K., Kaldayeva, S. S., Kondaurov, R. G., Erzhan, B., & Erzhet, B., (2013). Mutual activation and high selectivity of polymeric structures in intergel systems. *Abstract of Communications, 4*[th] *International Caucasian Symposium on Polymers and Advanced Materials*, 191.
3. Alimbekova, B. T., Korganbayeva, Z. K., Himersen, H., Kondaurov, R. G., & Jumadilov, T. K., (2014). *J. Chem. Chem. Eng., 8*, 265.
4. Jumadilov, T. K., (2014). Electrochemical and conformational behavior of intergel systems based on the rare crosslinked polyacid and polyvynilpyrydines. *Proceedings of the International Conference of Lithuanian Chemical Society "Chemistry and Chemical Technology,"* 226.
5. Jumadilov, T., Akimov, A., Eskalieva, G., & Kondaurov, R., (2016). Intergel system polyacrylic acid hydrogel and poly-4-vinylpyridine hydrogel sorption ability in relation to lanthanum ions. *Proceedings of VIII International Scientific-Technical Conference "Advance in Petroleum and Gas Industry and Petrochemistry,"* 68.
6. Jumadilov, T. K., Abilov, Z. A., Kaldayeva, S. S., Himersen, H., & Kondaurov, R. G., (2014). *J. Chem. Eng. Chem. Res., 1*, 253.
7. Alimbekova, B., Erzhet, B., Korganbayeva, Z., Himersen, H., Kaldaeva, S., Kondaurov, R., & Jumadilov, T., (2014). Electrochemical and conformational properties of intergel systems based on the crosslinked polyacrylic acid and vinylpyridines. *Proceedings of VII International Scientific-Technical Conference "Advance in Petroleum and Gas Industry and Petrochemistry"* (APGIP–7), 64.
8. Jumadilov, T. K., Himersen, H., Kaldayeva, S. S., & Kondaurov, R. G., (2014). *J. Mat. Sci. Eng. B., 4*, 147.

9. Jumadilov, T., Abilov, Z., Kondaurov, R., Himersen, H., Yeskalieva, G., Akylbekova, M., & Akimov, A., (2015). *J. Chem. Chem. Tech., 4*, 459.

10. Jumadilov, T. K., Kondaurov, R. G., Abilov, Z. A., Grazulevicius, J. V., & Akimov, A. A., (2017). *Pol. Bul., 74*, 4701.

11. Jumadilov, T. K., (2016). Hydrogels remote interaction–basis of new phenomena and technologies. *Proceedings of 10ᵗʰ Polyimides & High Performance Polymers Conference*, CIV–7.

12. Jumadilov, T., Abilov, Z., Grazulevicius, J., Zhunusbekova, N., Kondaurov, R., Agibayeva, L., & Akimov, A., (2017). *J. Chem. Chem. Tech., 11*, 188.

13. Jumadilov, T. K., Kondaurov, R. G., Khakimzhanov, S. A., Eskalieva, G. K., & Meiramgalieva, G. M., (2017). Activated structures of interpenetrating networks–new type of effective sorbents for different nature ions. *Proceedings of 5ᵗʰ International Caucasian Symposium on Polymers and Advanced Materials*, 33.

14. Ghaffarlou, M., Sutekin, S. D., & Guven, O., (2018). *Rad. Phys. Chem., 142*, 130.

15. Rumyantsev, M., & Savinova, M. V., (2018). *Pol. Bul., 75*, 17.

16. Swift, T., Seaton, C. C., & Rimmer, S., (2017). *Soft Matt., 13*, 8736.

17. Wang, Y. H., He, J., Aktas, S., Sukhishvili, S. A., & Kalyon, D. M., (2017). *J. Rhe., 61*, 1103.

18. Pandey, P. K., Kaushik, P., Rawat, K., Aswal, V. K., & Bohidar, H. B., (2017). *Soft Matt., 13*, 6784.

19. Izmailov, N. A., (1976). Electrochemistry of solutions. *M.: Chemistry*, 488 pp.

20. Dorfner, K., (1991). *Ion Exchangers.* Berlin: Walter de Gruyter, 1496 pp.

21. Harland, C. E., (1994). *Ion Exchange: Theory and Practice.* Cambridge: The Royal Society of Chemistry, 285 pp.

22. Muraviev, D., Gorshkov, V., & Warshawsky, A., (2000). *Ion Exchange.* New York: M. Dekker, 918 pp.

23. Zagorodni, A. A., (2006). *Ion Exchange Materials: Properties and Applications.* Amsterdam: Elsevier, 496 pp.

24. Shkolnikov, V., Bahga, S. S., & Santiago, J. G., (2012). *Phys. Chem. Chem Phys., 14*, 32.

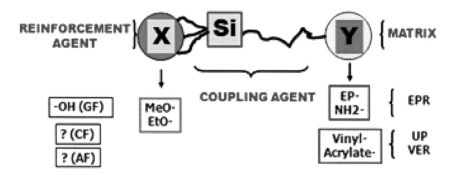

FIGURE 1.2 Interaction of the coupling agent with matrix and reinforcement agent.

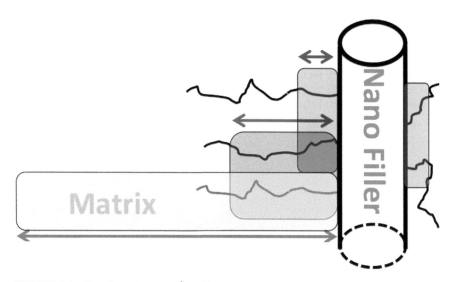

FIGURE 1.3 Interface (1 to 100 Å) – blue part, Interphase (up to 100 μm) – red part and the integration of the filler into the composite – yellow part. Filaments represent the coupling agent.

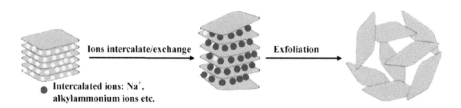

FIGURE 1.9 Intercalated *vs.* exfoliated structure.

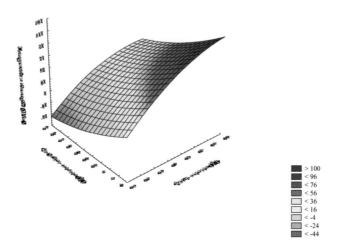

FIGURE 3.1   Graphical presentation of the material model of the polyester composite with glass reinforced polyester and nanofiller, showing the 'compressive strength' as a function of the component variables: Polyester resin and glass polyester recyclate.

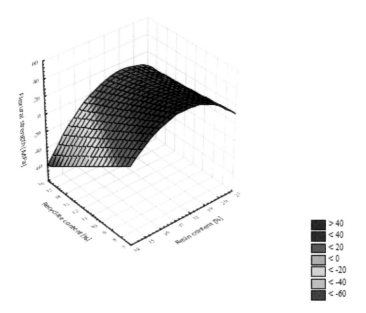

FIGURE 3.2   Graphical presentation of the material model of the polyester composite with glass reinforced polyester and nanofiller, showing the 'flexural strength' as a function of the component variables: Polyester resin and glass polyester recyclate.

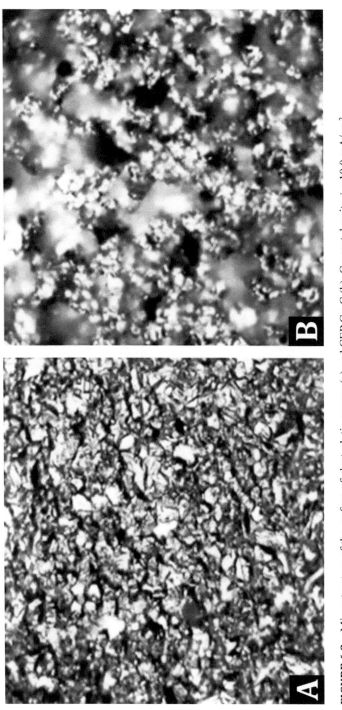

FIGURE 6.2 Microstructure of the surface of electrolytic copper (a) and CEP Cu-C (b). Current density $i = 10.0$ mA/cm^2.

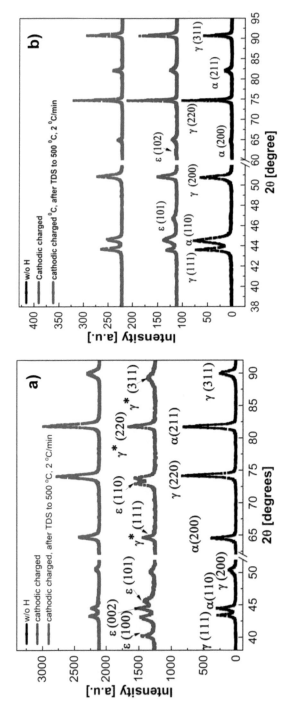

FIGURE 23.1 XRD pattern of the uncharged and 72 h hydrogenated samples (a) LDS and (b) DSS [11], [37].

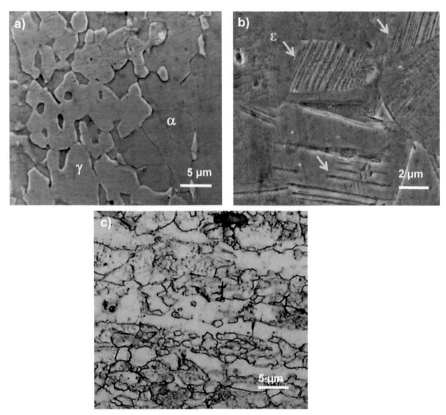

FIGURE 23.2 OM observations of (a) as received LDS and DSS and after 72 h cathodic charging of (b) LDS, showing the formation of needle shape martensite phase after one month at RT and (c) DSS [12], [37].

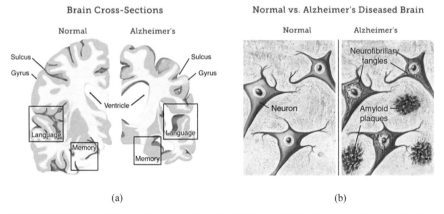

FIGURE 25.1 Alzheimer disease effect (a) [10], Amyloid plaques and neurofibrillary tangles (b) [10].

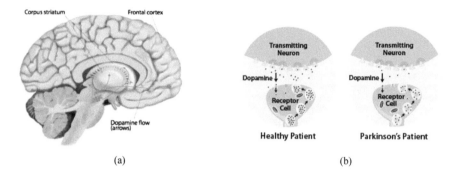

FIGURE 25.2 Involved region in Parkinson's disease (a) [12], dopamine transfer in brain neurons (b) [11].

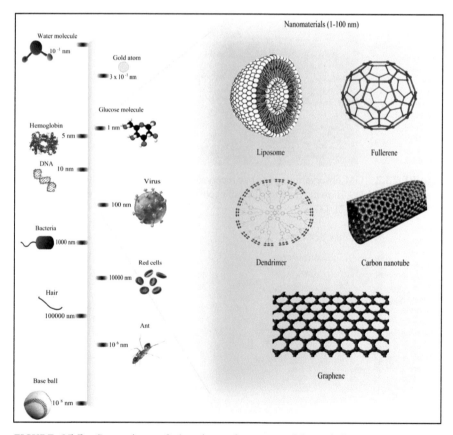

FIGURE 25.5 Comparison of the sizes of nanomaterials and famous structures in nanotechnology [17].

FIGURE 25.6   Fractal pattern in leaves of plants or trees (a), snowflakes (b), cauliflower (c), romanesco broccoli (d), Architectural patterns (e), historical buildings (f) [11].

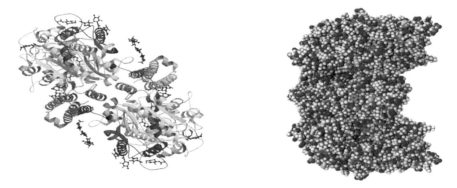

FIGURE 27.1   Chemical structure of prostate-specific membrane antigen (PSMA, Glutamate carboxypeptidase II).

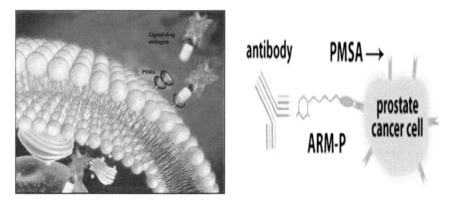

FIGURE 27.2 Transporter molecules carrying therapeutic drugs to PSMA targets on a prostate cancer cell.

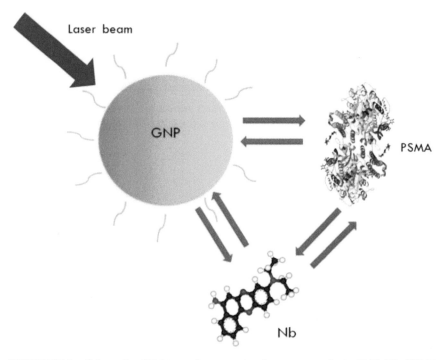

FIGURE 27.4 Schematic of light to enhancement and energy transfer in GNPs/Nb /PSMA nanocomposite.

CHAPTER 17

Some Chromatographic Peculiarities of Polymer-Mineral Packing Materials with Octadecyl- and Butyl-Functional Groups

S. S. HAYRAPETYAN, M. S. HAYRAPETYAN, G. P. PIRUMYAN, and H. G. KHACHATRYAN

Yerevan State University, 1 Alek Manoukian Str., 0025 Yerevan, Armenia, E-mail: haysers@ysu.am

ABSTRACT

Some peculiarities of behavior of columns packed by polymer-mineral packing materials in reversed-phase high-performance liquid chromatography (RP-HPLC) regime have been considered as compared with columns containing conventional C_{18}-phase by example of Waters Symmetry C_{18} column.

Poly-octadecyl methacrylate-butyl acrylate (poly-ODMA-BA) and poly-octadecyl methacrylate-bu-tylacrylate-maleic anhydride (poly-ODMA-BA-MA) have been used as polymer phases.

And micro-spherical silica gels (MS) having following characteristics: specific surface area ($S_{sp.}$) 80–100 m^2/g; total pore sorption volume (V_Σ) 0.56–1.00 cm^3/g; average pore diameter ($d_{av.}$) 30 nm have been applied as initial materials of inorganic origin.

It has been shown by means of NMR method that above mentioned functional groups are actually fixed on the surface of silica carrier.

17.1 INTRODUCTION

The most common type of packing materials for reversed-phase chromatography is fully porous silica, whose surface has been derived with hydrocarbon chains grafted via Si-O-Si-C bonds. Since the evidence showed that at least the presence and activity of surface silanol groups (further–simply "silanols") can be measured unequivocally using a single type of packing it became interesting to apply the technique to range of different packings. While the major focus was to establish a broad measure of silanol activity at neutral pH made it possible to assess other parameters as well. This allows for an additional classification of packings; for example, the hydrophobicity of packing can be measured via the retention of neutral analytes. In addition, the newer generation packings with embedded polar functional groups can be easily characterized chromatographically [1].

The results of the method obtained with over 50 different reversed-phase packings were discussed in Ref. [2]. For characterization of many different reversed-phase packings were used the following test mixture: uracil (16 mg/L) as a marker for column dead volume, naphthalene (60 mg/L) and acenaphthene (200 mg/L) as hydrophobic markers, butylparaben (20 mg/L) and dipropyl phthalate (340 mg/L) as polar probes, propranolol (400 mg/L) and amitriptyline as a basic probes (100 mg/L) [2].

Silanol groups left on the surface due to incomplete derivatization play an important role in retention mechanism. Therefore, studies were undertaken to understand the influence of silanols on retention and peak shape [3]. The retention of most compounds with basic functional groups depends not only on the bonded phase properties, but on the amount of silanol groups available for interaction with the analyte. Thus, complete characterization of the properties of reversed-phase packings should test not only hydrophobic properties of a packing, but also the "silanol" activity.

Recently, hybrid packings have become available that are stable within wider pH range (up to pH 12) but with retention characteristics identical to classical silica-based packings. This allows for broader pH range than was possible in the past [4].

Problems of the modification of the silica gel surface with polymer layers and the influence of the coated polymer quantity on chromatographic properties of the resulting sorbents have been considered. The polymer modification of the surface of wide-porous MS obtained by means of hydrothermal treatment of mesoporous silica gel under autoclave conditions was described. The polymer layer itself was formed by octadecyl methacrylate-methyl-methacrylate co-polymer. As a result, efficient packings for reversed-phase high-performance liquid chromatography (RP-HPLC) were obtained [5, 6].

The development of new methods for synthesis of packing materials for chromatography with the goal to improve their structure and chromatographic parameters always was and remains significant. Most often researchers turn to different means of surface modification for already existing chromato-graphic systems [7–10].

17.2   EXPERIMENTAL PART

17.2.1   *SYNTHESIS OF SORBENTS*

17.2.1.1   *PREPARATION OF OCTADECYLMETHACRYLATE-BUTHYLACRYLATE CO-POLYMER (PRE-POLYMER).*

Octadecylmethacrylate (ODMA) (30.0 g), solid at room temperature, was dissolved together with dicumenyl peroxide (DCP) (0.25 g) in butyl acrylate (BA) (10.0 g), heated under reflux for 3 hours and cooled. The resulting oligo-ODMA-MMA copolymer (75% ODMA, 25% BA) is soluble in *n*-pentane [5].

17.2.1.2   *COATING OF SILICA SUPPORTS*

MS possessing the following characteristics: $S_{sp.} = 80.0$ m^2/g; $V_{\Sigma} = 0.57$ cm^3/g; $d_{av.} = 30$ nm, particle size 5 mcm and symmetry type MS (Waters) with the following characteristics: $S_{sp.} = 100.0$ m^2/g; $V_{\Sigma} = 1.00$ cm^3/g; $d_{av.} = 30$ nm, particle size 5 mcm were used as porous silica.

Silica beads (5.0 g) were added to the solution of pre-polymer and DCP (1.25 g and 40 mg correspondingly) in 15 mL *n*-pentane. The subsequent evaporation of the solvent was carried out in a rotary evaporator. The dry powder thus obtained was placed into hermetically closed container and subjected thermal treatment. The temperature was initially increased up to 100°C during 1 hour then kept for another hour at this temperature, then heated up to 130°C over 1 hour and then kept at this temperature for another hour. The sorbent obtained was rinsed with dimethylformamide-toluene hot mixture, furthermore with ethanol, and then dried with acetone on the filter. Final drying of the samples was carried out in the drying oven at 120C over 2 hours.

Carboxyl groups were produced on the polymer surface layer by treatment with maleic anhydride (MA). First polymer layer comprising ODMA-BA copolymer and DCP was deposed on the silica as described above. An appropriate amount of MA dissolved in acetone was then added, and the solvent was again removed by rotary evaporation. Polymerization was then performed by thermal treatment as described above.

The ODMA: BA mass ratio in the olygomer was 3:1 and olygomer: MA mass ratio was 8:1.

The used monomers and polymerization initiator (DCP) was Aldrich produced.

It follows from Figures 17.1 and 17.2 that polymer film with the following functional groups was actually deposited to the silica surface: ODMA, butylacrylate, carboxylic (obtained as a result of MA hydrolysis).

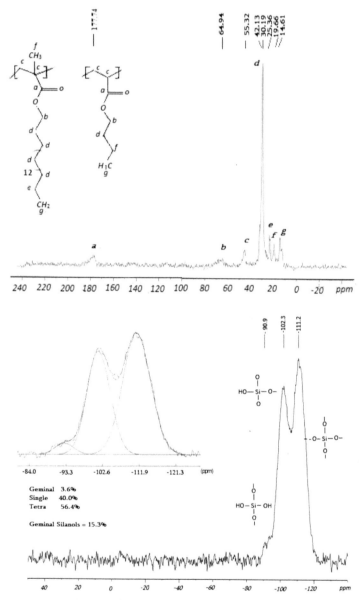

FIGURE 17.1 H-C CPNAS NMR spectrum of ODMA-BA coated symmetry 300A: (Symmetry 300A coated octadecyl methacrylate - bytilacrilate copolymer).

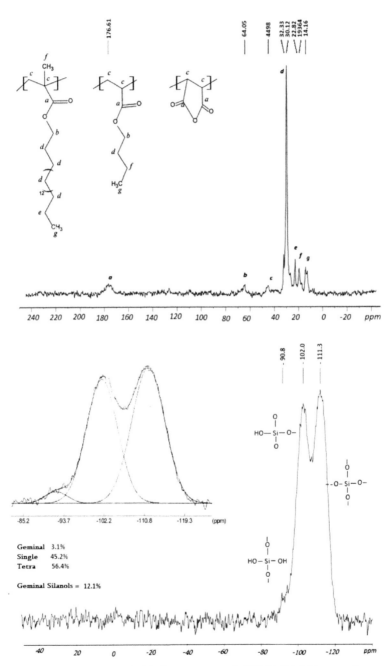

FIGURE 17.2 H-Si CPMAS NMR spectrum of ODMA-BA-MA coated symmetry 300A: (Symmetry 300A coated octadecyl methacrylate, bytilacrilate, and maleic anhydride copolymer).

It had been shown earlier by us [11] that hybrid polymer-mineral packing materials based on wide-porous silica ($S_{sp.}$ = 80.0 m²/g; V_Σ = 0.58 cm³/g; $d_{av.}$ = 30 nm) and ODMA-BA-MA polymer composition as deposited stationary phase possess weak cation-exchange properties (see Table 17.1).

Uracyl (1)–pyridine (2)–phenol (3)–N,N-DMA (4)–*para*-BBA (5)– toluene (6) mixture was used for testifying the columns packed by obtained polymer-containing sorbents (Table 17.1). The fact that functional groups on the surface of prepared sorbents will possess weak cation-exchange properties may be easily foreseen. One can conclude from the Table 17.1 that pyridine peak appears earlier than phenol peak, and it is logical to suppose that polymer deposition eliminates so-called "silanol" interaction.

It is evident when comparing chromatographic data obtained on the base of the indicated testing mixture on the column packed by the sorbent Si–300 + ODMA-BA-MA containing 20% mass. polymer that toluene peak shifted (from 3.7 min to 6.042 min. One can assume that the introduction of the third component to the polymer composition content leads to the secondary structure-formation whereupon the increase of toluene retention factor k^1 from 1.05 to 2.85 occurs.

Retention time for N,N-DMA peak after introducing MA into the polymer composition content is 6.48 min (without MA it was 3.33 min, Table 17.1). It is natural because after introducing MA carboxylic groups are present onto the surface (as a result of hydrolysis). A change of pyridine peak retention time also witnesses about their presence (rather their formation during the chromatography process). At the initial stages of the chromatography process, pyridine peak appears before phenol peak. Then it shifts to the higher times side and finally disappears at all, because pyridine may be tied completely by carboxylic groups.

When column packed by Si–300 + ODMA-BA-MA sorbent is washed by pH 12.0 phosphate eluent *p*-BBA, and uracyl peaks appear together at 1.57 min, pyridine peak appears at 1.69 min and before phenol peak, N,N-DMA peak appears at 3.45 min, and toluene peak–at 4,294 min (Table 17.1). It follows from the Table 17.1 that appearance of the pyridine peak at 1.69 min after localization of the carboxylic groups (alkaline hydrolysis at pH 12.0) indicate that the polymer coating screened completely silanol groups on the MS surface.

Further, after treatment by acidic eluent and restoration of carboxylic groups on the surface the following picture obtained (Table 17.1): pyridine peak disappears again, *p*-BBA, toluene, and N,N-DMA peaks shift to the higher times side (3.375 min, 5.55 min, and 6.10 min correspondingly).

TABLE 17.1 Chromatographic Parameters of Obtained Packing Materials (ACN–acetonitrile; PB–phosphate buffer; N,N-DMA–N,N-dimethylaniline; p-BBA–*para*-butyl benzoic acid)

Sample	Mobile phase	Peak retention time (min)						$k^I_{(toluene)}$
		Uracyl	Pyridine	Phenol	N,N-DMA	p-BBA	Toluene	
Si–300 + ОДС	ACN: water 50/50	1,65	–	2.23	–	–	3.75	1.27
Si–300 + ODMA-BA	ACN–50 µM PB pH 7.0	1.82	2.14	2.25	3.33	1.20	3.73	1.05
Si–300 + ODMA-BA-MA	ACN–50 µM PB pH 7.0	1,57	2.26	2.29	6.84	1.57	6.04	2.85
Si–300 + ODMA-BA-MA	ACN–50 µM PB pH 12.0	1,27	1.69	1.83	3.42	1.20	4.29	2.36
Si–300 + ODMA-BA-MA	ACN: (0.1% H$_3$PO$_4$) 50/50	1.57	–	2.09	6.10	3.32	5.55	2.52

The introduction of MA into the content of polymer composition provides the presence of carboxylic groups along with C_{18}-groups on to the surface. They attach cation-exchange properties for the surface of packing materials.

On the other hand separation of basic compounds on such sorbents is unexpected (Table 17.1), at that pyridine peak appears considerably earlier than phenol peak, and tailing factor value for amitriptyline peak at 34 min equals only 1.18 (Figure 17.3). How one can explain this phenomenon? Most probably the carboxylic groups formed as a result of MA hydrolysis do not possess the same affinity towards compounds of basic nature as compared with silanol groups.

Chromatograms of uracyl (1); pyridine (2); phenol (3); propranolol (4); butylparaben (5); naphthalene (6); dipropylphtalate (7); acenaphthene (8); amitriptyline (9) testing mixture are presented in Figures 17.3 and 17.4.

It is noteworthy that the tailing factor for those compounds which have considerably higher retention times is comparatively lower values. That is the conclusion about affinity of such compounds towards polymer phase or about some interactions between them is groundless. It especially relates to propranolol, because retention time of it at Symmetry C_{18} column is 2.46 min. Values 13.42 min and 24.58 min correspondingly for columns with ODMA-BA Symmetry 100 and ODMA-BA-MA Symmetry 100 seem as a result of certain interaction with the surface. However, on the other hand, tailing factor values equals only 1.26 and 1.36 correspondingly. It is strange and obscure for us.

The naphthalene and acenaphthene behavior as hydrophobic markers is perfectly inscribed to the RP-HPLC logics. If retention times of these substances for column with Symmetry C_{18} are 8.11 min and 19.76 min correspondingly, for ODMA-BA Symmetry 100 and ODMA-BA-MA Symmetry 100 retention times are 4.81 min, 5.44 min, and 11.35 min, 10.93 min correspondingly. This means that definite (corresponding) number of C_{18} groups was involved to the chromatography process. However, it is necessary to note that the introduction of the MA to the polymer phase content leads to the redistribution of C_{18} groups, which increases chromatographic accessibility and the amount of reversed-phase groups on the surface.

Certain increase in naphthalene peak retention time was observed (from 4.81 min to 5.44 min) when passing from ODMA-BA Symmetry 100 column to ODMA-BA-MA Symmetry100 column. In the same time, acenaphthene peak retention time decreases from 11.35 min to 10.93 min. We suppose that some other retention mechanism for these hydrophobic substances is obvious

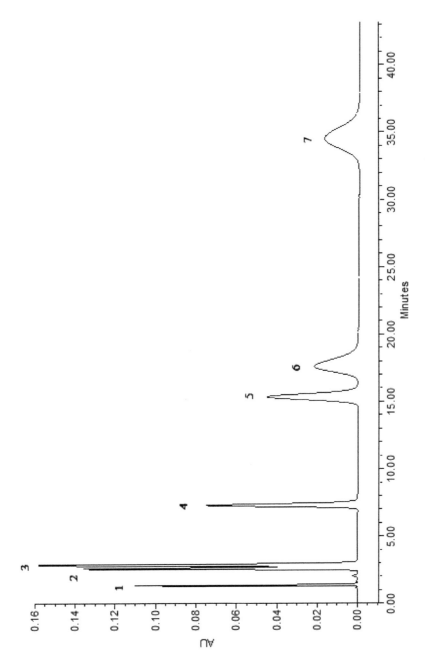

FIGURE 17.3 Chromatograms of uracyl (1); pyridine (2); phenol (3); propranolol (4); butylparaben (5); naphthalene (6); dipropylphtalate (7); acenaphtene (8); amitriptyline (9). Column–Symmetry - ODMA-BA.

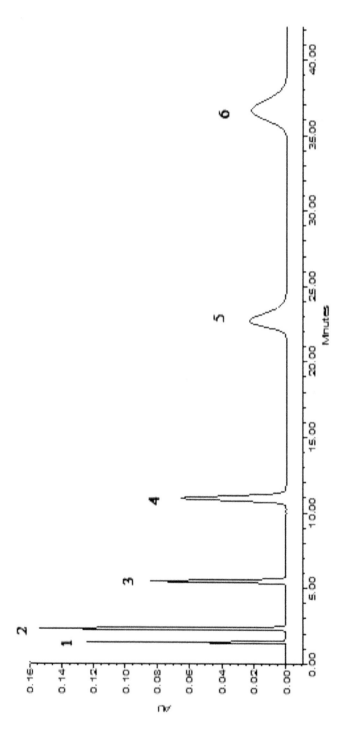

FIGURE 17.4 Chromatograms of uracyl (1); pyridine (2); phenol (3); propranolol (4); butylparaben (5); naphthalene (6); dipropylphtalate (7); acenaphtene (8); amitriptyline (9). Column–Symmetry - ODMA-BA-MA.

here. For substances with polar groups (butylparaben and dipropyl phthalate), the following picture was observed. On the column packed by ODMA-BA Symmetry 100, butylparaben and dipropyl phthalate appears virtually together with dead volume (0.95 min and 1.14 min correspondingly). On the column packed by the sorbent with carboxylic groups (ODMA-BA-MA Symmetry100 column), certain relay in retention times of the sea substances takes place (values correspond 2.3 min and 2.426 min). Apparently, this increase is stipulated by their affinity towards carboxylic groups. In case of ODMA-BA phase, the surface do not have polar groups and butylparaben retention time may serve as dead volume marker (Table 17.2).

Quite interesting is behavior of propranolol. Its peak retention time on the Symmetry C_{18} column is only 2.46 min, whereas on the ODMA-BA Symmetry 100 column it increases up to 13.42 min and reaches 24.58 min on the ODMA-BA-MA Symmetry 100 column, and at that tailing, factors have rather low values (1.26 and 1.36 correspondingly).

Increase occurs in uracyl peak retention time values in the C_{18} < ODMA-BA < ODMA-BA-MA row (Table 17.2). *Ex facte* it is rather strange, especially if considering that all columns are packed with the same packing material. In addition, the polymer component amount on the silica surface exceed the amount of ODS. This, in its turn, decreases the values of average pore volume of packing material, which also must to be a reason for decreasing of column dead volume values. However, for some reason the dead volume of column increases.

It is known that the dead volume value is composed from the following parameters of packing material and column: total pore volume of sorbent; inter-particle space in thoroughly packed column; out-column volume (inner volume of connecting capillaries).

The third component is constant, because one can always use capillaries of the same diameter and total length. The first component of the sum as we have already marked, must diminish in the C_{18} > ODMA-BA > ODMA-BA-MA row, inasmuch phase amount in this row increases and it will lead to the decrease of pore volume values.

The second component–inter-particle space in the column, It may influence on the dead volume value, but the influence is too small and impossible to measure it.

About the fact that the packing of polymer-coated micro-spherical (MS) particles is not thus much thorough as it observed for column packed by MS with ODS phase, witnesses values of pressure in the columns with polymer phases (correspondingly 1157 psi for C_{18} packed, 666 psi for ODMA-BA packed, and 782 psi for ODMA-BA-MA packed columns. Why the packing

TABLE 17.2 Eluent: MeOH/20 mM K$_2$HPO$_4$/20 mM K$_2$HPO$_4$ pH 7.0 (65/35)

	T$_r$			k$_{corr}$			T$_{usp}$			N$_{usp\ corr}$			RPH		
	1	2	3	1	2	3	1	2	3	1	2	3	1	2	3
Uracyl	1.015	1.212	1.405												
Propranolol	2.46	13.42	24.58	1.46 / 2.46	13.12 / 10.07	16.49	1.14	1.26	1.36	3962	1458	2743	7.44	20.57	10.94
Butylparaben	3.55	0.95	2.300	2.55 / 3.55	0.95	0.64	1.12	1.11	1.15	6080	1701	2631	4.85	17.64	11.40
Naphthalene	8.11	4.81	5.441	7.11 / 8.11	2.97	2.87	1.10	1.10	1.11	9583	4788	5753	3.08	6.27	5.21
Dipropylphthalate	6.65	1.14	2.426	5.65 / 6.65	−0.059	0.73	1.10	1.10	1.14	7483	1922	2883	3.94	15.61	10.41
Acenaphthene	19.76	11.35	10.925	18.76 / 19.76	8.36	6.78	1.05	1.07	1.06	9340	4580	5371	3.16	6.55	5.59
Amitriptyline	23.59	24.63	35.66	22.59 / 23.59	19.32	24.38	1.85	1.13	1.17	6328	2126	3058	4.66	14.11	9.81
Pyridine [single injection]	0.38	1.303	1.467	0.38	0.08	0.04	1.89	1.20	1.25	2519	4010	3745	11.70	7.48	8.01
Phenol [single injection]	0.59	1.542	1.630	0.59	0.27	0.16	1.20	0.00	1.26	8158	2741	3329	3.61	10.94	9.01

1. Symmetry C$_{18}$.
2. ODMA-BA Symmetry 100.
3. ODMA-BA-MA Symmetry 100.

of MS coated by polymer phases takes place стр. than in case of convenient reversed phases it is a subject for other discussion. Nobody can disclaim that poor packing of MS leads to the increase of dead volume owing to increase of inter-particle space. Nevertheless, such relatively great increase in dead volume values can't be explained only by this. It means that certain interactions exist between uracil molecules and polymer phases and uracyl may not serve as dead volume marker in this case. It follows from this that values of retention factor presented in table (calculated by uracil retention time values) are incorrect. It is necessary to solve the problem of dead volume marker. Apparently, it is expedient to take dead value of Symmetry C_{18} column, namely 1.015 min.

On the other hand, when passing from ODMA-BA to ODMA-BA-MA dead volume value increases from 1.212 to 1.405, although the increase of phase amount (as MA) is insignificant. Besides, packing quality becomes better. Pressure in column in the first case was 666 psi, and in the second– 782 psi. This, in its turn, one more time witnesses about certain interaction between polymer phase and uracyl molecules.

The increase of dead volume values by 20% and 40% for account of increase of inter-particle space would mean proportional decrease of packing material amount in the column. However, it is impossible inasmuch definite shrinkage must be observed in such cases when exploiting columns.

Butylparabene peak retention time on the ODMA-BA Symmetry 100 column is 0.95 min and may serve as dead volume index (marker). And it is likely true, inasmuch alkyl hydrophobic groups of polymer phase can't interact with polar groups of butylparabene.

Such interaction take place on the ODMA-BA-MA Symmetry 100 with polar groups on the polymer phase surface conditioned by the presence of MA in copolymer and its hydrolysis during the chromatography process. And retention time values of butylparabene witnesses about that. Similar picture we can observe also for polar dipropyl phthalate. If butylparabene retention time would taken as dead volume marker for polymer-containing columns, then values for retention factors in the Table 17.2 must to be changed (corrected).

17.3 CONCLUSION

It has been shown that polymer-containing packing materials for RP-HPLC possess some properties different from that for convenient C_{18} packing materials.

Uracyl may not serve as dead volume marker for columns packed by ODMA-BA, and ODMA-BA-MA polymer phases inasmuch retention of it takes place by these phases, i.e., interaction of uracil molecules with stationary phase surface take place.

It is expedient to use butylparabene retention time as dead volume marker for such columns. Chromatographic characteristics (behavior) of propranolol on columns containing polymer-coated MS also sharply differ from its chromatographic parameters on convenient C_{18} columns.

KEYWORDS

- **ODMA-BA**
- **ODMA-BA-MA**
- **RP-HPLC**

REFERENCES

1. Neue, U. D., Serowik, E., Iraneta, P., Alden, B. A., & Walter, T. H., (1999). *J. Chromatogr. A., 849*, 87–100.
2. Neue, U. D., Alden, B. A., & Walter, T. H., (1999). *J. Chromatogr. A., 849*, 101–116.
3. Tanaka, T., Goodel, H., & Karger, B. L., (1978). *J. Chromatogr., 158*, 233.
4. Neue, U. D., Phoebe, C. H., Tran, K., Cheng, Y. F., & Lu, Z., (2001). *J. Chromatogr. A., 925*, 49–67.
5. Hayrapetyan, S. S., Khachatryan, H. G., & Neue, U. D., (2006). *J. Separation Sci., 29*(6), 801–809.
6. Hayrapetyan, S. S., Neue, U. D., & Khachatryan, H. G., (2006). *J. Separation Sci., 29*(6), 810–819.
7. Neue, U. D., (1997). *HPLC Columns, Theory, Technology & Practice* (p. 393). Wiley-VCH Inc.
8. Hanson, M., Unger, K. K., & Schomburg, G., (1990). *J. Chromatogr., 517*, 269–284.
9. Hanson, M., & Unger, K. K., (1992). *TrAC Trends in Analytical Chemistry, 11*, 368–373.
10. Hanson, M., Unger, K. K., Mant, C. T., & Hodges, R. S., (1992). *J. Chromatogr. A., 599*, 65–75.
11. Hayrapetyan, S. S., & Khachatryan, H. G., (2005). *Chromatographia, 61*(1/2), 43–47.

CHAPTER 18

New Technical Access for Creation of Gradually Oriented Polymers

L. I. NADAREISHVILI, R. SH. BAKURADZE, M. G. ARESHIDZE,
I. I. PAVLENISHVILI, and L. K. SHARASHIDZE

*Georgian Technical University, Vl. Chavchanidze' Institute of Cybernetics,
5 Z. Anjaparidze St., 0186, Tbilisi, Georgia,
E-mail: levannadar@yahoo.com*

ABSTRACT

A new technical solution of zonal gradient and homogeneous stretching of linear polymers is described. The design features of the device allow to regulate the quantitative characteristics of oriented materials strictly. The version of zone heater, cooler, and their displacement mechanism in the form of an independent unit makes it possible to carry out zonal stretching practically on any tensile device.

18.1   INTRODUCTION

One of the actual directions of modern polymer science is the creation and research of materials with a gradient of different properties (mechanical, optical, electrical, magnetic, etc.) [1–3]. Gradient materials can be created by a gradient of chemical composition or material structure. Methods for creating polymer-based gradient materials include: spreading of chemical reaction front, the metal powders metallurgy methods, corona discharged, centrifugation, layering of thin films step by step method, ultraviolet irradiation, selective laser caking, die casting, etc.

We have developed a new approach to obtaining gradient materials, the essence of which is to create a gradient of the morphology of polymer, namely the degree of the preferential arrangement of structural elements (macromolecular chains and their parts, as well as permolecular structural fragments) in the chosen direction. Such structural transformation of polymers can be realized by stretching of isotropic polymer samples in a gradient mode. Technically this can be done in several ways:

- stretching in the inhomogeneous mechanical field [4];
- adjustable expansion of the stretch zone [5, 6];
- uniaxial controlled zonal gradient stretching [7, 8].

The design of new methods for obtaining of gradient polymeric materials, as well as the improvement of corresponding devices, will greatly contribute to the development of graded material science.

In this chapter, a new technical solution of obtaining gradient materials is considered; the distinctive features of the design are indicated, technical descriptions of the appropriate device and the principle of its operation are given.

18.2   UNIAXIAL CONTROLLED ZONAL GRADIENT STRETCHING DEVICE

Experience in the creation of polymeric gradient materials and the improvement of appropriate appliances for orientation drawing led to the design of a device that provides the following technical requirements:

- stretching with a given acceleration;
- regulation of the extent of the stretch zone; and
- inclusion in the construction of an active cooler and adjustment of its acceleration and motion direction.

These distinctive features of the technical design of the apparatus provide the possibility of creating gradient-oriented samples with specified quantitative parameters that characterize the gradually oriented state of the linear polymers (value, length, and profile of the distribution of the relative elongation/orientation degree) [9].

According to Ref. [8] uniaxial controlled zonal gradient stretching device consists of the three independent units–stretching machine (strength testing machine or its simplified analogue), add-on module (heater, cooler, and their displacement mechanisms) and stretch process control block. The version of a zone heater, a cooler and their displacement mechanism in the form of an independent unit – in the form of a functional module – makes it possible to carry out zonal stretching practically on any tensile device.

Figure 18.1 shows a scheme of one of the feasible technical solutions of the add-on module (*a* – front view, *b* – side view). Add-on module consists of supporting frame 1, in which the screws 2, 3, 4, 5 are inserted. Cross-arms 8, 9 are moved by rotating screws 2, 5 and 3, 4, respectively. The screws are rotated by electric motors 6, 7. In the holes 10, 11 of cross-arms 8, 9 the rods 12, 13 with grooves 14, 15 are inserted. The rods can move along own axes by screw mechanisms 16, 17. The arrow ↔ indicates the direction of movement of the rods.

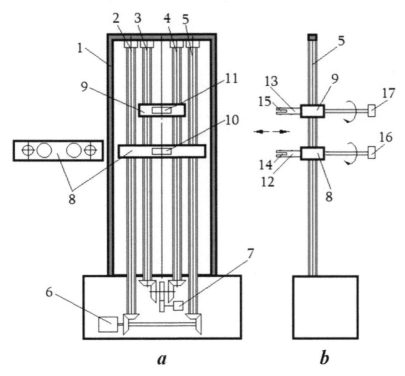

FIGURE 18.1   Add-on module.

Figure 18.2 presents the scheme of one of the possible solutions for the heater of the tested film (*a* – cut in a vertical plane; *b* – top view, the configuration for input into working zone of stretching machine; *c* – top view, operating state). The heaters 18 are placed in the thermal insulators 19, which are provided with covers 21 for regulating the height of the heat flow directed to the test sample 20.

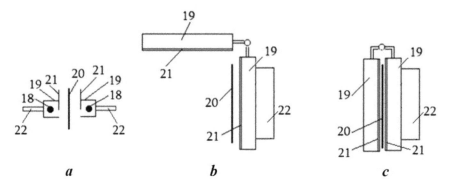

FIGURE 18.2   Scheme of the heater unit.

Before starting the test, the plate 22 of thermal insulator 19 is inserting into the groove 15 of the rod 13. The thermal insulator is positioning parallel to the plane of the film 20 at a selected distance from it ($\sim$ 0.5 ÷ 1 mm). The second thermal insulator is arranged in a similar manner, as a result of which the test film 20 is located between two equidistant heaters 18.

One of a possible scheme for the cooler (blowers) is presented in Figure 18.3 (*a*–cut in a vertical plane; the arrow ← indicates the direction of the air ejection; *b*–top view, the configuration for input into working zone of stretching machine; *c*–top view, operating state). Before starting the test, the plate 25 of the blower 23 is inserting into the groove 15 of the rod 13. The blower is positioning parallel to the plane of film 20 at a selected distance from it (~1 mm). The second blower is arranged in a similar manner, as a result of which the test film 20 is located between two equidistant blowers 23.

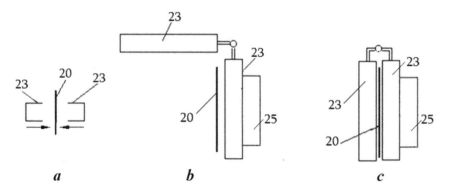

FIGURE 18.3   Scheme of the cooler unit–blower.

Cooling of the test sample is possible by liquid; for example, by water in a variable volume vessel. Figure 18.4 shows one of the possible options of using of variable volume vessel–the bellows 25 (*a*–front view; *b*–side view), in the inner bottom of which the clamp 26 is hermetically fixed. Bellows is equipped with pipes 27, 28 for input and output of the water. On the outer side of the bellows bottom the plate 29 is fixed. There is a ledge 30 on the perimeter of the upper open edge of the bellows 25. Before the beginning of the experiment, the ledge 30 is placed in the holder 31 (Figure 18.4, *c*) and the plate 32 of the holder 31 is inserted into the groove 14 of the rod 12.

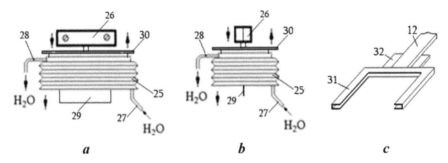

FIGURE 18.4   Scheme of the cooler unit–bellows.

One of the possible schemes of a stretching machine (in the form of a universal testing machine) is given in Figure 18.5 (*a*–front view, *b*–side view). It consists of a supporting frame 33, in which the leading screws 34, 35 are inserted. Cross-arm 36 is moved by leading screws. The screws 34, 35 are rotated by an electromotor 40. Clamp 37 is actually stationary and attached to load cell 38. Clamp 39 is fixed on the cross-arm 36 and moves with it. Depending on the cooling method, in the active clamp 39, either one end of the test sample 20 or the plate 29 of the bellows 25 is fixed.

In Figure 18.6, the scheme of the control block of the device is represented. The electromotor 6 moves the cooler (blower) 23 (or the ledge 30 of the bellows 25). The electromotor 6 rotation frequency is controlled by the driver 1. The electromotor 7 moves the rod 13 and at the same time the thermal insulator 19 with the heater 18. The electromotor 7 rotation frequency is adjusted by the driver 2. The electromotor 40 moves the active clamp 39. The electromotor 40 rotation frequency is regulated by driver 3.

We enter the selected program through the LCD (liquid crystal display, touch panel controller) operator panel in the PLC (programmable logic controller) and control the process. PLC gives signals to the drivers which determine the frequencies of electromotors rotation (i.e., the speed of movement of the device mobile components–the rods 12, 13 and active clamp 40) in accordance with the selected program. The temperature controller maintains the selected temperature within the set limits. Power sensors record the tensile stress required for stretching.

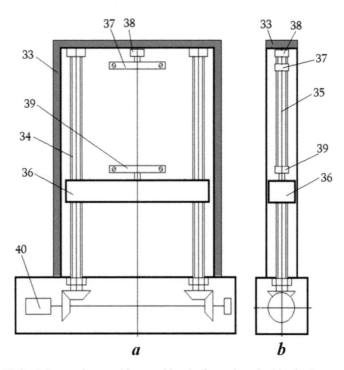

FIGURE 18.5 Scheme of a stretching machine (*a*–front view, *b*–side view).

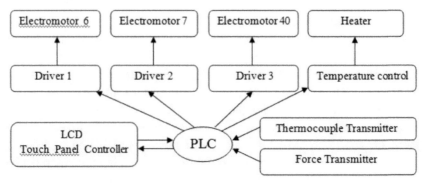

FIGURE 18.6 Scheme of the control block.

On the described device it is possible to stretch polymer samples (film, rod, plate) both in a homogeneous (non-gradient) and gradient modes.

Before the beginning of the experiment, the stretching mode is determined:

- gradient (or homogeneous) stretching;
- length of the stretched (obtained) sample;
- range of changes in the relative elongation;
- the distribution profile of the relative elongation in the stretched sample;
- changes in the speed of movement of the mobile components of the device (active clamp 40, heater 18, cooler 23) during the stretching process;
- initial size of the sample;
- stretching temperature.

Stretching mode is determined by the relative values of the speed of movement of the zonal heater 18, the cooler (blower 23 or ledge 30 of bellows 25) and the active clamp 39. Let us denote by V_1, V_2, V_3 the speed of movement of the zone heater, the cooler and the active clamp, respectively. Then the stretching conditions in the homogeneous and gradient modes can be written as follows:

$V_1 = V_2 = const$ $V_3 = const$ (homogeneous stretching)
$V_1 = const.$ $V_2 \neq const.$ $V_3 \neq const$ (gradient stretching)

All these parameters are set according to the patent [8] or on the basis of mathematical model developed by us [9]. The selected stretching program is introduced into the control block of the device (Figure 18.6).

Figure 18.7 shows the mutual arrangement of the test sample 20, thermal insulator 19 and cooler 23 prior to start of the experiment. The test film 20 is fixed in the clamps 37, 39 of the stretching machine. The cooler 23 and the thermal insulator 19 (with the heater 18) are fixed in the rods 12, 13, respectively. The cooler 23 and the heat insulator 19 are given the configurations depicted in Figures 18.3*b*, and 18.2*b*, respectively. After introduction them into the working zone of the stretching machine, they will be located near the active clamp 39 at a selected distance from each other and put into operation (Figures 18.3*c* and 18.2*c*).

The described device works as follows:

The heater 18 is turned on. After reaching the selected temperature, the electric motors 6, 7 and 40 are turned on. The arrow ↑ indicates the direction of the heater 18 and cooler 23 movements, and the arrow ↓ the direction of active clamp 7 and cross-arm 4 movement. Testing film 20 will start lengthening in the heating zone, where the liquid limit will be lowered. The stretched part of the sample is gradually coming out from the heating zone. The lengthening is stopped when this part enters into the cooler (air blower) zone. As a result of the movement of the active clamp 39, the heater 18 and the cooler 23, the process gradually spreads to the nonstretched region of the sample 20. The process terminates when the thermal insulator reaches the clamp 37. It is possible to change the stretching mode at any stretching step of the film 20. Figure 18.8 shows the mutual arrangement of the stretched sample 20, thermal insulator 19 and cooler 23 after the end of the stretching process.

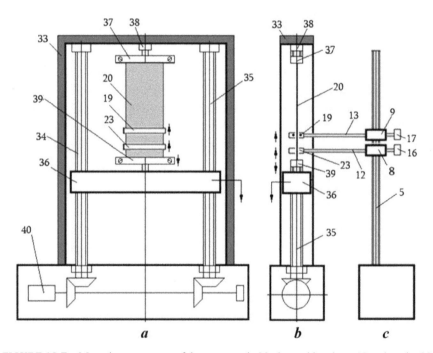

FIGURE 18.7   Mutual arrangement of the test sample 20, thermal insulator 19 and cooler 23 prior to start of experiment (stretching machine, *a*–front view, *b*–side view; *c*–add-on module, side view).

Figure 18.9 shows the mutual arrangement of the mobile components when using bellows 25 as a cooler. Bellows 25 is fixed in the active clamp 39 of the stretching machine by the plate 29 (Figure 18.4, a, b). The test film 20 is loaded into the clamps 26 and 37 (Figure 18.9, *a, b*). The bellows is placed in the holder nest 31 which is connected to the rod 12 (Figure 18.4, *c*). The thermal insulator 19 with the heater 18, as described above, is placed next to the clamp 26 at a selected distance. Water is supplied to the bellows 25 through the pipe 27. The arrow ↑ indicates the direction of movement of the upper open edge of the bellows 25 (ledge 33) and the thermal insulator 19 (with the heater 18), and the arrow ↓ indicates the direction of movement of the clamp 30, the cross-arm 36 and the clamp 29 embedded in the inner part of the bellows 25.

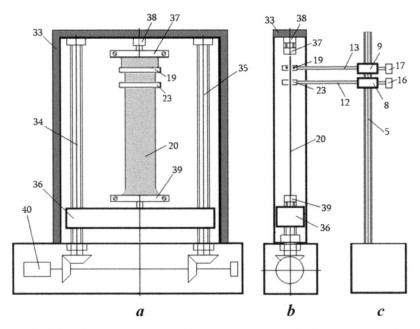

a   *b*   *c*

FIGURE 18.8 Mutual arrangement of the stretched sample 20, thermal insulator 19 and cooler 23 after the end of stretching (stretching machine, *a*–front view, *b*–side view; *c*–add-on module, side view).

Film 20 will start stretching in the heating zone, where the liquid limit is lower. The stretched part is gradually coming out from the heating zone. The lengthening is stopped when this part enters into the bellows 25 with water. As a result of the movement of the active clamp 39, the heater 18 and the upper open edge of the bellows 25 (ledge 33), the process gradually spreads to the nonstretched region of the sample 20. The process terminates when the thermal insulator reaches the clamp 37. It is possible to change the stretching mode at any stretching step of the film 20. Figure 18.10 shows the mutual arrangement of the stretched sample 20, thermal insulator 19 and bellows 25.

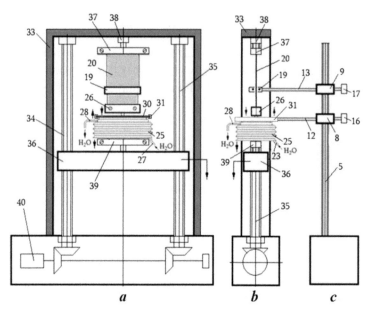

FIGURE 18.9 Mutual arrangement of the test sample 20, thermal insulator 19 and bellows 25 prior to the start of the experiment (stretching machine, *a*–front view, *b*–side view; *c*–add-on module, side view).

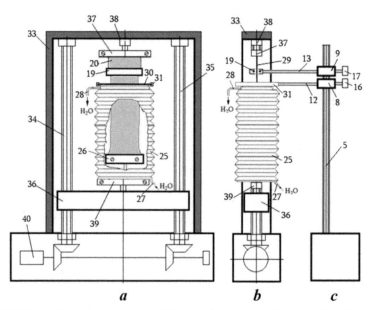

FIGURE 18.10 Mutual arrangement of the stretched sample 20, thermal insulator 19 and bellows 25 (stretching machine, *a*–front view, *b*–side view; *c*–add-on module, side view).

In the discussed examples the stretching of the sample occurs down from the top. At the same time, there are testing machines in which the stretching is realized up from the bottom (a fixed clamp is located on the lower part of the support frame 1 of the testing machine, while the active clamp moves up from the bottom). In this case, the thermal insulator 19 (with the heater 18) is fixed in the rod 12, and the cooler 23 is fixed in the core 13. Further operations are similar to the above. It is obvious that in this case the bellows as a cooler cannot be used.

Using this method, it is possible to create the gradient-oriented samples with predetermined values of parameters that characterize gradually oriented state of the linear polymers [9].

18.3   CONCLUSIONS

1.   A new technical solution of zonal gradient and homogeneous stretching of linear polymers is developed.
2.   Technical distinctive features of the device, in particular, inclusion in the construction of an active cooler and adjustment of its acceleration and motion direction as well regulation of the extent of the stretch zone provide the possibility of creating gradient-oriented samples with specified quantitative parameters that characterize the gradually oriented state of the linear polymers (value, length, and profile of the distribution of the relative elongation/orientation degree).
3.   The version of zone heater, cooler, and their displacement mechanism in the form of an independent unit makes it possible to carry out zonal stretching (gradient or homogeneous) practically on any tensile device.

KEYWORDS

- **controllable distribution of relative elongation**
- **controlled graded stretching**
- **orientation of linear polymers**
- **zonal stretching device**

REFERENCES

1. Kieback, B., Neubrand, A., & Riedel, H., (2003). Processing techniques for functionally graded materials. *Materials Science and Engineering, A362,* 81–105.
2. Miyamoto, Y., & Shiota, I., (1996). *"Functionally Graded Materials* (p. 792). *"* Elsevier Science, New York.
3. Nathan, J. R., (2011). *Functionally Graded Materials* (p. 324). Nova Science Publ., New York.
4. Nadareishvili, L., Bakuradze, R., Topuridze, N, et al., (2015). "Graded orientation of the linear polymers." *International Journal of Mechanical. Aerospace, Industrial and Mechatronics Engineering, 13*(02), 1352–1357.
5. Nadareishvili, L., Bakuradze, R., Topuridze, N., et al., (2015). In: Mukbaniani, O., Abadie, M., & Tatrishvili, T., (eds.), *"Method of Obtaining of Gradually Oriented Polymer Films."* High-Performance *Polymers for Engineering-Based Composites. Section 1. Applications of Polymer Chemistry and Promising Technologies. Ch. 13* (pp. 145–152). Apple Academic Press, Inc. USA.
6. Nadareishvili, L., Wardosanidze, Z., et al., (2007). *"Polymeric Films Deformation Method,"* (p. 4182). Georgian Patent.
7. Nadareishvili, L., (2016). *"Nadareishvili's Device for Stretching of Polymer Samples,"* (p. 6509). Georgian Patent.
8. Nadareishvili, L. (2018). *Nadareishvili's Device for Graded or Homogeneous Stretching of Polymer Films.* (P 6842 B). Georgian Patent.
9. Nadareishvili, L., Bakuradze, R., Aneli, J., Areshidze, M., Pavlenishvili, I., Sharashidze, L., & Basilaia, G. (in press/2019). *Mathematical modeling of the method of uniaxial zone controlled gradient and homogeneous stretching of the linear polymers.* eXPRESS Polymer Letters.

PART III
Materials and Properties

CHAPTER 19

Self-Organization and Sorption Properties in Relation to Lanthanum Ions of Polyacrylic Acid and Poly-2-Methyl-5-Vinylpyridine Hydrogels in Intergel System

T. K. JUMADILOV and R. G. KONDAUROV

JSC "Institute of Chemical Sciences After A.B. Bekturov," Almaty, Republic of Kazakhstan, E-mail: jumadilov@mail.ru

ABSTRACT

Hydrogels mutual activation was studied in aqueous solutions, particularly dependencies of the swelling coefficient, specific electric conductivity, and pH of aqueous solutions from hydrogels molar ratios in time were studied. Maximum activation of hydrogels occurs at gPAA:gP2M5VP = 5:1 and 4:2 ratio. Lanthanum ions sorption by intergel system polyacrylic acid hydrogel (gPAA)–poly-2-methyl-5-vinylpyridine hydrogel (gP2M5VP) was studied. Maximum sorption degree of the intergel system is observed at gPAA: gP2M5VP = 4:2 ratio, it is 91.09%, this is higher comparatively with individual hydrogels of gPAA and gP2M5VP (67.71% and 63.65%, respectively). At this hydrogels ratio polymer chain total binding degree is 75.83%, it is much higher than polymer chain binding degree of the initial hydrogels: binding degree of gPAA is 56.50%, of gP2M5VP is 53.00%. Desorption degree lanthanum ions by ethyl alcohol is 80.17%, by nitric acid–92.55%. Obtained results indicate that at gPAA:gP2M5VP = 4:2 ratio in intergel system there is a significant change of electrochemical, conformational, and sorption properties of the initial macromolecules.

19.1   INTRODUCTION

As a result of previous studies [1–5], it was found that remote interaction of polymer hydrogels leads to significant changes in their electrochemical and conformational properties. Subsequently, influence of different factors on polymer hydrogels remote interaction in the intergel systems was investigated [6–12].

At present, technologies of rare earth and other elements concentration and extraction in hydrometallurgy are based on use of ion-exchange resins. However, the ion exchangers does not have high degree of selective extraction of metals; their distribution coefficient has a low value. In addition, application of ion exchange resins is aimed to selective extraction of one metal. Whereas industrial solutions usually contain several valuable components. In this regard, the aim of this chapter is the study sorption capacity relatively to lanthanum on example of intergel system based on hydrogels of polyacrylic acid and poly-2-methyl-5-vinylpyridine ions and forecast of possibility of their use for rare and rare earth metals extraction.

19.2   EXPERIMENTAL PART

19.2.1   EQUIPMENT

For measurement of solutions specific electric conductivity conductometer "MARK–603" (Russia) was used, hydrogen ions concentration was measured on Metrohm 827 pH–meter (Switzerland). Samples weight was measured on analytic electronic scales Shimadzu AY220 (Japan). La^{3+} ions concentration in solutions was determined on spectrophotometers SF–46 (Russia) and Perkin Elmer Lambda 35 (USA).

19.2.2   MATERIALS

Studies were carried out in 0.005 M 6-water lanthanum nitrate solution. Polyacrylic acid hydrogel (gPAA) was synthesized in presence of crosslinking agent N,N-methylene-bis-acrylamide and redox system $K_2S_2O_8$-$Na_2S_2O_3$ in water medium. Synthesized hydrogels were crushed into small dispersions and washed with distilled water until constant conductivity value of aqueous solutions was reached. Poly-2-methyl-5-vinylpyridine (gP2M5VP) hydrogel of Sigma-Aldrich company (linear polymer cross-linked by divinylbenzene) was used as polybasis.

For investigation task from synthesized hydrogels an intergel pair polyacrylic acid hydrogel–poly-2-methyl-5-vinylpyridine hydrogel (gPAA-gP2M5VP) was created. Swelling coefficients of hydrogels are: $K_{sw(gPAA)}$ = 27.93 g/g; $K_{sw(gP2M5VP)}$ = 3.20 g/g. With gP2M5VP share increase in intergel system gPAA concentration decreased from 5.96 m mol/L to 0.97 m mol/L (gPAA:gP2M5VP ratios interval 6:0–1:5). Concentration of gP2M5VP with decrease of gPAA share increased from 1.125 m mol/L to 6.75 m mol/L (gPAA:gP2M5VP ratios interval 1:5–0:6).

19.2.3 ELECTROCHEMICAL INVESTIGATIONS

Experiments were carried out at room temperature. Studies of intergel system were made in following order: each hydrogel was put in separate glass filters, pores of which are permeable for low molecular ions and molecules, but impermeable for hydrogels dispersion. After that filters with hydrogels were put in glasses with lanthanum nitrate solution. Electric conductivity and pH of overgel liquid was determined in presence of hydrogels in solutions.

19.2.4 DETERMINATION OF HYDROGELS SWELLING

Swelling coefficient was calculated according to the equation:

$$K_{sw} = \frac{m_2 - m_1}{m_1}$$

where m_1 – weight of dry hydrogel; m_2 – weight of swollen hydrogel.

19.2.5 LANTHANUM IONS DESORPTION

After sorption, each hydrogel separately from another was subjected to lanthanum ions desorption in 96% ethanol solution and in 2M nitric acid [13].

Desorption degree was calculated according to this equation:

$$R_{des} = \frac{m_{des}}{m_{sorb}} * 100\%$$

19.2.6   METHODOLOGY OF LANTHANUM IONS DETERMINATION

Methodology of lanthanum ions determination in solution is based on formation of colored complex compound of organic analytic reagent arsenazo III with lanthanum ions [14].

Extraction (sorption) degree was calculated by equation:

$$\eta = \frac{C_{initial} - C_{residual}}{C_{initial}} * 100\%$$

where $C_{initial}$ – initial concentration of lanthanum in solution, g/L; $C_{residue}$ – residual concentration of lanthanum in solution, g/L.

Polymer chain binding degree was determined by calculations in accordance with equation:

$$\theta = \frac{V_{sorb}}{V} * 100\%$$

where v_{sorb} – quantity of polymer links with sorbed lanthanum, mol; v – total quantity of polymer links (if there are 2 hydrogels in solution, it is calculated as sum of each polymer hydrogel links), mol.

19.3   RESULTS AND DISCUSSIONS

19.3.1   STUDY OF GPAA AND GP2M5VP HYDROGELS MUTUAL ACTIVATION IN INTERGEL SYSTEM

Presence of intergel system based on PAA and P2M5VP hydrogels in an aqueous medium provides a variety of different processes which affect on electrochemical equilibrium in solution.

For polyacids, in this case for gPAA remote interaction with basic hydrogels occurs in three stages:

Stage 1: Ionization, formation of ionic pairs, dissociation;

Stage 2: After binding of a proton which was released into aqueous medium due to –COOH groups dissociation (dissociation takes place in accordance with the traditional representation [15]: firstly there is formation of ionic pairs, after that ionic pairs partially dissociates to individual ions), by nitrogen atoms equilibrium in solution in reaction ~COOH↔~COO⁻+ H⁺ shifts to right according to the principle of "Le Chatelier" (concentration of H⁺ decreases due to equation 2).

$$\sim\!COOH \leftrightarrow \sim\!COO^- + H^+ \tag{1}$$

$$\equiv\!N + H^+ \rightarrow \equiv\!NH^+ \tag{2}$$

Thus, –COOH groups undergo additional ionization and dissociation.

Stage 3: Inter-node links are in maximally unfolded state are forced to fold as a result of intramolecular and intermolecular interactions of the type $\sim\!COO^-\ldots H^+\ldots {}^-OOC\!\sim$ for maintaining of stable condition.

For polybases such as polyvinylpyridine remote interaction occurs in two stages:

Stage 1: The association of proton cleaved from carboxyl group:

$$\equiv\!N + H^+ \leftrightarrow \equiv\!NH^+ \tag{3}$$

Stage 2: Interaction of $\equiv\!NH^+$ with water molecules.

If water medium becomes acidic that the interaction occurs according to the reaction:

$$\equiv\!NH^+ + H.OH \leftrightarrow \equiv\!NH^+.OH^- + H^+ \tag{4}$$

If interaction of vinylpyridine links with water molecules leads to alkaline medium then the reaction will be:

$$\equiv\!N + H^+.OH^- \leftrightarrow \equiv\!NH^+ + OH^- \tag{5}$$

Result of these interactions is state in which some part of charged functional groups of hydrogels have no counterions. The concentration of ionized groups, which have no counterions, is in dependence from initial molar ratios of polymer networks and other factors.

Dependence of specific electric conductivity of solutions from hydrogels molar ratio in time is shown Figure 19.1. During remote interaction, there is an appearance of areas with minimum and maximum conductivity, while this parameter increases for almost all ratios of hydrogels with time. As it can be seen from figure minimum values of electric conductivity is observed at gPAA:gP2M5VP = 3:3 ratio for all time of hydrogels remote interaction. Also, low values of conductivity are observed at gPAA:gP2M5VP = 4:2 ratio. The area of maximum electric conductivity is hydrogels ratio 1:5. Maximum values of electric conductivity are achieved after 48 hours.

The minimum conductivity of intergel system gPAA-gP2M5VP is caused by binding of proton which was cleaved from carboxyl group by heteroatom of vinylpyridine. At the time when two hydrogels are put in aqueous medium, both polymers begin to swell due to interaction with water molecules. Carbonyl groups firstly ionize and then dissociate to carboxylate anion $–COO^-$ and hydrogen ions (protons) H^+.

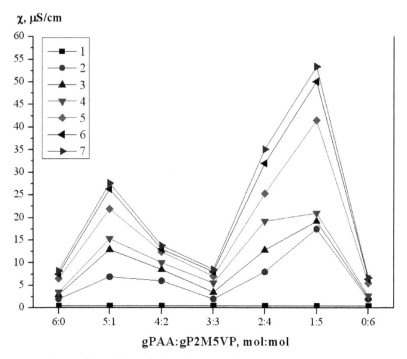

Curves' description (hydrogels remote interaction time): 1–0.5 h;
2–1 h; 3–2 h; 4–3 h; 5–6 h; 6–24 h; 7–48 h

FIGURE 19.1 Dependence of specific electric conductivity from hydrogels molar ratio in intergel system gPAA-gP2M5VP in time in an aqueous medium.

Ionization of cationic hydrogel (poly-2-methyl-5-vinylpyridine) in an aqueous medium occurs due to binding of hydrogen ions which were formed during dissociation of carboxylic groups and water molecules into ions H$^+$, OH$^-$. The long-range interaction effect of hydrogels leads to decrease of total content of positively charged ions in an aqueous medium.

High values of electric conductivity, in turn, indicate that for certain ratios of the two hydrogels dissociation of carboxyl groups prevails over the process of proton association. The reason of this phenomenon is conformational changes of inter-node links. At certain concentrations charged NH$^+$ groups can form intramolecular crosslinks $\geq$N.H$^+$.N$\equiv$, which lead to folding of macromolecular globes and to decrease of proton binding.

The dependence of aqueous solutions pH from gPAA:gP2M5VP molar ratios in time is shown on Figure 19.2. With prevalence of polyacid share in solution decrease of hydrogen ion concentration is observed. Distinct minimum is observed in presence of the polyacid (gPAA:gP2M5VP = 6:0).

The maximum pH value at 48 hours was observed at gPAA:gP2M5VP = 5:1 ratio. Gradual decrease of pH can be seen with the growth of polybasis share.

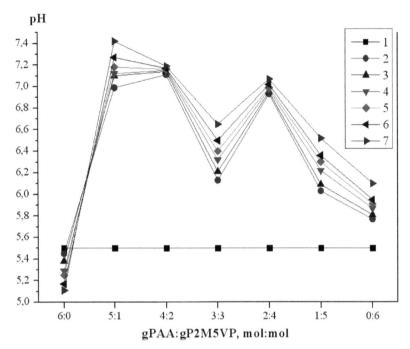

Curves' description (hydrogels remote interaction time): 1–0.5 h; 2–1 h; 3–2 h; 4–3 h; 5–6 h; 6–24 h; 7–48 h

FIGURE 19.2 Dependence of pH from hydrogels molar ratio in intergel system gPAA-gP2M5VP in time in an aqueous medium.

The appearance of H^+ ions excess is due to high speed of swelling and –COOH groups dissociation and also to an insufficient rate of basic groups swelling and their low concentration. The increase in the of OH^- ions content in aqueous medium is associated with low rate of swelling and low concentration of –COOH groups, as well as high speed of swelling and interaction of basic functional groups with H^+ ions. This is possible in case of the second reaction occurrence, in which hydroxyl anions are released in solution. In parallel, there is a third reaction, resulting in a free proton is association with pyridine ring and concentration of positively charged ions in solution dramatically decreases.

Comparing low values of hydrogen ions concentration and high values of electric conductivity in intergel system at ratios of gPAA:gP2M5VP =

4:2 and 3:3, it can be concluded that at these molar ratios there is maximum activation of hydrogels in intergel system.

The concentration of ions in the studied intergel system directly depends on swelling rate and hydrogels concentration in aqueous medium. Swelling rate and deprotonization is in dependence from nature, crosslinking degree, dispersion, and morphology of polymer hydrogels.

Dependencies of swelling coefficients of polyacrylic acid and poly-2-methyl-5-vinylpyridine hydrogels from hydrogels molar ratios in time are shown on Figures 19.3 and 19.4.

Dependence of swelling coefficient of acid hydrogel from hydrogels molar ratio is shown in Figure 19.3. As can be seen with increase of poly-basis share there is an increase of polyacid swelling. Maximum swelling of gPAA is observed at gPAA:gP2M5VP = 1:5 ratio at 48 hours hydrogels remote interaction.

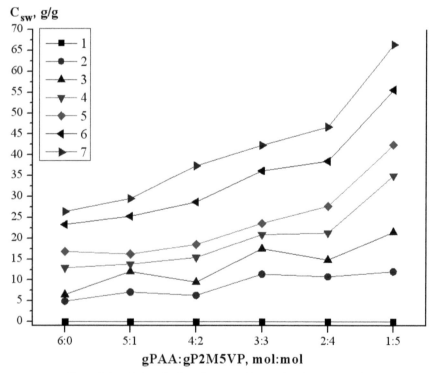

Curves' description (hydrogels remote interaction time): 1–0.5 h;
2–1 h; 3–2 h; 4–3 h; 5–6 h; 6–24 h; 7–48 h

FIGURE 19.3 Dependence of polyacrylic acid hydrogel swelling coefficient from gPAA:gP2M5VP hydrogels molar ratios in time.

It should also be noted that at a gPAA:gP2M5VP = 3:3 ratio up to 2 hours there are distinct maximums of polyacid swelling. This phenomenon can be explained that at this polymer hydrogels ratio ionization of polyacid's links occurs much faster than at other ratios. The area of minimum swelling of polyacrylic acid is observed in presence of only polyacid (gPAA:gP2M5VP = 6:0 ratio) due to the fact that there is no additional dissociation of carboxyl groups as a result of shift of equilibrium to right (towards proton formation).

Dependence of poly-2-methyl-5-vinylpyridine swelling coefficient from hydrogels molar ratios in time is presented in Figure 19.4. Minimal swelling of polybasis occurs when hydrogels ratio is 3:3 and 2:4. It occurs due to the fact that macromolecule tries to take the most advantageous form (globule), despite unfolidng due to repulsion of same charged units. Swelling of polybasis increases with growth of polyacid share. A distinct maximum of swelling is observed at a ratio 5:1 at 48 hours. In a region where polybasis dominates high values of swelling coefficient are caused by additional ionization of pyridine links under the influence of H^+ ions cleaved from –COOH groups. At intermediate ratios (3:3 and 2:4) concentration of H^+ and OH^- ions has almost similar values in intergel system. As a result, H^+ ions are cleaved from $\equiv N^+H$ and ionization degree of polybasis decreases. Accordingly, there is a contraction of polybasis. With further growth of –COOH groups ionization degree of nitrogen atoms increases. This is shown in values of poly-2-methyl-5-vinylpyridine swelling coefficient.

Comparing values of specific electric conductivity, pH, and swelling coefficient of the two hydrogels it can be concluded that in result of mutual activation PAA and P2M5VP hydrogels transfer into a highly ionized state. The area of maximum activation of polymer hydrogels is gPAA:gP2M5VP = 5:1 and 4:2 ratios. Highest ionization of gP2M5VP hydrogel occurs at 5:1 ratio. In turn, swelling of gPAA increases significantly when the polybasis prevails in solution (ratio 1:5).

19.3.2   STUDY OF LANTHANUM IONS SORPTION BY INTERGEL SYSTEM GPAA-GP2M5VP

From the results mentioned above, it can be expected that phenomenon of hydrogels mutual activation should be reflected in processes of metals ions extraction. To check this assumption, sorption properties of intergel system gPAA-gP2M5VP in relation to lanthanum ions were studied.

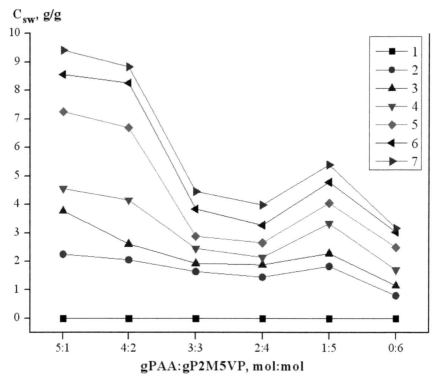

Curves' description (hydrogels remote interaction time): 1–0.5 h;
2–1 h; 3–2 h; 4–3 h; 5–6 h; 6–24 h; 7–48 h

FIGURE 19.4 Dependence of poly-2-methyl-5-vinylpyridine hydrogel swelling coefficient from gPAA:gP2M5VP hydrogels molar ratios in time.

There is a change in electrochemical and conformational properties of PAA and P2M5VP hydrogels during extraction of lanthanum ions by intergel system gPAA-gP2M5VP; however, the changes have are different comparatively to the situation when the hydrogels interact with each other in an aqueous medium.

Change of lanthanum nitrate electric conductivity in presence of gPAA-gP2M5VP intergel system in time is shown on Figure 19.5. The obtained data shows that lanthanum ions sorption causes significant changes of conductivity. As can be seen from figure, electric conductivity decreases for all gPAA:gP2M5VP ratios with time. Minimum values of conductivity are observed at hydrogels ratio 4:2 at 48 hours. In solution, there is a presence of ions formed in result of three-stage dissociation of lanthanum nitrate in addition to carboxylate anions and protons.

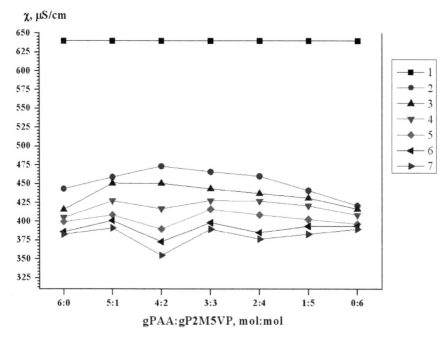

Curves' description (hydrogels remote interaction time): 1–0.5 h;
2–1 h; 3–2 h; 4–3 h; 5–6 h; 6–24 h; 7–48 h

FIGURE 19.5 Dependence of specific electric conductivity from hydrogels molar ratio in time in intergel system gPAA:gP2M5VP in 6-water lanthanum nitrate solution.

In solution the following chemical reactions occur:

1. Dissociation of lanthanum nitrate, along with carboxyl groups dissociation:

$$\sim COOH \leftrightarrow \sim COO^- + H^+ \tag{6}$$

$$La(NO_3)_3 * 6H_2O \leftrightarrow La^{3+} + 3NO_3^- + 6H_2O \tag{7}$$

2. Sorption of lanthanum ions by polymer hydrogels:

$$3\sim COO^- + La^{3+} \rightarrow \sim COO_3La \tag{8}$$

$$3\equiv N + La^{3+} \rightarrow \equiv N^+_3La \tag{9}$$

These reactions impact on electrochemical equilibrium in solution. Depending on predominance of one of them there will be changes in values of conductivity.

Figure 19.6 shows dependence of hydrogen ions concentration from polyacrylic acid and poly-2-methyl-5-vinylpyridine hydrogels molar ratios in time. Concentration of hydrogen ions increases with time. Minimum values of pH are observed at ratios gPAA:gP2M5VP = 6:0 and 5:1. This is a result of hydrogels ionization due to formation of coordination bonds with lanthanum ions.

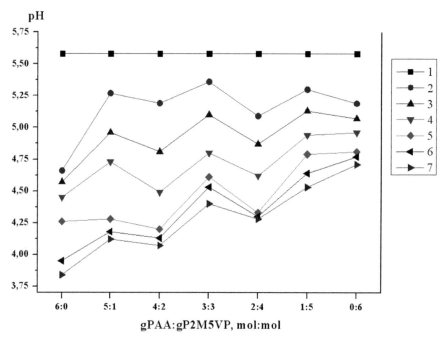

Curves' description (hydrogels remote interaction time): 1–0.5 h;
2–1 h; 3–2 h; 4–3 h; 5–6 h; 6–24 h; 7–48 h

FIGURE 19.6   Dependence of pH from hydrogels molar ratio in time in intergel system gPAA:gP2M5VP in 6-water lanthanum nitrate solution.

Swelling coefficient of gPAA behavior during lanthanum ions sorption is shown on Figure 19.7. Sharp increase of polyacid swelling is observed with polybasis share increase in solution. Maximum swelling of gPAA occurs at gPAA:gP2M5VP = 1:5 ratio at 1 hour of hydrogels remote interaction. There is a gradual decrease of swelling with time. Minimum value of swelling coefficient of gPAA is observed at gPAA:gP2M5VP = 5:1. Due to the fact that there is an occurrence of lanthanum ions sorption, links between node circuits do not have charges, contributing to unfolding of macromolecular globe and as a result, swelling of polyacrylic acid hydrogel decreases.

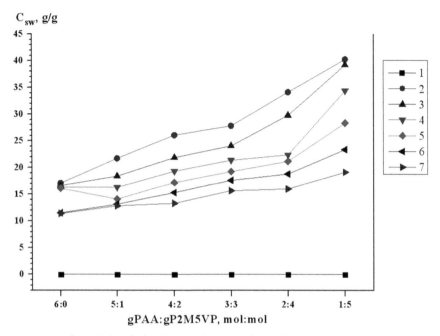

Curves' description (hydrogels remote interaction time): 1–0.5 h; 2–1 h; 3–2 h; 4–3 h; 5–6 h; 6–24 h; 7–48 h

FIGURE 19.7 Dependence of polyacrylic acid hydrogel swelling coefficient in presence of poly-2-methyl-5-vinylpyridine hydrogel in 6-water lanthanum nitrate solution.

Dependence of swelling coefficient of basic hydrogel poly-2-methyl-5-vinylpyridine from molar ratio gPAA:gP2M5VP in time is shown in Figure 19.8.

Changes of swelling coefficient of base hydrogel are similar to changes in polyacid swelling. Firstly there is a sharp increase of swelling, maximum value of swelling coefficient is reached at 1 hour of hydrogels remote interaction at gPAA:gP2M5VP = 5:1 ratio. Then there is a consequent gradual decrease of swelling as in the case with polyacrylic acid. The lowest swelling of poly-2-methyl-5-vinylpyridine is observed in presence of only polybasis (gPAA:gP2M5VP = 0:6 ratio) what is associated with the absence of hydrogels mutual activation phenomenon.

Extraction degree change in dependence of gPAA: gP2M5VP molar ratios is shown on Figure 19.9. As can be seen from the figure ratios at which there are 2 hydrogels in solution have higher sorption capacity comparatively to case when solution contains only polyacid or polybasis (ratios 6:0 and 0:6). However, at gPAA:gP2M5VP 5:1 and 4:2 ratios sorption of lanthanum ions is not very intense. A much higher of La^{3+} ions extraction degree is observed at ratios 2:4 and 1:5. Moreover, the maximum values of sorption degree are reached at gPAA:gP2M5VP = 1:5 ratio. Extraction degree at this hydrogels ratio is 90.34%. This is result of high ionization of gPAA and gP2M5VP due to hydrogels mutual activation in intergel system.

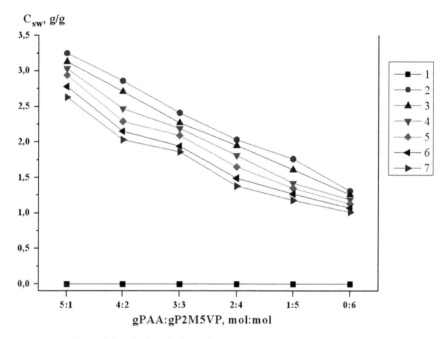

Curves' description (hydrogels remote interaction time): 1–0.5 h;
2–1 h; 3–2 h; 4–3 h; 5–6 h; 6–24 h; 7–48 h

FIGURE 19.8 Dependence of poly-2-methyl-5-vinylpyridine hydrogel swelling coefficient in presence of polyacrylic acid hydrogel in 6-water lanthanum nitrate solution.

Polymer chain binding degree is shown on Figure 19.10. Obtained data shows that intergel pairs (ratios of gPAA:gP2M5VP from 5:1 to 1:5) have higher binding degree than gPAA and gP2M5VP individual hydrogels (ratio 6:0 and 0:6). Similarly to sorption degree maximum values of polymer chain binding degree are observed at gPAA:gP2M5VP = 1:5 ratio, it is 73,13%.

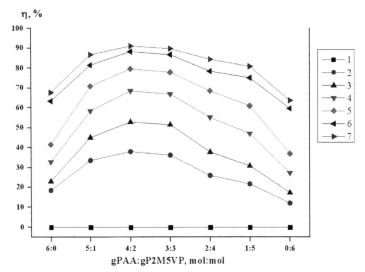

Curves' description (hydrogels remote interaction time): 1–0.5 h;
2–1 h; 3–2 h; 4–3 h; 5–6 h; 6–24 h; 7–48 h

FIGURE 19.9 Dependence of lanthanum ions extraction degree from hydrogels molar ratio in time in intergel system gPAA:gP2M5VP in 6-water lanthanum nitrate solution.

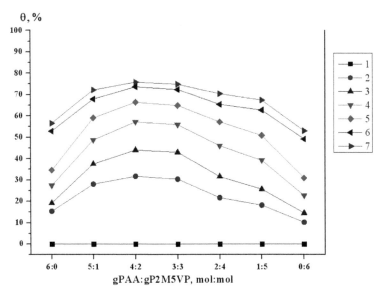

Curves' description (hydrogels remote interaction time): 1–0.5 h;
2–1 h; 3–2 h; 4–3 h; 5–6 h; 6–24 h; 7–48 h

FIGURE 19.10 Dependence of polymer chain binding degree from the molar ratio of hydrogels in time in intergel system gPAA:gP2M5VP 6-water lanthanum nitrate solution.

19.3.3   COMPARISON OF INTERGEL SYSTEM AND INDIVIDUAL HYDROGELS SORPTION ABILITIES

Lanthanum nitrate presents in solution in dissociated state. Dissociation of lanthanum nitrate occurs in 3 stages, in which dissociation constant of first stage is much higher than in second and third. In this regard, binding of dissociated ions in intergel system occurs according to different mechanisms. Lanthanum nitrate formed in first stage of dissociation is associated according to ionic mechanism. Products of second and third stages are bind due to coordination bonds formation.

It should be noted that in intergel system gPAA-gP2M5VP in result of hydrogels mutual activation their electrochemical and conformational properties are changed significantly. Also, there is an increase of extraction degree and polymer chain binding degree of each hydrogel. Sorption degree of lanthanum ions by gPAA and gP2M5VP individual hydrogels (ratios 6:0 and 0:6) is 67.71% 63.65%, respectively. As was mentioned above intergel system extracts 91.09% of lanthanum ions at gPAA:gP2M5VP = 1:5 ratio. Binding degree of gPAA and gP2M5VP individual hydrogels is 56.50% and 53.00%, respectively, when binding degree at gPAA:gP2M5VP = 4:2 ratio is 75.83%. Such increase in sorption properties of polymers is due to the fact that in result of ionization hydrogels undergo various conformational changes in macromolecular structure (e.g., globe unfolding due to repulsion of charged ions on internode links).

19.3.4   STUDY OF LANTHANUM IONS DESORPTION FROM INDIVIDUAL HYDROGELS MATRIX

Desorption kinetics of lanthanum ions ethyl alcohol and nitric acid from the individual hydrogels ratio (gPAA:gP2M5VP = 1:5) are shown in Figures 19.11 and 19.12.

Desorption kinetics of lanthanum ions by ethanol are presented on Figure 19.11. As can be seen, the overwhelming majority is desorbed after 6 hours. Then there is a further increase of lanthanum ions concentration, which becomes negligible after 24 hours. The total desorption degree of lanthanum is 80.17% at 48 hours.

Lanthanum ions desorption kinetics by nitric acid are shown on Figure 19.12.

The desorption process is similar to desorption with ethanol–the maximum desorption occurs within 6 hours. Consequent increase of lanthanum ions concentration in the desorbent reaches final values at 48 hours. Desorption degree is 92.55%.

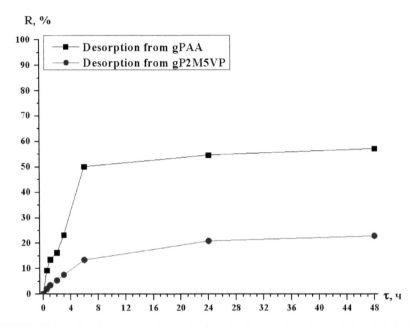

FIGURE 19.11 Lanthanum ions desorption kinetics from hydrogels matrix by ethyl alcohol (96%).

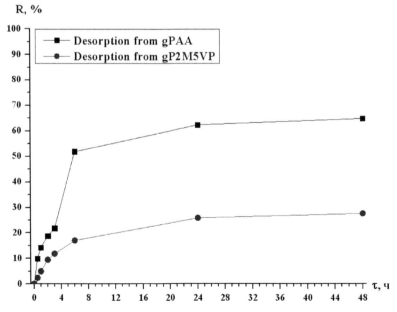

FIGURE 19.12 Lanthanum ions desorption kinetics from hydrogels matrix by 2M nitric acid.

19.4   CONCLUSION

The obtained results allow to make the following conclusions:

1. Result of polymer macromolecules mutual activation is significant change in electrochemical, conformational properties of hydrogels.
2. Based on obtained data on specific electric conductivity, pH, and swelling coefficient it can be concluded that there is an occurrence of lanthanum ions sorption by polymer hydrogels.
3. Maximum value of lanthanum ions extraction degree (91.09%) is observed at gPAA:gP2M5VP = 4:2 ratio at 48 hours of hydrogels remote interaction.
4. Maximum value polymer chain binding degree (75.83%) occurs at gPAA:gP2M5VP = 4:2 ratio at 48 hours of hydrogels remote interaction.
5. Desorption degree of lanthanum ions extraction from hydrogels matrix by ethanol is 80.17%. Desorption degree by nitric acid is 92.55%.
6. Mutual activation of gPAA and gP2M5VP hydrogels in intergel system provides significantly greater sorption degree of lanthanum ions in comparison with individual hydrogels.

ACKNOWLEDGMENTS

The work was financially supported by Committee of Science of Ministry of education and science of the Republic of Kazakhstan.

KEYWORDS

- desorption
- La^{3+} ions
- poly-2-methyl-5-vinylpyridine
- polyacrylic acid
- remote interaction
- sorption

REFERENCES

1. Alimbekova, B. T., Korganbayeva, Z.K., Himersen, H., Kondaurov, R. G., & Jumadilov, T. K., (2014). *J. Chem. Chem. Eng., 8*, 265.
2. Jumadilov, T. K., Kaldayeva, S. S., Kondaurov, R. G., Erzhan, B., & Erzhet, B., (2013). Mutual activation and high selectivity of polymeric structures in intergel systems. *Abstract of Communications, 4th International Caucasian Symposium on Polymers and Advanced Materials*, 191.
3. Jumadilov, T. K., (2014). Electrochemical and conformational behavior of intergel systems based on the rare crosslinked polyacid and polyvynilpyridines. *Proceedings of the International Conference of Lithuanian Chemical Society "Chemistry and Chemical Technology,"* 226.
4. Alimbekova, B. T., Jumadilov, T. K., Korganbayeva, Z. K., Erzhan, B., & Erzhet, B., (2013). *Bul.* d'EUROTALENT-FIDJIP, *5*, 28.
5. Erzhet, B., Jumadilov, T. K., & Korganbayeva, Z. K., (2013). *Bul. d'EUROTALENT-FIDJIP, 5*, 41.
6. Jumadilov, T. K., (2011). *Industry of Kazakhstan, 2*, 70.
7. Bekturov, E. A., & Suleimenov, I. E., (1998). *Polymer Hydrogels* (p. 133). Almaty.
8. Bekturov, E. A., & Jumadilov, T. K., (2009). News of National academy of sciences of RK, *Chem. Series, 1*, 86.
9. Bekturov, E. A., Jumadilov, T. K., & Korganbayeva, Z. K., (2010). KazNU herald, *Chem Series, 3*, 108.
10. Jumadilov, T., Shaltykova, D., & Suleimenov, I., (2013). Anomalous ion-exchange phenomenon. *Proceedings of Austrian-Slovenian Polymer Meeting*, p. 51.
11. Jumadilov, T., Akimov, A., Eskalieva, G., & Kondaurov, R., (2016). Intergel system polyacrylic acid hydrogel and poly-4-vinylpyridine hydrogel sorption ability in relation to lanthanum ions. *Proceedings of VIII International Scientific-Technical Conference "Advance in Petroleum and Gas Industry and Petrochemistry,"* p. 68.
12. Jumadilov, T. K., (2016). Hydrogels remote interaction–basis of new phenomena and technologies. *Proceedings of 10th Polyimides & High Performance Polymers Conference*, p. CIV–7.
13. Tereshenkova, A. A., Statkus, M. A., Tihomirova, T. I., & Tsizin, G. I., (2013). *Moscow Uni. Bul., Chem. Series, 54*, 203.
14. Petruhin, O. M., (1987). Methodology of physic-chemical methods of analysis. *M.: Chemistry*, pp. 77–80.
15. Izmailov, N. A., (1976). Electrochemistry of solutions. *M.: Chemistry*, p. 24.

CHAPTER 20

The Determination of Average Stability Constant of Zinc-Fulvate Complex by the Gel Filtration Method

T. MAKHARADZE

Rafiel Agladze Institute of Inorganic Chemistry and Electrochemistry,
E. Mindeli Street, 11, Tbilisi 0186, Georgia,
E-mail: makharadze_tako@yahoo.com

ABSTRACT

Natural geopolymers like fulvic acids (FAs) take an active part in complex formation processes proceeding in natural waters. In spite of researches, experimental data on average stability constants of complex compounds of FA with heavy metals (among them zinc) are heterogeneous, and they differ in several lines from each other. FA was isolated from Faravani Lake by the adsorption–chromatographic method.

Charcoal was used as a sorbent. The complex formation process of fulvate complexes of zinc was studied by using Sephadex G–25 at pH 8. Average value of the stability constant of zinc fulvate complexes was calculated.

20.1 INTRODUCTION

Fulvic acids (FAs) are one of the first geopolymers which were found in natural waters. FAs have functional groups, and that's why they take an active part in complex formation processes, proceeding in natural waters and form stable compounds with heavy metals and radionuclide elements [1–11]. Studying of FAs is an actual issue of macromolecular and environmental chemistry. The data on average stability constants of complex compounds of FAs with zinc (II) are heterogeneous, and they differ in several lines from each other [2, 3, 8, 9]. Therefore, it's difficult to investigate complex formation processes, taking place in natural waters, identify migration forms of heavy metals and evaluate and assess the chemical-ecological condition of natural waters.

Our objective was obtaining the pure samples of FAs, the investigation of complex formation processes between the pure samples of FAs and Zn(II) at pH = 8 by the gel-chromatographic method.

20.2   EXPERIMENTAL

After filtration through membrane filters (0.45 μm pore size), for obtaining pure samples of FAs, the water of Faravani lake was concentrated by the frozen method. The concentrated water samples were acidified with 6 M HCl to pH 2 and was put for 2 hours on water bath at 60°C for coagulation of humin acids. Then the solution was centrifuged for 10 min at 8000 rpm (Centrifuge T–23). For the isolation of FAs from centrifugate was used the adsorption-chromatographic method. Charcoal was used as a sorbent. Desorption of amino acids and carbohydrates was performed with 0.1 M HCl. For the desorption of polyphenols was used the 90% acetone-water solution. The evaluation of the fraction of FAs was performed with 0.1 N NaOH solution [11, 12]. Obtained alkalic solution of FAs, for the purification, was passed through a cation-exchanger (KU–2–8). For determination, the concentration of FAs in obtained solution was used gravimetric method, the part of the solution was dried under vacuum until the constant weight was obtained. Then, model solutions of FAs were prepared. The main solution of fulvic complexes were obtained by the solubility method. The same quantity of standard solution of FAs and suspensions of ZnO were placed in fluoroplastic cylinders. The constant ionic strength was made by adding 0.1M potassium nitrate. The final volume was 40 ml.

Concentration of hydrogen ions was regulated by 0.01M potassium hydroxide and 0.01M nitric acid, in model solutions at pH = 8 (pH meter pH 2006). The solution was put on a mechanical mixer for 60 hours (until the balance was achieved) and then suspension was filtered through the membrane filters (Sinpor N6). In filtrates, the concentration of zinc was measured by Atomic Absorption Spectrophotometer (Perkin Elmer 200).

Formation of fulvate complexes was studied by the gel filtration method. For the investigation of complex formation process, taking account the associates of FAs, various pH and the average molecular weight, for the optimal determination was used Sephadex G–25 (the limits of Fractionating 100–5000).

The parameters of Sephadex G–25: mass of dry gel – 17 g, height of swelled layer of gel–42 sm, inner diameter of column–1,6 sm. For the calibration of Sephadex G–25 was used blue dextran, polyethylene glycols with molecular weights 300, 600, 1000 and glucose. The titer of standard substances – 1 mg/ml, transmission speed – 3 ml/min, apply volume of solution – 2 ml.

Results of calibration: free volume 31–36 mL (I fraction), releasing volume of standard substances according to molecular weights (Mw) 1000 = 36–46 mL (II fraction), (Mw) 600 = 46–51 mL (III fraction), (600) Mw 300 = 51–56 (IV fraction), Mw 300 = 56–61 mL (V fraction) Mw 180 = 61–67 mL (VI fraction).

The aliquots of different solutions with different pH (2–2 ml) were placed in the top part of the column. The elution process was done by bidistilled water that has the same pH as the aliquots of solution.

The quantity of metals connected with FAs was determined in I-V fractions. These are the fractions, which releasing volume fits substances with molecular weight $300 \leq Mw > 5000$. For this reason, I–V fraction were all gathered, then were concentrated up to 10 mL and the quantity of metals was measured by Atomic Absorption Spectrophotometer.

20.3 RESULTS AND DISCUSSION

If it is not taken into consideration charges of ions, the reaction of formation of zinc fulvate complexes could be written in the following way $Zn(II) + mFA = ZnFA_m$ (1). The numeral value (m) of the stechiometral coefficient or the number of ligands in the inner coordination sphere of complex equals to tangens of tilt angle of straight line built in coordinates $\log([Zn(II)_{total}]-[Zn(II)_{free}])-m \log[FA_{total}]$ (2).

For the calculation, the exact value of tangens tilt angle of straight line, for this purpose was used the least square method. After the calculation, was obtained the numeral value of stechiometral coefficient (m), which equals to 1,2. So in $ZnO(solid)-Zn(II)(solution)-FA-H_2O$ system at pH = 8.0, dominates zinc fulvate complex with the structure 1:1. Therefore, we can be written as:

$$\beta = [ZnFA]/\{[Zn(II)_{free}][FA_{free}]\}$$

During the gel chromatographic investigation of fulvate complexes, these characteristics will be calculated very easy: [ZnFA] is the quantity of metals connected with FAs determined in the fractions $300 \leq Mw > 5000$, mol/l; $[FA_{free}] = [FA_{total}]-[ZnFA]$, mol/l; $[Zn(II)_{free}] = [Zn(II)_{total}]-[ZnFA]$, mol/l; $[Zn(II)_{total}]$ is the total quantity in main solution, mol/l.

The results of investigation fulvate complexes by the gel filtration method are given in Table 20.1.

TABLE 20.1 The Conditional Stability Constants Calculated on the Basis of the Results, Obtained by the Gel Filtration Method $\beta = [ZnFA]/\{[Zn(II)_{free}][FA_{free}]\}$

FA_{total}	Zn_{total}	ZnFA	FA_{free}	Zn_{free}	β
9.10×10^{-5}	9.03×10^{-5}	4.50×10^{-5}	4.60×10^{-5}	4.53×10^{-5}	2.2×10^{4}
1.09×10^{-4}	1.10×10^{-4}	6.19×10^{-5}	4.71×10^{-5}	4.81×10^{-5}	2.7×10^{4}
1.27×10^{-4}	1.28×10^{-4}	7.89×10^{-5}	4.81×10^{-5}	4.91×10^{-5}	3.3×10^{4}
1.45×10^{-4}	1.37×10^{-4}	8.79×10^{-5}	5.71×10^{-5}	4.91×10^{-5}	3.1×10^{4}
182×10^{-4}	1.65×10^{-4}	1.15×10^{-4}	6.70×10^{-5}	5.00×10^{-5}	3.4×10^{4}
2.18×10^{-4}	1.93×10^{-4}	1.42×10^{-4}	7.6050×10^{-5}	5.10×10^{-5}	3.7×10^{4}
2.54×10^{-4}	2.13×10^{-4}	1.60×10^{-4}	9.40×10^{-5}	5.30×10^{-5}	3.2×10^{4}

Average meaning of the stability constant of zinc fulvate equals to 3.1×10^{4} ($\beta = 3.1 \times 10^{4}$).

20.4 CONCLUSIONS

The pure samples of FAs were separated from the water of the Faravani Lake r by the adsorption-chromatographic method. By using the Sephadex G–25 was studied the complex formation process between zinc and FAs at pH = 8. It was established, that average value of the stability constant of zinc fulvate $\beta = 3.1 \times 10^{4}$.

ACKNOWLEDGMENTS

The work was done by supporting CRDF Global, Shota Rustaveli National Science Foundation (SRNSF) and Georgian Research and Development Foundation (GRDF).

KEYWORDS

- **average stability constant**
- **fulvic acids**
- **sephadex G–25**
- **zinc**

REFERENCES

1. Bertoli, A. C., Garcia, J. S., Trevisan, M. G., & Ramalho, T. C., (2016). *Biometals, 29*(2), 275.
2. Castro, R. T., Cunha, E. F., Alencastro, R. B., & Espınola, A. Y., (2007). *Water Air Soil Poll., 183*(2), 467.
3. Ephraim, (1992). *J. Anal. Chim. Acta., 267*(1), 39.
4. Kirishima, A., Ohnishi, T., Sato, N., & Tochiyama, O., (2014). *J. Nucl. Sci. Technol., 47*(11), *1044.*
5. Rey-Castro, C., Mongin, S., Huidobro, C., David, C., Salvador, J., Garces, J., Galceran, J., Mas, F., & Puy, J., (2009). *Environ. Sci. Technol., 43*(19), 7184.
6. Sasaki, T., Yoshida, H., Kobayashi, T., Takagi, I., Moriyama, H., & Americ. J., (2012). *Analytic. Chemis., 3*(7), 462.
7. Schnitzer, M., & Skinner, S. I. M., (1966). Organo-metallic interactions in soils: 5. Stability constants of Cu^{++}, Fe^{++}, and Zn^{++} fulvic acid complexes. *Soil Sci. 102*(6), 361.
8. Shizuko, H., (1981). *Talanta, 28*(11), 809.
9. Town, R. M., Van Leeuwen, H. P., & Buffle, J., (2012). *Environ. Sci. Technol., 46*(19), 10487.
10. Varshal, G. M., Kosheeva I. Y., Sirotkina, I. S., Velukhanova, T. K., Intskirveli, L. N., & Zamokina, N. C., (1979). *Geochemistry (Russia), 4*, 598.
11. Revia, R., & Makharadze, G., (1999). *Talanta, 48*(2), 409.

CHAPTER 21

Zeolites as Micro-Pore System and Their Usage Prospects

N. KIKNADZE[1] and N. MEGRELIDZE[2]

[1]Chemistry Department, Batumi Shota Rustaveli State University (BSU), Ninoshvili/Rustaveli Str. 35/32, 6010 Batumi, Georgia, E-mail: nino-kiknadze@mail.ru

[2]BSU Agrarian and Membrane Technologies Institute, Grishsashvili Str. 5, 601 Batumi, Georgia

ABSTRACT

Zeolites belong to the natural rock the constant usage of which in science and practice has the following basis: its high ion-exchanging capacity, prolonged activity of water and salt changing mode in soil (after effect); selectivity towards NH_4^+ and K^+, which causes increasing the coefficient of their usage by plants and enhancing efficacy of mineral fertilizers.

Taking growing dosages (3–6–9–12 T/Hectares) of zeolites in red soils of West Georgia at the background of mineral fertilizers (NPK) caused increasing of P, K, Ca, Mg in tea leaves. Nitrogen decreasing tendency must be caused by selective absorption of NH_4^+ by zeolites, which increases nitrogen usage coefficient by tea plant. Taking zeolites in the red soils with mineral fertilizers improves tea leave quality. Compared to the background option usage of zeolites by 12 tonnes/hectare dosage, caused increasing tannin and extractive substances by 3.8–3.3%. Zeolites improve physical-chemical properties of the soil, in particular: strong acid reaction of red soils was decreased up to pH 4.5–5.8, it was caused by increasing Calcium and Magnesium easily soluble in water, which caused increasing pH of soil. By taking 12 tonnes/hectare zeolites in soils P_2O_5 and K_2O concentration increase is indicated (P_2O_5 – 78.5 mg/100 g; K_2O – 14.6 mg/100 g) as well as hummus (4.22%) and total nitrogen consistency (0.29%) was increased in soils.

Usage of zeolites as mineral fertilizers improves productivity qualities of the tea plant. Compared to *NPK* background taking zeolites by 12 tonnes/ hectare in soils caused increasing normal shoots by 34.2%- and decreasing of stupid shoots by 51.3%.

21.1 INTRODUCTION

In recent period in Georgia and abroad as well using natural agro-ores – zeolites has intensively started in agriculture, chemistry, physics, mineralogy, and other fields as well for fundamental examinations. The world of this mineral is newly opened for the scientists which has great practical importance. Natural fertility of the soil in modern conditions cannot provide high-quality production of agricultural products. Organic-mineral fertilizers, as well as rationale usage of ecologically pure local natural ores, have got important role in solving this problem: peat, sapropel, zeolites. To this direction important role has covering local ores of minerals – zeolites and their usage in agriculture.

Georgia is a republic with small amount of lands where it is determined the reserves of further increase of agricultural lands and the population is densely populated. Hence, the main reserves for increasing agriculture is products intensification – at the basis of mechanization, chemicalization, irrigation, and waste-free technologies. One of the most important fields of agriculture of Georgia was tea-growing, which with its difficulties was connected to spiritual and material changes of Georgian nation. The nega- tive events happening during the last decade caused massive deforestation of tea plantations and substitution of them with other agricultural crops for which Georgia has deficit of native tea. The soils at which tea is cultivated are humid subtropical zone red soils of west of Georgia. Since 90s of 20th century, the fields of tea-growing and citrus-growing appeared to be in quiet non-beneficial conditions. From the same period agrochemical research of red soils and their regular, planned, scientifically grounded fertilization have totally canceled. This cannot provide high product quality at present competitive market conditions, and for this reason, our market depends on imported production almost totally – by 90% and including on imported tea.

The natural zeolites have been discovered 200 years ago by Swede scien- tist – Kronstville who called rock to zeolite which literally means "boiling stone." Zeolite crystal while heating in fire omits water and starts boiling. Zeolite is called "molecular sieve" as well. At present, about 50 species of zeolites are known from which each of them has independent crystal struc- ture, and they are met at various regions of the globe: Japan, Canada, Island,

India, New Zealand, Bulgaria, Hungary, Georgia, and Russia. Practical value of zeolites is determined by the substances of their consisting minerals. The richest ores are zeolites which consist of 8 minerals: clinoptilolite, mordenite, chabazite, erionite, phillipsite, limonite, analcime, and ferrierite.

G. Gvakharia has a great role in studying zeolites of Georgia. According to his examinations, the zeolites distributed to our country are changed volcanic tuffs which are enriched with analcimes, clinoptilolites, and phillipsites. Besides, they include montmorillonite, hydro micas, quartz, feldspar, and volcanic glass mixture. Analcimes ores are discovered at volcanic sediment rocks of Kutaisi-Tkhibuli, Ambrolauri-Oni, Guria (Lanchkhuti-Chokhatauri) and Imereti range (Vani-Baghdadi). Manufacturing ores of clinoptilolites zeolites are discovered at volcanic layers of Mtskheta (Armazi ravine-Dzegvi) and Akhaltsikhe-Aspindza regions. Phillipsite rocks are discovered are Akhaltsikhe and Guria range at north branch (village Shukhuti, Lanchkhuti region). These rocks are distributed at 1200–1500 meters' height and the consistency of phillipsites in them is up to 60–90%. At micro-porous structure of zeolites, the "entrance windows" diameter is 0.3–0.6 mm. This windows cause the activity of "molecular sieve" of zeolites: some small molecules selectively enter porous structure of zeolite from "entrance window" and happens inside the crystals and fills adsorptive space. The big molecules cannot enter "entrance windows" and stay at the surface of seeds.

Because of high ion-changing capacity and ability to maintain the moist for long period, natural zeolites are the regime of changing the water and salt in soil so-called long-term "conditioner." Zeolites promote long-term post-influence of fertilizers taken in agricultural lands. They are used for increasing ability of cationic change of soil, and they provide maintenance of NH_4^+ and K^+ ions in soil; besides, they improve quality of tea leaves, and they increase usage of nitrogen and potassium coefficients. The zeolites have high selectivity towards toxic elements for which they are ecologically pure fertilizer different from mineral and organic fertilizers. The main aim of zeolites is to increase soil absorption capacity and efficacy of mineral fertilizers which can be explained by ion-changing and absorptive substances of it. By the action of nitrogen fertilizers, zeolites can dismantle and release out Na^+, Ca^{2+}, Mg^{2+} ions which selectively are changed to NH_4^+ and K^+ ions by which they provide long-term detention-maintenance of potassium and nitrogen in soil.

It is identified that taking zeolites and mineral fertilizers in soil jointly increases the harvest of agricultural crops: carrot by 63%, eggplant by 55%, wheat by 15%, apple by 28%, corn by 15%, tomato, and pepper by 33%. At potato bulbs, increase of starch is noted by 1%. Taking 200 kg/ha zeolite in soil increases radish bulb mass by 13–16% and the upper soil mass by 13%. Using zeolites in carrot increases: protein consistency by 1.4%; general potassium by 0.44%; general phosphorus by 0.88%; at pepper product proteins are increased by 1.7%; potassium by 0.63% phosphorus by 0.09%. Consistency of nitrates of carrot is decreased by 76.6% and at pepper product by 41.2%. At the background of taking fertilizers in soil, zeolites increase consistency of sugars in carrot by 2.9%, vitamin C by 4.3% and carotenes by 34.5%. The Zeolites have long-term activity as a result of which the detention ability of cationic ions (Ca^{2+}, Mg^{2+}) absorbed by soil is increased. It causes decrease of acidity of soil. While taking into soil zeolites with clinoptilolities pH of soil, humidity, aeration is increased, and soil solidity is decreased, the consistency of P^{32} is increased in plants. By taking great dosages of zeolites in black soils of east of Georgia omission of phosphorus from the plants has been increased by 14% [1–11].

21.2   EXPERIMENTAL

21.2.1   MATERIALS

Foreseeing the huge importance of zeolites, as natural fertilizer, the goal of our research was to study their influence at chemical indicators of red soils, chemical consistency of tea leaves and quality of it, in order to identify the role of zeolites in enhancing productivity of tea plant. The tea plantation based on Institute of Tea, Subtropical Crops and Tea Production in Ozurgeti region, borough Anaseuli (Georgia) was selected as research subject at which sample was made for studying the efficacy of zeolites in 1987 at tea variety "Kimin." The tea plantation was built in 1965, and it was fertilized by the following: NH_4NO_3 – annually 300 kg/ha dosage; the simple super-phosphate – in 4 years once 600 kg/ha dosage; KCl – once per to years 200 kg/ha dosage. At present soils are fertilized mostly with nitrogen fertilizers. Philiphsite zeolites revealed in Lanchkhuti region village Shukhuti were taken into in soils twice: in 1987 and 1994 years according to the schemes given under the Tables 21.1–21.5. Since 1994 up to present the post-actions of zeolites are studied at the mentioned soils.

21.2.2   MEASUREMENT

Plant samples (young tea leaves) have been taken in the period of vegetation, at each variant from all repetitions. The taken samples were made fixation for 6 minutes in Kokh Apparatus, and we were drying it. For the purpose of identifying tea leaves chemical consistency and qualitative indicators the following indicators were determined in them:

1. General nitrogen with Keldali method;
2. General phosphorus and potassium with Ginzburg-Shcheglova method;
3. Calcium and magnesium – with Trinolometric method;
4. Tannin – with Levental-Neibauer method;
5. Extract substances – with weight method.

The soils samples have been taken from 0–45 cm depth with borer from four points of each variant of experiment. We were drying soil samples at the room temperature, loosening, and we were sifting at 1 mm diameter space. The following chemical indicators have been determined in soil samples:

1. General nitrogen with Keldali method;
2. General humus with Tiurin method;
3. Changing acidity with Sokolov method;
4. Hydrolyzed acidity with Sokolov method;
5. pH in water and KCl suspension with electro potentiometer method;
6. Calcium and magnesium with Trinolometric method;
7. Movable phosphorus and potassium with Oniani method [12–14].

21.3 RESULTS AND DISCUSSION

Before making the experiment at the land plot chemical indicator shave been determined in red soils (Table 21.1). The hummus and general nitrogen consistency is average in soil as for movable potassium, calcium, magnesium, and phosphorus are low. The soil reaction is strongly acid. Therefore, before making the experiment red soil is less full of nutritious elements.

21.3.1 ZEOLITE INFLUENCE AT CHEMICAL AND QUALITATIVE INDICATORS OF RED SOIL AND TEA LEAF

According to experiment data (Table 21.2) received by us, it is identified that general nitrogen consistency in tea leaves at *NPK* background is 4.2%. At zeolite taking variants, unimportant decrease tendency of nitrogen is noted 0.74–1.1%. The decrease of nitrogen consistency in tea leaves at zeolite taking variant must be caused by these rocks selective absorption of NH_4^+ ions. It enhances nitrogen usage coefficient by tea plant, and it decreases loss of nitrogen fertilizers by the way of their washing off and evaporation.

TABLE 21.1   Chemical Indicators of Experimental Red Soil Before Making Experiment at Zeolites

Chemical parameters of red soils	Soil sample depth 0–45 cm
Humus, %	3.16
Nitrogen, %	0.22
P_2O_5, mg/100 g soil	12.1
K_2O, mg/100 g soil	11.2
CaO, mg/100 g soil	34.2
MgO, mg/100 g soil	14.6
Changing acidity, mg. equivalent/100 g soil	5.3
pH in water suspension	5.0
pH in KCl suspension	4.2

TABLE 21.2   Zeolites Influence at Chemical Consistency of Tea Leaf

№	Variant	The consistency of nutritious elements, %				
		N	P	K	Ca	Mg
1	*NPK* – background	4,2	1,0	0,60	0,65	0,03
2	*NPK* + zeolite 3 t/ha	3,8	1,2	0,64	0,72	0,08
3	*NPK* + zeolite 6 t/ha	3,5	1,7	0,82	0,76	0,12
4	*NPK* + zeolite 9 t/ha	3,2	1,7	0,86	0,80	0,12
5	*NPK* + zeolite 12 t/ha	3,1	1,9	0,95	0,89	0,15

General phosphorus consistency at background variant is 1.0%, and at zeolites variants, the increase tendency is noted by 0.2–0.9%. Potassium consistency at background variant is 0.60% and at variants with zeolites by 6–9–12 tonnes/ha, the quantity of it is increased at tea shoots by 0.22–0.35%. The same regularity we have for the magnesium, the increase compared to the background variant is noted with 0.09–0.12%. The zeolites have the ability to selectively absorb and detain in their micropores $K^=$ ions. Herewith, zeolites are enriched with K_2O (4.6% K_2O). Increase of Calcium quantity at 12 t/ha zeolite variant compared to background equals to 0.24%. In the aspect of nitrogen consistency, unimportant decrease tendency is noted. It is caused by selective absorption of NH_4^+ ions by zeolites, which increases nitrogen usage coefficient by tea.

Taking zeolites at the background of full mineral fertilizers makes influence at consistency of Tannin and extract substances, in particular: if at the background of *NPK* tannin quantity was 18.2%, at zeolite 12 t/ha variant this indicator is increased by 3.8% (Table 21.3). At the same variant, the quantity of extract substances is 39.0%, which is more by 3.3% compared to background. Hence, zeolites improve qualitative indicators of tea leaf.

TABLE 21.3   The Influence of Doses Zeolites at Qualitative Indicators of Tea Leaf

№	Variant	By moving to% absolutely dry substance	
		Tannin	**Extract substances**
1	*NPK* – background	18.2	35.7
2	*NPK* + zeolite 3 t/ha	18.7	36.0
3	*NPK* + zeolite 6 t/ha	19.3	37.4
4	*NPK* + zeolite 9 t/ha	20.8	38.9
5	*NPK* + zeolite 12 t/ha	22.0	39.0

21.3.2   INFLUENCE OF ZEOLITES AT CHEMICAL CONSISTENCY OF RED SOILS

We have carries chemical analysis of experimental plot at 0–45 cm depth layer where mostly tea root system is located (Table 21.4). The soils have been fertilized during long period of time (since 1965) with physiologically acid fertilizers, and therefore their reaction at *NPK*-background variant is strongly acid: in H_2O suspension pH is 3.7, and in KCl suspension pH is 3.0. Taking zeolites with a large amounts in soils (12 t/ha) has move the soil reaction to average acidity: pH in KCl suspension is equal to 3.9. Therefore, changing and hydrolyzed acidity of the soils have been decreased. In particular, compared to the background variant the decrease of changing acidity was 2.03 units, and as for hydrolyzed acidity, it was – 2.8 units.

By the influence of zeolites the consistency of movable phosphorus is increased in soil compared to background variant: taking zeolite with 12 t/ha dosage caused increase of movable phosphorus by 18.0 mg per 100 g soil. At the 5-th variant increase of CaO compared to the background equaled to 25.7 mg/100 g and increase of MgO – by 7.2 mg/100 g and therefore, it can be considered regular the decrease of pH soil from strongly acid to average acidity.

Despite the selective absorption of K^+ ions, the consistency of movable potassium is increased by taking zeolites in the soil. This can be explained

TABLE 21.4 The Influence of Different Dosages of Zeolites at Chemical Indicator of Red Soils (0–45 cm)

Variant	pH		Acidity, mg equivalent/100 g soil		Mg/100 g soil				%	
	H_2O	KCl	Changing	Hydrolyzed	P_2O_5	K_2O	CaO	MgO	Humus	Nitrogen
NPK – background	3.7	3.0	7.15	19.8	60.5	10.2	17.0	2.9	2.35	0.12
NPK + zeolite 3 t/ha	3.9	3.3	6.23	19.2	62.3	11.0	17.6	4.0	2.96	0.16
NPK + zeolite 6 t/ha	4.0	3.4	5.89	18.5	71.4	12.3	22.5	5.4	3.15	0.20
NPK + zeolite 9 t/ha	4.3	3.5	5.45	17.3	82.6	14.1	33.8	7.3	3.56	0.24
NPK + zeolite 12 t/ha	4.7	3.9	5.12	17.0	78.5	14.6	42.7	10.1	4.22	0.29

by high consistency of K_2O in zeolite structure which is moved to soil solution after rock dismantle. The received results are in regular Attitude with general potassium consistency in tea leaves. Herewith, the consistency of movable phosphorus is increased in soil at zeolite variants. Humus quantity at the 5-th variant equals to 4.22% which is more by 1.87% to background variant. Hence nitrogen amount of organic admixtures in humus is increased by 0.17%.

21.3.3   ZEOLITE INFLUENCE AT TEA PLANT PRODUCTIVITY

The main goal of our research was to study influence of zeolites at shoot producing process if tea plant for the purpose of which the quantity of normal and deaf shoots has been determined (Table 21.5). From the experiment data it is identified that if in case – background the quantity of normal shoots is 175 units, at zeolite variant 3 t/ha variant it equals 192 units and the zeolite at 12 t/ha variant it is – 235 units, so it is increased by 34.2% compared to background variant. The quantity of deaf shoots at background variant is 102 units, and at zeolite 12 t/ha variant it is 31 units, so it is less by 69.2% compared to background variant.

TABLE 21.5   The Influence of Zeolite at Tea Plant Productivity

№	Variant	The quantity of normal shoots		The quantity of deaf shoots		totally	
		Unit	%	Unit	%	Unit	%
1	*NPK* – background	175	100	102	100	277	100
2	*NPK* + zeolite 3 t/ha	192	109.7	63	61.8	255	92.1
3	*NPK* + zeolite 6 t/ha	202	115.4	54	52.9	256	92.4
4	*NPK* + zeolite 9 t/ha	226	129.1	34	33.3	260	93.9
5	*NPK* + zeolite 12 t/ha	235	134.2	31	30.8	266	96.0

21.4   CONCLUSIONS

Taking increasing dosages of zeolites in soils at the background of mineral fertilizers (*NPK*) caused increase of phosphorus, potassium, calcium, magnesium consistency. The unimportant nitrogen decrease tendency was caused by selective absorption of NH_4^+ ions which to its part enhances nitrogen usage by tea plant. At *NPK*-background with 12 t/ha dosage of zeolites the increase of tannin and extract substances was fixed with 3.8 and 3.3%. The zeolites caused to weaken strongly acid reaction of red soils which is stipulated by increase of calcium and magnesium in the mentioned soils. By taking zeolites in the soils the increase of movable forms of phosphorus and potassium is noted, the consistency of important characteristics of soil fertility humus and general nitrogen was enhanced. Using zeolites improves productivity indicators of tea leaves, taking them by 12 t/ha in soils caused increase of normal shoots by 34.2% and the decrease of deaf shoots by 69.2% compared to background.

KEYWORDS

- **fertility**
- **microporous system**
- **mineral fertilizers**
- **productivity**
- **red soils**
- **tea**
- **zeolites**

REFERENCES

1. Marshania, E. I. I. G. V., (1980). *Natural Zeolites in Agriculture* (pp. 125–131). Tbilisi.
2. Erkvania, G. V., (1964). The influence of natural zeolites at the effectiveness of nitrogen fertilizer norm and form under essential oil geranium on alluvial soils, Abkhazia, *Auto Thesis for the Academic Degree of Candidate of Agricultural Sciences* (p. 24).-Sokhumi.
3. Tsitsishvili, G. V., (1978). Natural zeolites–New perspective materials for modern industry and agriculture.–*Geology, Genesis, and Consumption of Natural Zeolites, Thesis of Abstracts from Zvenigorod* (pp. 4–6).
4. Zardalishvili, O. U., (1986). Some biochemical indicators of carrot and pepper cultivated at soil containing clinoptilolites, *Thesis of Scientific Conference* (pp. 36–41). Tbilisi.
5. Brek. D., (1980). *Zeolite Molecular Seieve (Translation from English)* (p. 778). Moscow: "MIR."
6. Chelishev, N. F., & Chelisheva, R. V., (1980). *Natural zeolites in Agriculture* (pp. 217–226.). Tbilisi: "Metsniereba."
7. Eliseeva, I. S., (2003). *Manufacturing and Development of Synthetic Zeolites* (p. 23). Auto thesis for Academic Degree of Candidate of Technical Sciences. Ufa.

8. Shkhapatsev, A. K., (2000). *Agro-Ecological Efficacy of Zeolite Consumption in Rice Growing* (p. 134). Dissertation for academic degree of candidate of agricultural sciences. Maikop.

9. Kiknadze, N., Kiknadze, M., & Bolkvadze, S., (2009). *Influence of Zeolites at Qualitative Indicators of Tea Leaf / Herald of Scientific National Academy of Georgia (chemistry serial)* (Vol. 35, No. 1, pp. 80–81). Tbilisi.

10. Kiknadze, N., (2009). *Influence of Zeolites at Chemical Consistency of Tea Leaf / Herald of Scientific National Academy of Georgia (Chemistry Serial)* (Vol. 35, No. 1, pp. 78–79). Tbilisi.

11. Kiknadze, N., Kiknadze, M., Lomtatidze, N., & Alasania, N., (2008). *Influence of Zeolite on Red Soils in Southwestern Georgia Artvin Çoruh University Faculty of Forestry Journal* (Vol. 9. No. 1 & 2, pp. 77–80). Orman Fakültesi Dergisi.

12. *Agrochemical Method for Studying the Soils* (1975). Moscow, Publishing House "Science," p. 645.

13. Oniani, O., & Margvelashvili, G., (1975). *Chemical Analysis of Soil* (p. 507). Tbilisi, Publishing House "Ganatleba."

14. Oniani, O., & Margvelashvili, G., (1978). *Chemical Analysis of Plant* (p. 417). Tbilisi, Publishing House "Ganatleba."

CHAPTER 22

Water-Proofing Materials, Luminophore, and Other Deficient Products

G. KHITIRI[1], I. CHIKVAIDZE[2], and R. KOKILASHVILI[3]

[1]P. Melikishvili Institute of Physical and Organic Chemistry,
Iv. Javakhishvili Tbilisi State University Tbilisi, Georgia

[2]Iv. Javakhishvili Tbilisi State University Tbilisi, Georgia,
E-mail: iosebc@yahoo.com

[3]Georgian Technical University Tbilisi, Georgia

ABSTRACT

Water-proofing of concrete, ferroconcrete, metal, wood, and other construction structures, pipelines of various purpose, including oil- and gas pipelines, as well as agglutination of metal-concrete, metal-wood, and other combinations is one of the most important and actual problems. Materials of such types are not produced in Georgia. Import, transportation, customs duty of those products and other expenses significantly increase the value of structures. Thus, the use of these materials is unprofitable. Analogous situation is in the field of separation of paraffin and ceresin, petrolatum, luminophore, and different types of bitumen's from oil, on the basis of which it is possible to produce small-scale deficient inexpensive products.

The production of the above-mentioned materials from the local raw materials and various types of waste is highly relevant in Georgia.

To this end, the first step is to create laboratory and experimental modeling methods as a basis for the future production technologies.

The strategic goal of the project is to create zero-discharge treatment schemes for organic and mineral wastes to produce various types of deficient products in Georgia and, in addition, to avoid pollution of the environment.

The project proposal deals with the preparation of various compositions of hydro insulation materials from inorganic components and oil waste in Georgia. It is also taken into account the separation of luminophore, paraffin-ceresin, and bitumen masses of various types from the wastes in pipelines. The latter will be used as one of the components in waterproofing compositions.

The purpose of the project is to create the basis for the production of hydro insulating materials and small-scale deficient, high-quality products–paraffin and ceresin, petrolatum, luminophore, lubricants, and bitumen's for various purposes.

22.1 INTRODUCTION

The novelty is the use of sedimentary waste in the oil pipeline to produce small-scale deficient and expensive products such as [1, 2]:

1. Paraffin (C_{25}–C_{35}; melt temperature 50–65°C and more) and ceresin (C_{36}–C_{53}; melt temperature is within the range 50–65°C), which are of great demand for industrial and domestic purposes.
2. Luminophore component is needed to detect invisible cracks with the methods of luminescent marking and luminescence microscopy.
3. High-melting tar mass is needed to obtain raw insulation materials (water-proofing and anticorrosive ones), electrode coke and high-melting bitumen.

This is also important to avoid pollution of the environment.

22.2 EXPERIMENTAL

Calculations showed that the use of waste is reasonable and economically profitable if the content of the oil components is more than 30%. The Baku-Tbilisi-Ceyhan oil pipeline contains 60–75% and more oil components. Therefore, separation of the above-mentioned products from the waste for the further use is prospective to produce cheap materials.

Operating characteristics of the proposed luminophore satisfy requirements of international standards. In particular, the solutions made from it have an ability to detect invisible cracks of the thick <1 mm. The light intensity in its greenish-yellow area is 4.5–5 times more than that of standard–uranyl nitrate. It is stable with high quantum efficiency (40–45%), (measured with a luxometer b–116) [3, 4].

22.3   RESULTS AND DISCUSSION

One of the main objectives of the project is to develop a methodology for obtaining high-quality water-proofing materials. The above-mentioned high-melting tar mass will be used as the main raw material. Especial treatment and further homogenization of the mass–mixture of high-dispersive quartz sand, silicates, and other wastes–gives possibility to make various universal and inexpensive compositions. With the aim to improve hydro-insulating and other specific characteristics of the compositions they will be varied by changing the ratio of ingredients. The elasticity of these compositions is conditioned by macromolecular paraffin-ceresin and the polymeric (rubber) components; varying the strength and adhesion is possible by soluble and insoluble silicate-quartz components. The mentioned components can be got in a large amount in region. Therefore, products obtained from cheap raw materials will be inexpensive.

The use of the mentioned waste is also important to avoid the pollution of nature (Figures 22.1–22.3).

The project plan provides fractioning of sedimentary waste in the oil pipeline and clearing fractions without using traditional, expensive stages and adsorbents, catalysts, and reagents. Specifically, melting of sediment and fractionation of viscous mass is provided by rectification (under atmospheric pressure) and molecular distillation (1×10^{-3} mm Hg column residual pressure) up to the temperature of 80–550°C; obtaining the following five fractions as a result of the rectification:

1. Fraction 80–190°C–a): Solvent for industrial-technical purposes, b) Petroleum component;
2. Fraction 190–300°C: Diesel fuel component;
3. Fraction 300–350°C: Diesel fuel component with high Cetane number;
4. Fraction 350–450°C: Paraffin-ceresin component to produce candles, lubricating grease, petrolatum, ointments, and creams, mastics, impregnates, etc.;
5. Fraction > 450°C: Raw material to obtain electrode coke, water-proofing materials, bitumen's and luminescence components, etc.;
6. Two more fractions: 450–500°C (for obtaining high-quality luminophore) and 500–550°C (for obtaining high-quality electrode coke) will be got by distillation of remaining viscous mass.

The residue left after molecular distillation is the best raw material for insulation products as in addition to the oil components it contains a large amount of fine-grained sand.

FIGURE 22.1   Waterproofing materials.

FIGURE 22.2   Waterproofing of bridges.

22.4 CONCLUSIONS

The 5th fraction (>450°C), and the residue left after the molecular distillation are to be used as the main component for preparing the test waterproofing compositions. Then the highly dispersed secondary organic components (polymeric and/or rubber) and also highly dispersed inorganic components (sand, gravel), then soluble glass (silicate solution) and other special additives will be added. Finally, these mixtures will be homogenized at the temperature 60–75°C and intensive stirring.

FIGURE 22.3   Waterproofing–flat roofing.

Obtaining of different samples of test hydro-insulation compositions is planned by varying the component mass ratio and the type and quantity of specific additives.

The study of mechanical and the physical and chemical parameters of each sample will allow us to select the optimal composition of the insulation material according to the specifics of its use.

The preparation of insulation materials for the following purposes is provided:

- Waterproofing of concrete, ferroconcrete, metal, wood, and other construction structures; filling and bonding of cracks;
- Flat roofing;

- Isolation of pipelines, including oil-and gas pipelines (hydro- and anti-corrosive isolation). For the implementation of the project, the following works are provided:

  1. Patent search on the case of the use of residuals in materials used in the same direction and an analysis of these experiences. To work out recommendations for the use in the project;
  2. Selection of raw materials from the local wastes and working out processing program;
  3. Purchasing of laboratory equipment, chemical reagents, waste, and raw materials and other necessary supplies;
  4. Preparation and homogenization of different compositions;
  5. Study of technical and physical and chemical parameters and properties of different compositions;
  6. Testing of compositions having the best indices and working out the recommendation for their practical use;
  7. Preparation of a patent application.

The expected outcomes of the project are of great importance for the Georgia and other countries of the region since the above-mentioned materials are not produced in these countries and they are dependent on import of expensive materials.

Luminophore has great potential for use in biology, medicine, analytical chemistry, criminology, polygraph, luminescent print manufacturing, and nanomaterial techniques

As a result of the project implementation, cheap materials will be created, which actually are products of utilization of secondary raw materials and waste. After the project is completed, it is prospective development of the corresponding technologies on the basis of the worked out experimental modeling techniques and their implementation in manufacturing. Production of the mentioned materials in required quantity will be profitable and will facilitate the development of small- and medium-scale industry; also, ecological problems in the region will be reduced.

ACKNOWLEDGMENTS

Financial support by the Shota Rustaveli Georgian National Science Foundation, Tbilisi, Grant # 30–295055, is gratefully acknowledged.

KEYWORDS

- **bitumen**
- **luminophore**
- **paraffin**
- **utilization**
- **waterproofing**

REFERENCES

1. Tiagunova, G. V., & Iaroshenko, I. G., (2005). *Ecology. M.*, p. 504. (in Russian).
2. Khitiri, G., Chikvaidze, I., Gabunia, T., & TsurTsumia, M., (2015). *Bulletin of Georgian National Academy of Sciences, 9*, 3.
3. Khitiri, G. S., & Akhobadze, I., (2005). *Method of obtaining Luminophor P4054*, (In English). Georgia Patent (2005).
4. Khitiri, S. G., & Chikvaidze, I. S. (2018). *AP2017 13972A the Method of Utilization of the Sludge of the Pipeline*. Georgia Patent (2018).

CHAPTER 23

Hydrogen Embrittlement of Duplex Stainless Steels

R. SILVERSTEIN[1] and D. ELIEZER[2]

[1]Materials Department, University of California, Santa Barbara, California, USA
[2]Department of Materials Engineering, Ben-Gurion University of the Negev, Beer-Sheva, Israel, E-mail: barrav@post.bgu.ac.il

ABSTRACT

Hydrogen trapping state and desorption behavior in hydrogen charged duplex stainless steels (DSS) alloys have been investigated. The susceptibility of steels to hydrogen fracture mechanism is directly related to the interaction between traps (defects) and hydrogen; therefore, it is being affected the most by the deformation process, and hydrogen induced-phase transition. The effect of strain rates and hydrogen fugacity on hydrogen embrittlement in DSS is discussed in details. The common effect of hydrogen on phase transformation has been shown. The impact of hydrogen fugacity and strain rate on DSS have been demonstrated with thermal desorption spectrometry (TDS), X-ray diffraction (XRD), and microstructural observations.

23.1   INTRODUCTION

The ferritic-austenitic duplex stainless steels (DSS) combining good strength with corrosion properties are under increasing development; these grades include the lean duplex stainless steel (LDS) and the super duplex stainless steel (SDSS). The LDS is a recently developed low alloyed duplex stainless steel. LDS alloys are being used in many architectural applications due to their high strength, good corrosion resistance and lower cost [1–8]. The LDS microstructure is very similar to that of the SDSS; it has a mixed crystal structure of ferrite (BCC, α) and austenite (FCC, γ), but its uniqueness is in the different quantity of the alloying elements, granting it better mechanical properties with a reduced cost. These alloys have the advantage of lower nickel content than austenitic alloys, while providing similar or better corrosion resistance in many environments. Therefore, they are candidates for undersea oil and gas applications such as flow lines [7–9]. The DSS alloys have been proposed to many applications where high strength and ductility are desired for hydrogen resistance and resources savings. These are prominent demands in pressure vessels and underwater pipelines industries and in hydrogen fuel productions which uses material for storing and transporting of hydrogen. Those stainless steels have been extensively studied with hydrogen due to their attractive properties for the above-mentioned industries [5–8].

One of the main problems in maximizing the service life of metals in industry is their low resistance for hydrogen. Since hydrogen is common in most manufacturing processes and services, the common used DSS are subjected to a deleterious effect known as hydrogen embrittlement. The susceptibility of DSS to the hydrogen-assisted cracking (HAC) phenomenon is related to some factors, among them is the presence of the ferrite phase and the hydrogen-trap interaction [4–7]. In order to initiate a crack, a critical hydrogen concentration must be reached at potential crack sites. Therefore, factors affecting the hydrogen diffusivity into a crack site are of great importance to the analysis of DSS behavior in hydrogen environment.

Hydrogen has different effects on the α and γ phases. In addition, it is known that the stability of the γ phase, which can have a significant effect on the material's properties, is affected both by the steel composition and by the presence of hydrogen in solid solution [10–14]. It is known from earlier publications [15–17] that irreversible traps act only as sinks, however reversible traps, which can act both as sinks or sources, will have a major influence on the material's susceptibility to HAC. Since, γ stability can affect both phase transition and hydrogen's diffusion, it will have a major effect on the steel's HAC characteristics and on the trap's activation energy for hydrogen. A 'good' trapping site for hydrogen has to possess an optimum combination of high binding energy to hydrogen and hydrogen capturing kinetics due to the dynamic nature of temperature and stress involved in the operated steel [16]. Therefore, the stability of the γ phase and the activation energy of traps are measured and compared for the studied steels.

In this work, hydrogen embrittlement mechanisms were examined on hydrogen charged at different strain rates applied by quasi-static loading in tensile test (10^{-7} s^{-1}) and dynamic loading by shock waves (10^5 s^{-1}) [1–3], were examined.

The dynamic experimental part was performed by accelerating a projectile in a gas gun towards the analyzed sample. Strength information was achieved by wave calculation which is recorded by the velocity interferometer for any reflector (VISAR) [26, 27].

Hydrogen effect on strength was studied by means of thermal desorption analysis (TDA). This analysis is used to assess the hydrogen evolution and trapping parameters within the context of the various microstructures of the studied titanium alloys, and to explain the correlation between prior microstructure, hydrogen concentration, potential trapping states and the alloy's response to hydrogen embrittlement.

23.2 EXPERIMENTAL

23.2.1 MATERIALS

The chemical composition of different duplex stainless steel: SDSS, SAF 2507, and LDS, LDX 2101, are presented in Table 23.1. LDS is a low alloyed duplex stainless steel.

DSS alloys were delivered as a 1.5 mm thick plate in the fully annealed condition. The microstructure of the sheets consists of equiaxed grains with an average grain size of about 20 μm as was describes by us in previous studies [8, 29, 30].

TABLE 23.1 Chemical Composition of the Studied LDS and SDSS (wt%)

Sample's commercial name	C	S	P	Mn	Si	Ni	Cr	Mo	N	Cu
LDX 2101	0.026	0.001	0.025	4.9	0.63	1.53	21.53	0.2	0.22	0.28
SAF 2507	0.025	0.004	0.022	0.411	0.288	6.745	24.012	4.63	0.3	0.072

The samples are cathodically precharged with hydrogen or precharged at high temperature and high pressure in a hydrogen gas chamber. The cathodic charging was performed in a 0.5N H_2SO_4 (sulfuric acid) water solution and 0.25 g l^{-1} of $NaAsO_2$ (sodium arsenide) powder with a current density of 50 mA cm^{-2} at 25°C for 24 and 72 hours (h). The sodium arsenide acts as a surface recombination inhibitor that increases the solute hydrogen concentration in the material by an order of magnitude. The gaseous charging was performed at 300°C and 60 MPa pressure for 3 hr in a gas chamber. These charging conditions have been carefully selected by diffusion calculations [30, 31] in order to receive an homogenous hydrogen content of 54 wt ppm along the sample bulk.

23.2.2   MEASUREMENT

23.2.2.1   MICROSTRUCTURE ANALYSIS

The microstructure, phase composition, and lattice parameter of DSS were examined both before and after hydrogenation, using the following micro-structural observations: transmission electron microscopy (TEM), scanning electron microscopy (SEM), energy dispersive spectrometry (EDS), optical microscopy (OM), and X-ray diffraction (XRD). XRD patterns were measured at RT using a Philips PW 1050/70 diffractometer, scanned at steps of 0.02° in a 2θ range between 30° and 100°, using Ni-filtered Cu-K$_\alpha$ radiation (1.54 Å) and a scan rate of ~0.3°/min. TEM images were taken by JEM 2100F. Samples were cut using Focus Ion Beam Scanning Electron Microscopes (FIBS).

23.2.2.2   DYNAMIC EXPERIMENTS

Dynamic experiments were conducted by plate impact at room temperature (RT) and strain rate of ~10^5 s^{-1} using a 25 mm and 6 m long gas gun. The impact was performed by accelerating an LDS or Al impactor towards an LDS target. The target's surface is being illuminated by a coherent 532 nm Nd-Yag. The scattered light from the surface is collimated by mirrors and lens and returned to the VISAR [26, 27]. After impact, the targets are softly caught by a special made soft catching cell and are taking to microstruc-tural observation. This procedure gives another analysis perspective to the deformed targets. The experiments were performed at dynamic pressure of ~0.5–2 GPa which was achieved by impact velocity of ~50–200 m/s.

23.2.2.3   THERMAL DESORPTION ANALYSIS (TDA)

The characteristics of hydrogen desorption were investigated by means of TDA. This technique involves accurate measurement of the desorption rate of hydrogen atoms, as solute or trapped in the material, while heating the sample a non- isothermal heating at a known rate under UHV ~10 µpa. In this work, the samples were heated from RT to 500°C at constant heating rates of 2°C/min, 4°C/min and 6°C/min. The mass spectrometer was operated under the fast multiple mode detection; the measured intensity channel was set to 2 amu in order to detect hydrogen desorption. The working procedure, as described by Lee and Lee [30] and others [31], allowed for the identification of different types of traps coexisting in the specimen.

23.2.2.4   *LECO HYDROGEN DETERMINATOR ANALYSIS*

In order to determine the quantity of absorbed hydrogen in the sample, vacuum extraction using a LECO RH–404 hydrogen determinator was applied after the charging process; measurement accuracy was estimated for ± 0.05 wt ppm (~2% of the result). This method includes heating the sample to the sample's melting temperature by an electrode furnace. The emitted hydrogen is detected by a thermal conductive cell, which has the ability to detect differences in thermal conductivity of different gases.

23.2.2.5   *TIME OF FLIGHT-SECONDARY ION MASS SPECTROMETER (TOF-SIMS)*

For the comparison of the TDA results and with those of the microanalysis technique ToF-SIMS samples were also charged by deuterium. The use of deuterium instead of hydrogen for the ToF-SIMS analyzes is due to the need to distinguish between the hydrogen that is intentionally in the microstructure of DSS to the one contained in the residual gas species adsorbed thereon. Considering the differences in diffusibility and solubility in the microstructure, the properties of the two isotopes differ only within the same order of magnitude. Before charging the plates were polished up to a 0.25 μm diamond suspension to reach a surface quality sufficient for imaging ToF-SIMS. The charging with deuterium was performed at RT in D_2O containing 0.05M D_2SO_4 and 1.3 g l^{-1} of $NaAsO_2$ (sodium arsenite) with a constant current density of 5 mA cm^{-2} for 96 hours. Immediately after charging samples were cleaned in an ultrasonic bath with Acetone for 5 minutes and were introduced into the ToF-SIMS UHV analysis chamber. The samples were cooled down while purging with dry nitrogen to a temperature of –130°C. Samples were transferred and analyzed at a temperature below –100°C. Further details on sample preparation, analysis, data, and data fusion are given elsewhere [32, 33].

23.3 RESULTS AND DISCUSSION

23.3.1 *MICROSTRUCTURAL OBSERVATIONS IN THE PRESENCE OF HYDROGEN*

23.3.1.1 *THE EFFECT OF ELECTROCHEMICAL HYDROGEN CHARGING (HIGH FUGACITY HYDROGEN)*

The XRD diffraction patterns (Figure 23.1a) show a comparison between LDS without hydrogen (black line) and LDS with hydrogen (blue line) aged for different times at RT. On exposure to 72 h, electrochemical hydrogen environment hydrogen induces strain in the LDS sample. The γ phase reflections exhibit a decrease in intensity and increased lattice parameter of ~2%. Additional reflection observed was ascribed to the ε_H-martensite, a hydrogen-containing HCP solid solution, which appears as a result of the combination between hydrogen induced in the γ phase and plastic deformation [3, 22–25]. These observations were already seen in cathodic charged DSS alloys [3, 10, 11, 35]. A similar result is also seen in hydrogen cathodic charged Ti-based alloys, where hydrogen increases the lattice parameter and can lead to second phase's precipitation [36]. The plastic deformation is created during cathodic charging, since the lower hydrogen diffusivity inside the γ phase induces severe stresses on the sample's surface. During aging, it can be seen that the ε_H-martensite peak becomes more stable. In addition, when the aging continues the next transformations can appear $\varepsilon_H \rightarrow \varepsilon$, or $\varepsilon_H \rightarrow \alpha'$. In the first transition, the ε is the HCP martensite phase free of hydrogen, while in the second the α'- martensite phase represents martensite with body-centered tetragonal structure (BCT) [11, 22, 28, 29]. The α'-martensite might appear in the XRD spectra; however, it is hard to identify since its peaks may overlap with those of α-bcc. Figure 23.2 compares micrographs of non-charged LDS (Figure 23.2a) and hydrogen charged for 72 h (Figure 23.2b). The ε-martensite in the shape of lath martensite, appears in Figure 23.2b, indicating the stability of ε-martensite within electrochemical charged LDS after a long time at RT.

In order to compare the microstructural changes of hydrogen on LDS, XRD analysis was performed on DSS (black line in XRD spectra) with the same charging condition of LDS (72 h charging and aging for 1 month at RT-blue line in XRD spectra). The comparison of the XRD spectra, Figure 23.1a and 1b, showed that the γ^* and the ε_H-martensite phase's reflections in LDS demonstrate higher intensity (~80% higher), meaning greater stability of these phases in LDS. In order to estimate the ε-martensite quantities after

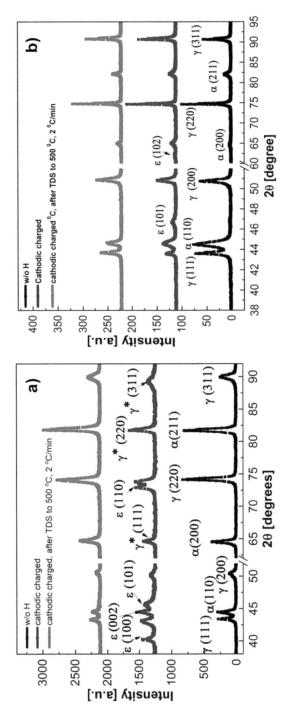

FIGURE 23.1 **(See color insert.)** XRD pattern of the uncharged and 72 h hydrogenated samples (a) LDS and (b) DSS [11], [37].

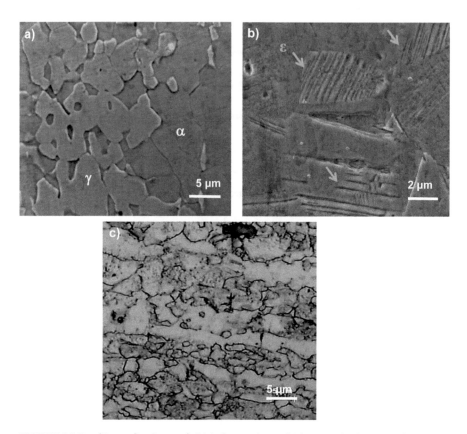

FIGURE 23.2   (See color insert.) OM observations of (a) as received LDS and DSS and after 72 h cathodic charging of (b) LDS, showing the formation of needle shape martensite phase after one month at RT and (c) DSS [12], [37].

1 month aging in the studied steels, we conducted refinement of the XRD pattern using Rietveld method. Rietveld method is one of the most popular approaches to quantitative phase analysis in XRD, for it provides information about the quantity of the crystalline by a simple Gaussian model for the different intensities [24, 25]. The quantitative analysis is based on the fact that intensity of different peaks on XRD depends on concentration of this phase in the studied material. According to the Rietveld method, the quantities of ε-martensite were 30 and 0% of the whole sample in LDS and SDSS, respectively. This phenomenon can be explained due to the different quantities of the alloying element in those steels. Although the matrix of both alloys is composed of ~50–50% α and γ phases, the γ phase in LDS is less steady in the presence of hydrogen due to the lower nickel content in it. In order to maintain an equal quantity of the phases, the phase is

compensated by an increased quantity of manganese and nitrogen to assure a balanced microstructure with approximately equal quantities of α and γ. Consequently, the manganese increment can cause a decrease in the stacking fault energy of the material, which induces the martensite transformation. In addition, it is known that high hydrogen concentrations inside the lattice, which are affected by the lattice's chemical composition, decrease the γ-phase stability and may induce transformations of the γ-phase to α' or ε-martensite [12–14, 22, 26, 27]. These indications are supported by earlier publications of Floreen and Mihalisim [43] and Rozenak and Eliezer [12–14, 22]. According to their work, internal stresses or strains, which accompany the absorption of hydrogen, might provide a significant driving force for γ decomposition. The γ phase stability of stainless steels is dependent on the alloy composition and is given by [12–14, 22, 28]:

$$S = Ni + 0.68Cr + 0.55Mn + 0.45Si + 27\,(C + N), \tag{1}$$

where S is the austenite stability factor and *Ni, Cr, Mn, Si, C, and N* are the stabilizing elements in weight percentage. Higher values of S indicate a greater value of the γ stability.

According to the above equation (Eq. 1) and the alloys' compositions as referenced in Table 23.1, the stability factor and the phase quantities, as estimated by the Rietveld method, were calculated for LDS and SDSS. The results are presented in Table 23.2.

TABLE 23.2 Phase Quantities of 72 h Cathodic Charged DSS and LDS After One Month at RT [11]

Sample	Phase concentration (wt%)			S stability factor
	α	γ	ε/α	
LDS	46	26	30	25
DSS	50	50	0	32

It can be seen from Table 23.2 that the LDS presents 28% lower value of S compared with the value of SDSS (25 and 32, respectively). Furthermore, the extent of hydrogen induced formation of α'- and ε martensite which are still stable in LDS after 1 month aging at RT correlates well with the calculated stability of the γ-austenite phase. The larger amounts of α' and ε martensite, also exhibited lower values of the stability factor. This phenomenon generally means that the dominant phase transformation in LDS is the γ→ε→α' rather than the γ→γ*. Nevertheless, the second phases appeared in the XRD spectra on both, LDS, and DSS, disappear after thermal desorption to 500°C (appears as red line in XRD spectra). These results were also supported by OM observations of SDSS (Figure 23.2c) showing no structural changes after a long period of aging of 1 month. These observations indicate the increased quantity of α' and ε-martensite phase in LDS compared with SDSS. Nevertheless, the OM observation of SDSS reveals the same changes in α phase as with LDS, showing significant grain refinement (a decrease of 10 μm to ~5 μm). The larger amount of second phases as will be show by TDA analysis plays a very important role on the hydrogen trapping mechanisms and HAC. In order to better understand the behavior of the second phases, the stability factor of DSS and LDS was compared to other stainless steels. Figure 23.3 presents the position of stability factor (and the content of second phases as a result) of DSS and LDS compared to other stainless steels.

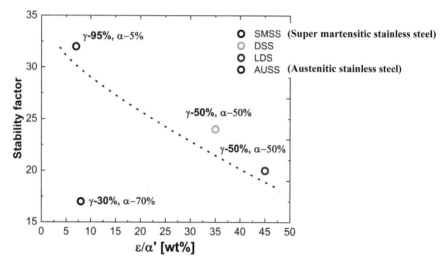

FIGURE 23.3   Phase stability factor of different stainless steels compared to those of DSS and LDS.

It was already mentioned in our previous works [11, 37, 44] that high stability factor leads to lower amount of second phases. Susceptibility to hydrogen embrittlement will be the highest in austenitic stainless steels and the lowest in supermartensitic stainless steels (SMSS), where γ-phase stability is lower and trapping states are higher. Based on our former works, the location of LDS compared with DSS, Figure 23.3 indicates on lower susceptibility of LDS to hydrogen embrittlement.

23.3.1.2 THE EFFECT OF GASEOUS HYDROGEN CHARGING (LOW FUGACITY HYDROGEN) ON LDS SAMPLE

Due to the larger amount of second phases in LDS, further experiments with hydrogen interactions were examined on this sample. The microstructure of the gas-phase hydrogen charged LDS, Figure 23.4, as was showed in previously published works by us [8, 29, 30] revealed for the first time the appearance of needle-shaped sigma (σ) phase which is an intermetallic compound with the Fe(CrMo) composition. The appearance of that phase was confirmed using XRD and energy dispersive diffraction (EDS) measurements which indicate on σ phase with a composition of 1:1:1 for Fe, Cr, and Mo, respectively [12]. This phase plays an important role in the hydrogen trapping mechanism in LDS.

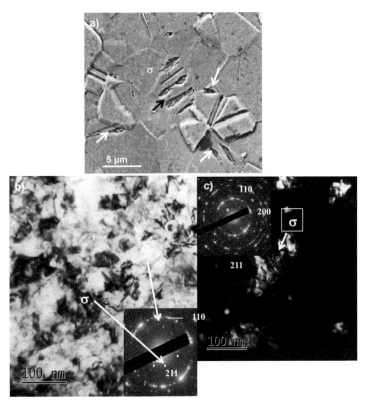

FIGURE 23.4 Micrographs of (a) LDS after gas-phase charging for 3 h at 60 MPa and 300°C, showing the formation of needle-shaped σ phase after one month at RT, (b) bright field TEM image, the inset shows the [100] zone axis from the σ-phase grain and (c) dark field TEM image [35, 45].

23.3.1.3   DEFORMATION RESPONSE AT LOW AND HIGH STRAIN RATES ($\sim 10^{-7}\,S^{-1}$ AND $10^5 S^{-1}$)

Tensile testing was performed at RT on gas-phase hydrogen charged LDS with length of 136 mm, a gauge length of 32 mm and a cross-section of 6 × 1.5 mm². The applied deformation rate was set to 0.02 mm/min which corresponds to a train rate of ~ 10^{-7} s^{-1}.

It can be seen from Figure 23.5 that hydrogen causes an increment in yield strength. This increment was calculated to be 20% higher in gas-phase hydrogen charged sample compared with the non-charged sample. The dynamic experiments, Figure 23.6, did not show the same tendency. Therefore, it can be said that only when hydrogen has enough time for diffusion, as in the case of lower strain rate experiments, solid solution hardening will occur. This hardening can be related to the pinning of dislocations by the attached solute atoms [46].

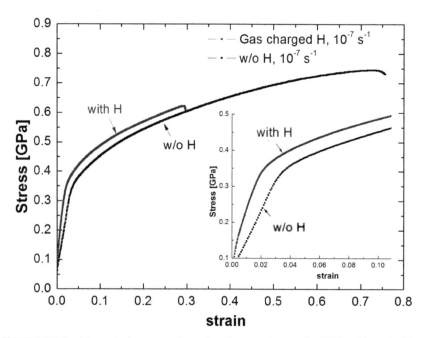

FIGURE 23.5   Stress-strain curve of quasi-static experiments for LDS with and without hydrogen. The inset presents the enlargement of the elastic-plastic transition [47, 48].

In order to investigate the plastic deformation behavior of LDS at higher strain rate, shock wave experiments were performed by accelerating an Al

1100 impactor towards different LDS targets at dynamic pressures of 1.2–2.3 GPa. We also performed metallurgical analysis of the dynamic deformed LDS in order to investigate its deformation mechanism after impact.

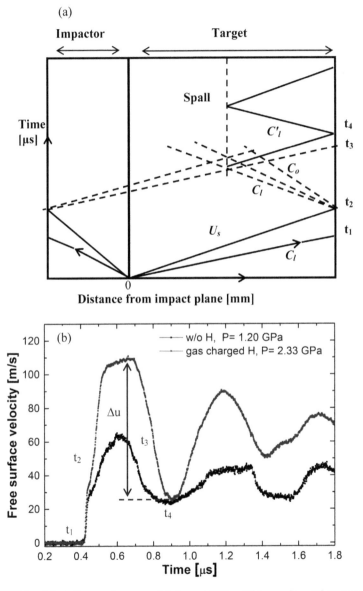

FIGURE 23.6 (a) x-t diagram for symmetric impact [49] and (b) experimental measurements of the free surface velocity of LDS with and without hydrogen at dynamic pressures 1.2–2.3 GPa [50].

The traveling waves during impact are presented in the 'time versus distance' diagram, Figure 23.6a. This diagram is converted to 'free surface velocity versus time' diagram as recorded by VISAR, Figure 23.6b. At impact two waves are produced; an elastic wave with longitudinal sound velocity (c_l), and a plastic wave propagating with shock velocity (U_s), Figure 23.6a. The elastic-plastic transition also known as the elastic precursor amplitude is called the Hugoniot elastic limit (HEL) [1, 34]. When a shock wave reaches the back surface of the impactor and the target it is reflected as a rarefaction waves fan (release waves), these waves' velocities are between the longitudinal sound (C_l) velocity and the bulk sound velocity (C_o). When these two rarefactions waves meet together, they produce tension, leading to micro-cracks or micro-voids nucleation and growth. If that tension exceeds the material's strength, a spall will occur by coalescence of cracks or voids, (can be seen in Figure 23.6a, as Spall) and the new wave velocity of the shocked metal is C'_1 [51].

The deformation behavior at high strain rate was investigated by calculating the HEL [1, 34] and rapture (spall) strength from the VISAR profile. The dynamic yield stress under uniaxial strain loading was extracted from HEL, according to the follow relation [52]:

$$\sigma_{HEL} = \rho_0 c_l u_{HEL} / 2 \tag{2}$$

$$\sigma_y = (1 - 2v)\sigma_{HEL} / (1 - v), \tag{3}$$

where c_l is the longitudinal sound velocity (propagation velocity of the elastic precursor), ρ_0 is the initial density, u_{HEL} is the free surface velocity at the precursor front, and v is the Poisson's ratio which was taken as 0.33.

The spall strength, also known as dynamic tensile stress σ_{spall}, is proportional to the fracture stress. The σ_{spall} is determined using the measured pull-back velocity Δu, where $\Delta u = u_{max} - u_{min}$ is the difference between the first maximum (u_{max}) and the first minimum (u_{min}) of the target's free surface velocity. Within the acoustic approach, the spall strength is given by the following relationship [52]:

$$\sigma_{spall} = \rho_0 c_0 (\Delta u + \delta) / 2 \tag{4}$$

$$\delta = h(1 / c_0 - 1 / c_l) \times (|\dot{u}_1| \cdot \dot{u}_2 / (|\dot{u}_1| + \dot{u}_2)), \tag{5}$$

where c_0 is the bulk sound velocity, ρ_0 is the initial density, δ is the correction for profile distortion, h is the distance between the small plane and the free surface, and $\dot{u}_1$ and $\dot{u}_2$ are the left and right slopes of the first minimum in the free surface velocity profile, respectively.

In all the experiments, with, and without hydrogen, the HEL velocity is 0.75±0.05 GPa, meaning that the presence of hydrogen in LDS at low dynamic pressures did not affect the HEL, Figure 23.6b. These results are summarized in Table 23.3.

TABLE 23.3 Dynamic Experiments Results for LDS With and Without H [50]

Sample	Dynamic pressure [GPa]	σ_{HEL} ±0.05 [GPa]	σ_y ±0.02 [GPa]	P_{Spall} ±0.20 [GPa]
LDS w/o H	1.2	0.75	0.38	0.89
LDS with H	2.3	0.75	0.38	1.68

Spall strength calculation of the gas-phase hydrogen charged sample indicates on 1.68 GPa spall strength. The spall strength of non-charged LDS was calculated to be 0.89 GPa. From that profile, it is hard to estimate the spall strength value since the first minimum curve is broadened. Microstructural observations after dynamic loading, Figure 23.7, support those finding by showing no cracking coalescence, meaning partial spall.

23.3.1.4   *MICROSTRUCTURE OBSERVATIONS AFTER DYNAMIC LOADING*

The impacted samples were softly caught and collected for metallurgical analysis (Figure 23.7). From these experiments one can see that the spall was not completed, i.e., the cracks did not coalesce and did not create a continuous tearing of the target. In order to observe the dynamic pressure effect on LDS with hydrogen, an additional post-mortem experiment has been conducted on hydrogen charged LDS at 60 m/s impact velocity (0.5 GPa dynamic pressure). The microstructural result can be seen in Figure 23.7c. Hydrogen seems to propagate crack motion at both dynamic pressures (0.5 GPa and 2.3 GPa). In addition, it can be seen that the cracks' are developed inside the grains; trans-granular cracking. No change has occurred in grain size in all samples; average grain size was estimated at about 20 μm.

The microstructural result of 0.5 GPa dynamic pressure shows a multiplication of cracks Figure 23.7c. At this lower dynamic pressure hydrogen has enough time to propagate a crack. The microstructural observations, in that case, indicate on higher and denser cracks compared with the non-charged sample. This demonstration clearly represents the HAC phenomenon; when hydrogen is presented in a bulk metal, it will promote cracking.

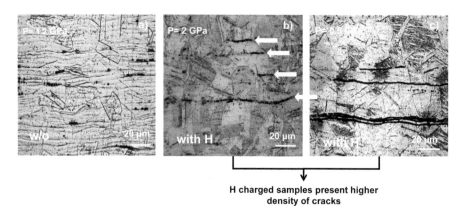

**H charged samples present higher
density of cracks**

FIGURE 23.7   OM analysis of the impacted cross-section targets: (a) non-charged LDS at dynamic pressure of 1.2 GPa, (b) gas-phase hydrogen charged LDS at dynamic pressure of 2.3 GPa and (c) gas-phase hydrogen charged LDS at dynamic pressure of 0.5 GPa [50, 53].

23.3.1.5 HYDROGEN TRAPPING MECHANISM

Hydrogen trapping mechanism of gas-phase hydrogen charged LDS samples non-deformed and deformed were studied by means of TDA analysis. The deformed samples include gas-phase hydrogen charged LDS after quasi-static (10^{-7} s^{-1}) tensile loading (quasi-static loaded), and gas-phase hydrogen charged LDS after dynamic (10^{5} s^{-1}) loading (dynamic loaded) at dynamic pressures of 0.5 GPa and 2.3 GPa.

The trapping activation's energies were estimated, according to Lee and Lee's model [30, 54], from ramped TDA which produce different desorption temperature.

Hydrogen evolution rate from a trap site is given by [30, 54]:

$$dX / dt = A\left(1 - X\right)exp\left(-E_a / RT_c\right), \tag{6}$$

where X is the hydrogen content that escapes a trap, A is the reaction rate constant, E_a is the activation energy for releasing hydrogen from its trapping site, R is the gas constant, and T is the absolute temperature.

The desorption peak temperature increases with heating rate, as demonstrated in the TDA spectra, and the activation energy (Ea) depends on the critical temperature for desorption (Tc) and heating rate (φ):

$$\partial ln\left(\varphi / T_c^2\right) / \partial\left(1 / T_c\right) = -E_a / R, \tag{7}$$

where R is the gas constant.

By applying Lee and Lee's model [30], the activation energy for hydrogen release may be calculated from the slope of $\ln(\varphi/T_c^2)$ versus $1/T_c$. According to their model an activation energy value, E_a, equal to or higher than 60 kJ/mol will be ascribed as an irreversible trap, else it will be ascribed as a reversible trap. This notion is highly important since only diffusible hydrogen through lattice sites or hydrogen residing at traps with the lowest binding energy contributes to a metal's embrittlement [55].

Hydrogen desorption rate as a function of temperature for non-loaded, quasi-static loaded, and dynamic loaded at P = 2.33 GPa and P = 0.50 GPa at a 2°C/min heating rate are shown in Figure 23.8. In order to evaluate the trap's activation energies, three different heating rates in the range of 2–6°C/min were applied to the samples. By referring to the shift of a peak temperature with different heating rates in Figure 23.8a, the activation energy for hydrogen desorption from a certain trap can be calculated, Figure 23.8b. Calculations are based on the behavior of the peaks' movement towards higher temperature with increasing heating rate, as expected for a thermally controlled activation desorption process [11, 12].

Hydrogen trapping states are seen in Table 23.4. All samples present about the same lower trapping energy values of ~20 kJ/mol and ~40 kJ/mol, which relates to the same trapping sites- elastic stress field [37, 56–58] and, at higher activation energies, core dislocation or grain boundary. The great differences are seen in the quasi-static loaded sample. It can be seen from Table 23.4 that the activation energy measurements for the quasi-static loaded LDS are 19 kJ/mol, 22 kJ/mol, and 40 kJ/mol. These values are lower than the value of 60 kJ/mol, and they can be classified as reversible traps. The maximal activation energy value in quasi-static loaded is ~40% lower than the non-loaded sample and ~50% lower than the dynamic loaded sample. The non-loaded sample presents two traps with lower values than 60 kJ/mol (19 kJ/mol and 39 kJ/mol) and one higher value of 66 kJ/mol which was ascribed to the formation of σ phase.

Regarding the dynamic loaded samples, they present two lower values than 60 kJ/mole, which can be ascribed as reversible trapping sites, and one higher value – 72 kJ/mol and 80 kJ/mol for the 0.50 GPa and 2.33 GPa dynamic pressures, respectively. Those higher activation energy value can be classified as irreversible trapping sites and ascribed to σ phase. The divergence in σ-phase's activation energy values belongs to its density.

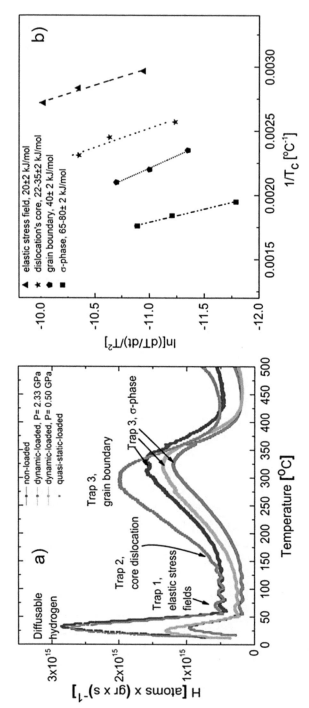

FIGURE 23.8 (a) TDS spectra at a 2°C/min heating rate of gas-phase hydrogen charged at different loading states: non-loaded, quasi-static loaded and dynamic loaded at different dynamic pressures (0.5 GPa and 2.33 GPa). (b) Determination of the activation energies for four-fitted hydrogen desorption peaks of TDS spectra in Figure 23.8a.

TABLE 23.4 Calculated Activation Energies for Non-Loaded, Quasi-Static Loaded and Dynamic Loaded at Different Dynamic Pressures (0.50 GPa and 2.33 GPa) [50]

Non-loaded		Quasi-static loaded		Dynamic loaded at 0.50 GPa		Dynamic loaded at 2.33 GPa	
Temp ± 1 [°C]	Ea [kJ/mol]	Temp ± 1 [°C]	Ea [kJ/mol]	Temp ± 1 [°C]	Ea [kJ/mol]	Tempe ± 1 [°C]	Ea [kJ/mol]
60–90	19±2	64–95	19±2	67–107	21±2	80–115	20±2
143–160	39±1	117–140	22.0±0.5	122–160	34±2	147–165	41±2
300–340	66±2	210–300	40.0±0.5	314–353	72.0±0.5	324–360	80±0.5

The higher trapping energy value of the dynamic loaded sample (80 kJ/mol) was ~18% higher than the non-loaded sample (60 kJ/mol), meaning that the dynamic loaded samples created deeper traps (higher trapping energy values) compared with the non-loaded sample. This phenomenon was related to the dynamic loading, which creates higher energy trapping sites. In addition, differences can also be seen between the dynamic loaded samples; at higher pressure (2.33 GPa) the sample has ~10% higher trapping energy value compared with the sample subjected to lower dynamic pressure (0.50 GPa). These differences were explained due to the greater deformation which caused hydrogen to be trapped deeper in trapping sites.

These results were in line with the post-metallographic cross sections of the impacted samples, Figure 23.7, which showed massive cracking for the low-pressure sample, due to hydrogen's ability to escape at lower energy from trapping sites and promote cracking. We explain this phenomenon using the HELP theory since trapping affects the metal's diffusivity it has a major influence on its validity to potential cracking sites. Therefore, when the activation energy values are lower, it will be easier for hydrogen to escape and, therefore, cracking will be massive. This phenomenon is clearly seen when comparing between dynamic loaded at different dynamic pressures. The low dynamic pressure at ~0.50 GPa presented a multiple cracking phenomenon, Figure 23.7c, compared with dynamic loaded at ~2.33 GPa, Figure 23.7b.

In order to confirm the hydrogen trapping energies' effect on hydrogen evaluation at different targets, integration of the TDA spectra (giving the desorbed hydrogen content within a range temperature) in addition to LECO measurement (giving the total absorbed hydrogen content by melting sample procedure) were applied to the samples, as can be seen in Figure 23.9. Additional information on the LECO procedure can be found in different works [8, 30, 41]. A simultaneous decrease in hydrogen content can be seen in all charging condition when increasing heating rate. The quasi-static loaded sample shows the greatest hydrogen desorption values, ~50% higher than the non-loaded sample, due to its lower activation energy's values.

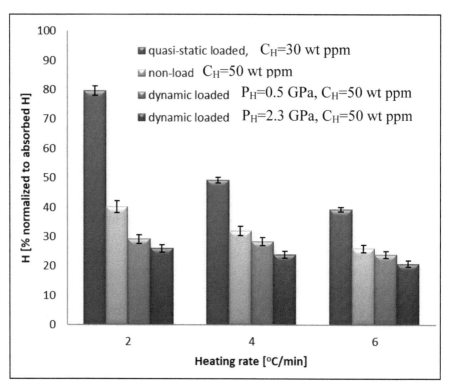

FIGURE 23.9 Influence of different heating rates on hydrogen evaluation, achieved by TDA's integration, of different gas-phase hydrogen charged LDS samples: non-loaded, quasi-static loaded and dynamic loaded at different dynamic pressures (0.5 GPa and 2.3 GPa). The C_H value near each name represents hydrogen total amount in each sample achieved by LECO [50].

In order to relate hydrogen trapping to hydrogen absorption during the thermal desorption process (TDA), relative hydrogen desorption as a function of temperatures were examined for all samples, Figure 23.10. Our results show substantial difference between the quasi-static loaded sample (Figure 23.10b) to the rest of the samples (Figure 23.10a, 10c, and 10d). It can be seen that in all temperature ranges (trapping sites) the quasi-static loaded sample shows the highest desorbed hydrogen fraction. The lower desorbed hydrogen fraction in all temperature range was from the dynamic loaded sample at 2.3 GPa, which showed the highest trapping energies values.

Additional observations were the same for all samples; in all heating rates, an increment in hydrogen evaluation content was seen as the temperature increased. Each heating rate has a certain temperature in which a maximal amount of hydrogen is desorbed.

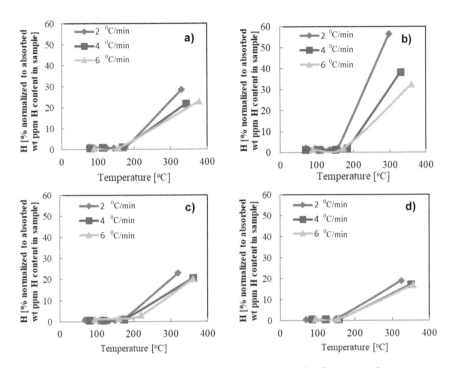

FIGURE 23.10 Influence of heating rate on hydrogen evaluation from a certain temperature in different gas-phase hydrogen charged LDS: (a) non-loaded, (b) quasi-static loaded, (c) dynamic loaded at P = 2.3 GPa and (d) dynamic loaded at P = 0.5 GPa [50].

In similar heating rates performed during TDA- Figures 23.9 and 23.10, the quasi-static loaded sample shows the greatest hydrogen desorption values, ~50% higher than the non-loaded sample, due to its lower activation energy's values. In addition, in all temperature range of hydrogen desorption (trapping sites), the quasi-static loaded sample shows the highest desorbed hydrogen content. The lower desorbed hydrogen content in all temperature range was from the dynamic loaded sample at 2.3 GPa, which showed the highest trapping energies values. These results are explained by the calculated trapping energies, which showed the following order of trapping energy: dynamic loaded at P = 2.3 GPa > dynamic loaded at P = 0.5 GPa > non-loaded > quasi-static loaded.

In Figure 23.11, we compare the influence of heating rate on hydrogen desorption rate in dynamic loaded at P = 2.3 GPa, dynamic loaded at P = 0.5 GPa and non-loaded samples. In that comparison, we ignore from the quasi-static loaded sample since it absorbs lower hydrogen content in charging (30 wt ppm compared with 50 wt ppm in the others) and it was impossible to refer to relative hydrogen desorption rate.

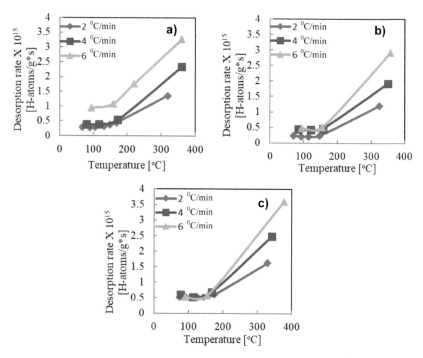

FIGURE 23.11   Influence of heating rate on hydrogen desorption rate vs. temperature in dynamic loaded at (a) 0.5 GPa gas-phase hydrogen charged LDS, (b) dynamic loaded at 2.3 GPa and (c) non-loaded [50].

Significant changes are at different heating rates; increasing heating rate amplifies the hydrogen desorption rate, since hydrogen needs to desorbs faster from traps along with short time for diffusion. The results in Figure 23.11 are in line with the trapping energies calculation (Table 23.4). The non-loaded sample shows the highest hydrogen desorption rate capability in all heating rates, due to its lower trapping energies for hydrogen. The dynamic loaded sample at P = 2.3 GPa shows the lowest desorption rate capability in all heating rates, presumably due to its higher trapping energies values compared with the rest of the samples.

Figure 23.12 [33] compares between ToF-SIMS images and surface cracking. The correlation between the results in Figure 23.11 to desorbed hydrogen versus temperature (Figures 23.9 and 23.10) can shine their light on the hydrogen embrittlement model of DSS and LDS and proves the role of second phases.

ToF-SIMS

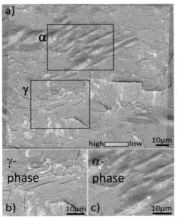

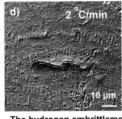

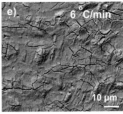

Surface cracking

The hydrogen embrittlement model in austenitic-ferritic steel was proved to be effected as follows:

heating rate↓ cracking↓ trap↑

The presence of hydrogen around cracks correlates with the existence of different trapping sites in strained regions.

FIGURE 23.12 (a) ToF-SIMS image data representing deuterium in the microstructure (colored) [32, 60]. The images show the local differences in the deuterium concentration in the (b) austenite and (c) in the ferrite. Surface cracking after TDA are seen at (a) 2°C/min and (e) 6°C/min heating rates [33].

The distribution of deuterium within γ-phase is heterogeneous can be attributed to the presence of hydrogen traps in this phase (such as hydrogen induced martensite formation [20] or sigma phase in the case of gas-phase charging [44]).

Sigma phase in the hydrogen embrittlement mechanism plays two important, but contradicting roles. On the one hand, it creates a higher diffusion passage for hydrogen, about six orders of magnitude higher than the austenite phase. On the other hand, according to TDS analysis, sigma phase acts as a high trapping energy site for hydrogen (~80 kJ/mol; Table 23.4), which can stop hydrogen progress, and can thus stop cracking. This effect can clearly be seen in different cracking regions, which can be ascribed to different potential trapping sites, such as strained regions (Figure 23.12). In order to settle these two roles, we approached the heating rate effect on trapping and diffusion. From Figures 23.10 and 23.11, it can be seen that when time for diffusion is low, such as in the case of 6°C/min, cracking would be intensive. This means that sigma effect is inclining towards diffusion effect rather than trapping effect. As for the case of low heating rate, such as in the case of 2°C/min, less crackings are seen due to sigma acting as a trapping site for hydrogen, hence preventing cracking.

In order to better understand the role of second phases in the hydrogen embrittlement mechanism, we explored the hydrogen trapping mechanisms at different strain rates by applying theoretical modeling based on diffusion calculations [61]. Activation's energies calculations indicate on great differences between quasi-static loaded sample to the rest of the sample. In order to better understand those changes, diffusion calculations were applied to quasi-static and dynamic loaded samples. The effective diffusion for the gas-phase hydrogen charged LDS target was calculated by using the given Turnbull et al.'s equation [62]:

$$D_{eff} = 1.2x10^{-6} exp\left(-50.7\left[kJ/mol\right]/RT\right), \qquad (8)$$

where R is the gas constant [J/mol K^{-1}] and T the temperature [K].

The following calculations for dynamic and quasi-static experiments were made according to the experiment's loading time duration. The experiments' times were 240 sec and 10^{-6} sec for the quasi-static and dynamic loaded samples, respectively. According to these values, the diffusion distances in the samples were: 6.8×10^{4} µm^2/sec and 5.72×10^{-4} µm^2/sec for the quasi-static and dynamic loaded samples, respectively. It can be seen that hydrogen diffusion in the quasi-static experiment is eight orders of magnitude higher than that in the dynamic experiment. Therefore, during deformation process created in quasi-static loading, it can be said that hydrogen has enough time to migrate from deeper potential trap site to less deep (low potential energy) trapping sites.

The diffusion calculation results were demonstrated in an analytical model described in Figure 23.13, and showed in this work for the first time. This model predicts hydrogen trapping mechanisms based on the diffusion time. Applying this model to the quasi-static and dynamic loaded LDS samples enabled us to predict the hydrogen diffusion distance and to relate these results to the hydrogen trapping phenomenon.

During deformation processes created in the quasi-static loading, it can be seen that hydrogen has enough time to migrate with dislocations, created during the quasi-static loading, to different trapping sites. During its migration hydrogen can transform from deeper potential trapping sites (irreversible trapping site) to lower potential trapping sites (reversible trapping).

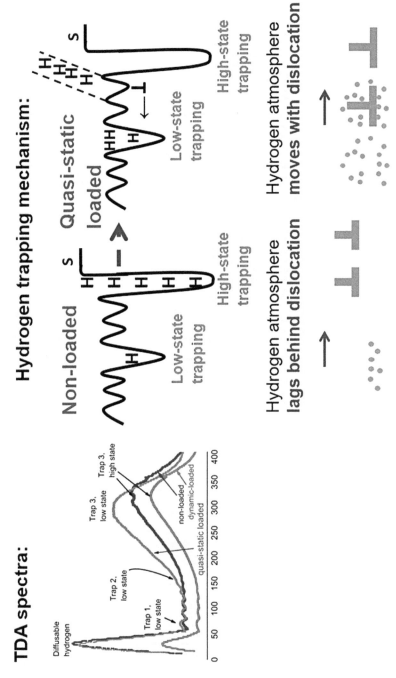

FIGURE 23.13 Hydrogen trapping model for non-loaded LDS and quasi-static loaded LDS.

Therefore, in the quasi-static loading, which provides a significantly higher time for diffusion, all trapping sites are characterized as reversible trapping sites only. The dislocation illustration is based on the works of Sofronis et al. [13, 16, 46].

The flowchart in Figure 23.14 compares different loadings and their effect on hydrogen trapping characteristics.

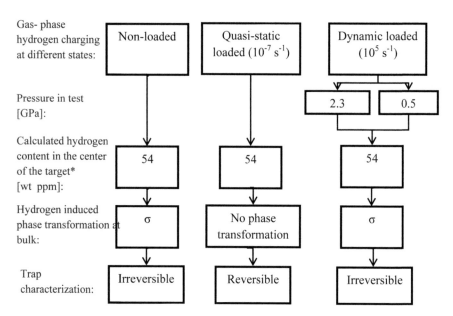

FIGURE 23.14 Flowchart showing the effect of deformation rates, created at low strain rate and high strain at different pressures, on trapping mechanisms and phase transformations [50].

23.4 CONCLUSIONS

This research study examines the effect stain rates and hydrogen-induced phase transition on the hydrogen embrittlement mechanism. The results showed that high stability factor leads to lower amount of second phases, and that high activation energies are the result of second phases.

Quasi-static experiments showed an increment of ~20% in yield strength which was related to the experiment deformation rate; higher deformation rate will not allow for enough time for hydrogen diffusion. This statement is well pronounced at dynamic experiments which did not show any effect of hydrogen on mechanical properties.

Hydrogen trapping energies values of the quasi-static loaded ($\sim 10^{-7}$ s^{-1}) sample reaches to ~40% lower values than the non-loaded LDS sample and ~50% lower values than the dynamic loaded LDS sample. At quasi-static loading, hydrogen had enough time to escape from the irreversible trapping site – σ phase, created during gas-phase hydrogen charging. The differences between the activations energies of the rest of the samples were belonged to σ phase density and deformation response which is responsible for hydrogen velocity or escape.

The greater deformation which was formed at the higher deformation at $P = 2.3$ GPa caused hydrogen to be trapped deeper in trapping sites (high energy trapping sites). This explanation is also supported by the microstructural observations results after dynamic loading, which showed massive cracking for the low-pressure sample (P = 0.5 GPa), due to hydrogen's ability to escape at lower energy from trapping site and promote cracking.

In this chapter, we demonstrate that the developed analytical model for hydrogen trapping based on diffusion calculation is suitable to predict the hydrogen trapping behavior in a large range of strain rates. Based on this behavior it was shown that hydrogen can escape from deeper potential trapping sites to less deeper (low potential energy) trapping sites during quasi-static deformation.

This assumption was explained due to longer time for diffusion compared with dynamic deformation 240 sec and 10^{-6} sec for the quasi-static and dynamic loaded samples, respectively. The model takes into consideration the stability of hydrogen induced second phases which affect the trapping phenomenon. This model shows that the susceptibility to hydrogen embrittlement decreases with second phases. Moreover, at high strain rates hydrogen atmosphere lags behind dislocation, only at low strain rate hydrogen atmosphere moves with dislocation. The model associated with DSS alloys was proved to be HELP model. In this, we proved a direct relationship between heating rate and cracking and opposite one to high trapping states (such as second phases).

KEYWORDS

- **duplex stainless steels**
- **quasi-static experiments**
- **thermal desorption spectrometry**
- **x-ray diffraction**

REFERENCES

1. Głowacka, A., & ŚwiąTnicki, W. A., (2003). "Effect of hydrogen charging on the microstructure of duplex stainless steel," *J. Alloys Compd.*, *356–357*, 701–704.
2. Zakroczymski, T., Glowacka, A., & Swiatnicki, W., (2005). "Effect of hydrogen concentration on the embrittlement of a duplex stainless steel," *Corros. Sci.*, *47*(6), 1403–1414.
3. Dabah, E., Lisitsyn, V., & Eliezer, D., (2010). "Performance of hydrogen trapping and phase transformation in hydrogenated duplex stainless steels," *Mater. Sci. Eng. A*, *527*(18 & 19), 4851–4857.
4. Oriani, R. A., (1970). "The diffusion and trapping of hydrogen in steel," *Acta. Metall.*, *18*, 147–157.
5. Turnbull, A., & Hutchings, R. B., (1994). "Analysis of hydrogen atom transport in a two-phase alloy," *Mater. Sci. Eng. A.*, *177*, 161–171.
6. Owczarek, E., & Zakroczymski, T., (2000). "Hydrogen transport in a duplex stainless steel," *Acta Mater.*, *48*(12), 3059–3070.
7. Davis, J. R., (1999). *Stainless Steels* (3rd edn). Ohaio: ASM international.
8. Sieurin, H., Sandstrom, R., & Westin, E. M., (2006). "Fracture toughness of the lean duplex stainless steel," *Metall. Mater. Trans. A*, *27A*, 2975–2981.
9. Lacombe, P., Baroux, B., & Beranger, G., (1993). *Stainless Steel*, Les Edition, 1–50.
10. Ravit Bar, Eithan Dabah, Dan Eliezer, Thomas Kannengiesser, & Thomas Boellinghaus, (2011). The influence of hydrogen on thermal desorption processes in structural materials. *Procedia Engineering 10,* 3668–3676.
11. Silverstein, R., & Eliezer, D., (2015). "Hydrogen trapping mechanism of different duplex stainless steels alloys," *J. Alloys Compd.*, *644*, 280–286.
12. Silverstein, R., Eliezer, D., Glam, B., Eliezer, S., & Moreno, D., (2015). "Evaluation of hydrogen trapping mechanisms during performance of different hydrogen fugacity in a lean duplex stainless steel," *J. Alloys Compd.*, *648*, 601–608.
13. Birnbaum, H. K., & Sofronis, P., (1994). "Hydrogen-enhanced localized plasticity- a mechanism for hydrogen-related fracture," *Mater. Sci. Eng. A*, *176*, 191–202.
14. Robertson, I. M., (2001). "The effect of hydrogen on dislocation dynamics," *Eng. Fract. Mech.*, *68*, 671–692.
15. Oltra, R., Bouillot, C., & Magnin, T., (1996). "Localized hydrogen cracking in the austenitic phase of a duplex stainless steel," *Scr. Mater.*, *35*(9), 1101–1105.
16. Maroef, I., Olson, D. L., Eberhart, M., & Edwards, G. R., (2002). "Hydrogen trapping in ferritic steel weld metal," *Int. Mater. Rev.*, *47*, 191–223.
17. Teus, S., Shyvanyuk, V., & Gavriljuk, V., (2008). "Hydrogen-induced $\gamma\rightarrow\varepsilon$ transformation and the role of ε-martensite in hydrogen embrittlement of austenitic steels," *Mater. Sci. Eng. A*, *497*(1 & 2), 290–294.
18. Chen, S., Gao, M., & Wei, R. P., (1996). "Hydride formation and decomposition in electrolytically charged metastable austenitic stainless steels," *Metall. Mater. Trans. A.*, *27*, 29–40.
19. Rozenak, P., Zevin, L., & Eliezer, D., (1984). "Hydrogen effects on phase transformations in austenitic stainless steels," *Mater. Sci.*, *19*, 567–573.
20. Rozenak, P., & Eliezer, D., (1987). "Phase changes related to hydrogen- induced cracking in austenitic stainless steel," *Acta Met.*, *35*, 2329–2340.

21. Rozenak, P., & Eliezer, D., (1984). "Effects of aging after cathodic charging in austenitic stainless steels," *J. Mater. Sci., 19*, 3873–3879.
22. Presouyre, G. M., (1980). "Trap theory of hydrogen embrittlement," *Acta Metall., 28*, 895–911.
23. Mcnabb, A., & Foster, P. K., (1963). "A new analysis of the diffusion of hydrogen in iron and ferritic steels," *Trans. Metall. Soc. AIME, 227*, 618–627.
24. Asay, J. R., (1993). *High-Pressure Shock Compression of Solids*. Springer. pp. 100–200.
25. Nellis, W. J., (2006). "Dynamic compression of materials: Metallization of fluid hydrogen at high pressures," *Reports Prog. Phys., 69*, 1479–1580.
26. Bourne, N. K., Millett, J. C. F., & Gray, G. T. III, (2009). "On the shock compression of polycrystalline metals," *J. Mater. Sci., 44*, 3319–3343.
27. Hemsing, W. F., (1979). "Velocity sensing interferometer (VISAR) modification," *Rev. Sci. Instrum., 50*, p. 73.
28. Barker, L. M., (1972). "Laser interferometer for measuring high velocities of any reflecting surface," *J. Appl. Phys., 43*, p. 4669.
29. Ravit Silverstein, Dan Eliezer, Benny Glam, Daniel Moreno, & Shalom Eliezer (2014). Influence of hydrogen on the microstructure and dynamic strength of lean duplex stainless steel. *Journal of Materials Science, 49*, 4025–4031.
30. Lee, S., & Lee, J., (1986). "The trapping and transport phenomena of hydrogen in nickel," *Metall. Trans. A, 17*, 181–187.
31. Tal-Gutelmacher, E., Eliezer, D., & Abramov, E., (2007). "Thermal desorption spectroscopy (TDS): Application in quantitative study of hydrogen evolution and trapping in crystalline and non-crystalline materials," *Mater. Sci. Eng. A, 445–446*, 625–631.
32. Sobol, O., Holzlechner, G., Holzweber, M., Lohninger, H., Boellinghaus, Th., & Unger, W. E. S., (2016). "First use of data fusion and multivariate analysis of ToF-SIMS and SEM image data for studying deuterium-assisted degradation processes in duplex steels," *Surf. Interface Anal. 48*, 474–478.
33. Silverstein, R., Sobol, O., Boellinghaus, T., Unger, W., & Eliezer, D., (2017). "Hydrogen behavior in SAF 2205 duplex stainless steel," *J. Alloys Compd., 695*, 2689–2695.
34. Rozenak, P., & Bergman, R., (2006). "X-ray phase analysis of martensitic transformations in austenitic stainless steels electrochemically charged with hydrogen," *Mater. Sci. Eng. A, 437*, 366–378.
35. Silverstein, R., Eliezer, D., Glam, B., Moreno, D., & Eliezer, S., (2014). "Influence of hydrogen on microstructure and dynamic strength of lean duplex stainless steel," *J. Mater. Sci., 49*, 4025–4031.
36. Tal-Gutelmacher, E., & Eliezer, D., (2005). "High fugacity hydrogen effects at room temperature in titanium-based alloys," *J. Alloys Compd., 404–406*, 613–616.
37. Silverstein, R., & Eliezer, D., (2017). "Mechanisms of hydrogen trapping in austenitic, duplex, and super martensitic stainless steels," *J. Alloys Compd., 720*, 451–459.
38. Rietveld, H. M., (1969). "A profile refinement method for nuclear and magnetic structures," *J. Appl. Crystallogr., 2*, 65–71.
39. McCusker, L. B., Von Dreele, R. B., Cox, D. E., Louër, D., & Scardi, P., (1999). "Rietveld refinement guidelines," *J. Appl. Crystallogr., 32*, 36–50.
40. Rozenak, P., & Eliezer, D., (1988). "Nature of the γ and γ* phases in austenitic stainless steels cathodically charged with hydrogen," *Metall. Trans. A, 19*, 2860–2862.

41. Eliezer, D., Chakrapani, D. G., Altstetter, C. J., & Pugh, E. N., (1979). "The influence of austenite stability on the hydrogen embrittlement and stress- corrosion cracking of stainless steel," *Metall. Trans. A, 10*, 935–941.

42. Minkovitz, E., Talianker, M., & Eliezer, D., (1981). "TEM investigation of hydrogen induced ε-hcp-martensite in 316L-type stainless steel," *J. Mater. Sci., 16*, 3506–3508.

43. Floreen, S., & Mihalisim, J. R., (1965). "High strength stainless steel by deformation in low temperature," In: Advances in the Technology of Stainless Steels and Related Alloys. *American Society for Testing Materials, 17*–25.

44. Silverstein, R., & Eliezer, D., (2017). "Effects of residual stresses on hydrogen trapping in duplex stainless steels," *Mater. Sci. Eng. A, 684*, 64–70.

45. Silverstein, R., Eliezer, D., & Tal-Gutelmacher, E., (2018). "Hydrogen trapping in alloys studied by thermal desorption spectrometry," *J. Alloys Compd. 747*, 511–522.

46. Robertson, I. M., Sofronis, P., Nagao, A., Martin, M. L., Wang, S., Gross, D. W., & Nygren, K. E., (2015). "Hydrogen embrittlement understood," *Metall. Mater. Trans. B, 46*, 1085–1103.

47. Silverstein, R., Glam, B., Eliezer, D., Moreno, D., & Eliezer, S., (2015). "The influence of inclusions and hydrogen on the microstructure and dynamic strength of materials," *AIP SCCM (Shock Compression Condens. Matter). 1793*, 110009.

48. Silverstein, R., Eliezer, D., Glam, B., Moreno, D., Eliezer, S., & Ben-Gurion, W. Q., (2014). "Dynamic strength of duplex Steel in the presence of hydrogen," *Steely Hydrog. Conf. Proc.*, 662–666.

49. Silverstein, R., & Eliezer, D., (2016). "Influences of hydrogen and textural anisotropy on the microstructure and mechanical properties of duplex stainless steel at high strain rate (~10^5 s^{-1})," *J. Mater. Sci., 51*, 10442–10451.

50. Silverstein, R., & Eliezer, D., (2016). "Hydrogen trapping energy levels and hydrogen diffusion at high and low strain rates (~10^5 s^{-1} and 10^{-7} s^{-1}) in lean duplex stainless steel," *Mater. Sci. Eng. A, 674*, 419–427.

51. Tarabay, A., Seaman, L., Curran, D. R., Kanel, G. I., Razorenov, S. V., & Utkin, A. V., (2003). *Spall Fracture* (1st edn.) New York: Springer. pp. 40–100.

52. Kanel, G. I., Razorenov, S. V., & Fortov, V. E., (2004). *Shock-Wave Phenomena and the Properties of Condensed Matter*. New York: Springer. pp. 100–200.

53. Silverstein, R., Eliezer, D., & Glam, B., (2017). "Hydrogen effect on duplex stainless steels at very High strain rates," *Energy Procedia, 107*, 199–204.

54. Turnbull, A., Hutchings, R. B., & Ferriss, D. H., (1997). "Modelling of thermal desorption of hydrogen from metals," *Mater. Sci. Eng., 238*, 317–328.

55. Sofronis, P., Dadfarnia, M., Novak, P., Yuan, R., Somerday, B., Robertson, I. M., Ritchie, R. O., Kanezaki, T., & Murakami, Y., (2009). "A combined applied mechanics/materials science approach toward quantifying the role of hydrogen on material degradation," In: *Proceedings of the 12th International Conference on Fracture (ICF–12)* (pp. 1–10).

56. Turnbull, A., (2015). "Perspectives on hydrogen uptake, diffusion, and trapping," *Int. J. Hydrogen Energy, 40*, 16961–16970.

57. Yamabe, J., Yoshikawa, M., Matsunaga, H., & Matsuoka, S., (2017). "Hydrogen trapping and fatigue crack growth property of low-carbon steel in hydrogen-gas environment," *Int. J. Fatigue, 102*, 202–213.

58. Kamilyan, M., Silverstein, R., & Eliezer, D., (2017). "Hydrogen trapping and hydrogen embrittlement of Mg alloys," *J. Mater. Sci., 52*, 11091–11100.

59. Eliezer, D., Tal-Gutelmacher, E., Cross, C. E., & Boellinghaus, T., (2006). "Hydrogen absorption and desorption in a duplex-annealed Ti–6Al–4V alloy during exposure to different hydrogen-containing environments," *Mater. Sci. Eng. A, 433,* 298–304.
60. Sobol, O., Straub, F., Wirth, T., Holzlechner, G., Boellinghaus, T., & Unger, W. E. S., (2016). "Real-time imaging of deuterium in a duplex stainless steel microstructure by time-of-flight SIMS," *Sci. Rep., 6,* 1–7.
61. Silverstein, R., Glam, B., Eliezer, D., Moreno, D., & Eliezer, S., (2018). "Dynamic deformation of hydrogen charged austenitic-ferritic steels: Hydrogen trapping mechanisms, and simulations," *J. Alloys Compd., 731,* 1238–1246.
62. Turnbull, A., Beylegaard, E., & Hutchings, R., (1994). "Hydrogen transport in SAF 2205 and SAF 2507 duplex stainless steels," In: *Hydrogen Transport and Cracking in Metals* (pp. 268–279).

PART IV
Constitutional Systems for Medicine

CHAPTER 24

The Influence of PEG on Morphology of Polyurethane Tissue Scaffold

J. KUCINSKA-LIPKA, H. JANIK, A. SULOWSKA, A. PRZYBYTEK, and P. SZARLEJ

Gdansk University of Technology, Chemical Faculty, Polymer Technological Department, 11/12 Narutowicza Street, 80–232 Gdansk, Poland, E-mail: helena.janik@pg.edu.pl, juskucin@pg.edu.pl

ABSTRACT

In this study, polyurethanes (PU) were synthesized from oligomeric α,ω-dihydroxy(ethylene-butylene adipate), poly(ethylene glycol) (PEG), hexamethylene diisocyanate (HDI), 1,4-butanediol (BDO) as chain extender and stannous octoate as catalyst. PEG due to its hydrophilic character influences physical and chemical properties of PU. For testing were used PU having the following weight contents of PEG: 0%, 7%, and 14%. Porous scaffolds were prepared by the method of solvent casting and particulate leaching. The materials were subjected to microscopic analysis. Porosity and pore morphology were evaluated. Based on the results from experimental data, it was found that PEG influences the pore's morphology of the obtained scaffold. The scaffold containing 7% wt. of PEG showed optimal properties for tissue engineering.

24.1 INTRODUCTION

Polyurethanes (PU) have a number of properties, thanks to which they are currently the object of many scientific studies on bone scaffolding. Their most important advantage is biocompatibility [1]. They demonstrate good biological properties, allowing cell adhesion on the scaffold surface and bone cell proliferation [2]. Research on the use of PU in bone regenerative medicine is based in particular on the synthesis of materials with osteoconductive properties, which are responsible for promoting bone formation to an extent that exceeds the rate of biodegradation.

Many biomedical applications require PU to have the form of a three-dimensional structure (scaffolds). Particularly in tissue engineering, PU should have a highly porous structure (>90%) with open and interconnected pores. This structure allows the penetration of cells and nutrients into the scaffold, which allows the proliferation of osteoblasts and mesenchymal stem cells and the formation of new bone tissue. Optimal porosity and pore size depend on the purpose of the scaffolding. Scaffolds biodegradable products must not have cytotoxic activity. In addition, an appropriate degradation rate is required, allowing the gradual absorption of mechanical loads by the regenerating tissue.

Poly (ε-caprolactone) (PCL) and aliphatic diisocyanates such as: 1,6-hexamethylene diisocyanate (HDI), 4,4'-dicyclohexylmethane diisocyanate (HMDI) or 1,4-butanediisocyanate (BDI) (polyethylenes) are the most commonly used substrates for PU syntheses. Aromatic diisocyanates are not used due to the carcinogenic nature of biodegradation products [3]. Most PUs are synthesized using suitable catalysts, such as dibutyl tin dilaurate (DBTDL) or tin 2-ethylhexanoate, commonly referred to as tin octoate ($Sn(Oct)_2$). These compounds catalyze the reaction of isocyanate groups with hydroxyl groups. The innovative polyurethane systems are presented in Table 24.1. And differ primarily by used chain extenders and the presence of some compounds, in order to improve the physicochemical, mechanical, and osteoconductive properties.

TABLE 24.1 Innovative Polyurethane Systems for Application in Bone Tissue Engineering

No.	Polyurethane system	References
1	PCL | HDI | D-idosorbidol | DBTDL	Gorna I Gogolewski [3]
2	PCL+PEG | HDI | benzoic acid	Shokrolahi et al. [6]
3	GCO | IPDI | BDO | $Sn(Oct)_2$ | + HA	Du et al. [7]

24.1.1   SYSTEM 1: PCL/HDI/D-IDOSORBIDOL/DBTDL

Gorna and Gogolewski received a biodegradable porous polyurethane with enhanced cell affinity [4]. PCL, HDI were used for the synthesis. The biologically active 1,4:3,6-dianhydro-D-sorbitol (D-isosorbidol) was used as a chain-extender, as it is a bone-building agent [5]. The three-dimensional scaffold was obtained as a result of a combination of two methods:

salt leaching and phase separation in a solvent/non-solvent system. PU after the synthesis was dissolved in the following three solvents: dimethyl sulfoxide (DMSO), dimethylformamide (DMF), N-methyl-2-pyrrolidone (NMP). The 15 g polymers were dissolved in 105 mL of the appropriate solvent. The porogen agent was sodium hydrogen phosphate heptahydrate $Na_2HPO_4 \cdot 7H_2O$. The following non-solvents were used: acetone, isopropanol, ethanol, and tetrahydrofuran (THF). It has been shown that the best solvent for obtaining a polyurethane scaffold was DMF. This scaffold had a more open pore character compared to other solvents. Furthermore, the type of non-solvent added to the PU solution had a significant effect on the pore structure. DMF-THF system allows obtaining a porous scaffold with the best porosity. The effect of the size of the porous agent and its concentration on the formation of the scaffold was also determined. For this purpose, salt crystals with the following fraction sizes were used: 90–140, 140–300 and 300–400 μm. By selecting the optimum conditions for the polyurethane foam formation process, a scaffolding with a porosity of 90%, regular, and interconnected pores and high water permeability was obtained [3]. In further research on this system, it was also shown that the scaffolding based on D-isosorbidol is not cytotoxic [5]. The samples used in the measurements were in the form of a film with a thickness of 0.1 mm. Part of the film was kept together with the scaffold during the salt washing-in procedure in the water. They had a contact angle of 69.3°. However, the films being a control sample (not in common with scaffolds during leaching of salt) had a contact angle of 74.4°.

24.1.2   SYSTEM 2: PCL+PEG/HDI/BENZOIC ACID

Shokrolahi et al. [6] obtained PU from mixture of PCL and PEG as polyols, HDI, and benzoic acid as chain extender. Porous scaffolds were obtained using three different techniques: freeze-drying, phase separation in the solvent/nonsolvent system and modified method of leaching solid particles. The PU solutions at 5% wt. and 8% wt. in DMSO were used. The scaffolds obtained by freeze-drying were characterized by two small pore size, which would prevent proper in growth of osteoblasts. The use of phase separation technology also proved to be an ineffective method, as they obtained scaffolds had mechanical properties well below the allowable limits for scaffolds intended for bone regeneration. The best results were obtained using a method based on a combination of compression molding, heating, and leaching of solid particles. The best mechanical properties and morphology of the pores were characterized by scaffolding obtained by using a porogen agent with a NaCl/PEG ratio equal to 65/20 (porosity equal to 87.6%, compression strength equal to about 1 MPa, Young's modulus equal to approx. 3.5 MPa). At a later stage of work, they were used for the cell culture test. The interaction of G292 cells with the scaffold was investigated. Based on the analysis of SEM photos, it was shown that after 3 days of incubation the scaffolding surface was covered with cells. Moreover, the cells freely penetrated the scaffolding pores. In connection with the above, it was found that the scaffolding was characterized by an appropriate pore structure, enabling the in growth of osteoblast cells [6].

24.1.3   SYSTEM 3: GCO/IPDI/BDO/SN(OCT)$_2$/+ HA

Du et al. received a porous scaffold consisting of 40% of hydroxyapatite (HA) powder and aliphatic polyurethane [7]. The use of natural compounds in the production of PU, applicable in bone tissue engineering, is a new area of research in this field. For the synthesis of PU castor oil glycerides (GCO), isophorone diisocyanate (IPDI), BDO as a chain extender and Sn(Oct)$_2$ as a catalyst were used. The molar ratio of OH to NCO groups was 1.1:1. The porous structure was obtained using the gas foaming technique by adding distilled water to the prepolymer. The scaffolding had an average pore size of 500 μm and a compressive strength of 4.6 MPa. A study of rat osteoblast culture in vitro was carried out. Effective anchoring of cells on scaffolding and high affinity of osteoblast cells for scaffolding was observed after 7 days. It was also observed that after seven days on the scaffold there was a cell layer with a morphology characteristic for osteoblasts. Moreover, the cells were spread throughout the pores of the scaffold. The obtained scaffold allowed high cell adhesion and proliferation. The study also examined the effect of bone regeneration using this scaffold. For this purpose, in vivo studies were carried out. The scaffolds covered with osteoblast cells and uncovered with these cells were implanted in the defects of the femoral rats. No inflammation was observed during the experiment. The regenerated bone was created both on scaffolding covered with cells and on scaffolding without cells. The newly created bone appeared around the scaffolding and inside the pores. The scaffold covered with osteoblast was characterized by a larger amount of regenerated bone [7].

The aim of the work was to determine the effect of polyethylene glycol content on mechanical properties, and pore scaffold morphology for PU obtained from α,ω-dihydroxyoligo (ethylene-butylene adipate), poly (ethylene glycol), HDI, 1,4-butanediol, and tin 2-ethylhexanoate.

24.2 EXPERIMENTAL PART

24.2.1 MATERIALS

Oligomeric α,ω-dihydroxy(ethylene-butylene adipate) (dHEBA) was purchased from Purinova. Poly(ethylene glycol) (PEG), HDI and stannous octoate (Sn(Oct)$_2$) were purchased from Sigma Aldrich. 1,4-butanediol (BDO) was purchased from POCH S.A.

24.2.1.1 PU SYNTHESIS

The synthesis of PU was carried out using the prepolymer method [8, 9]. The isocyanate index for each PU was 0.9. Three PUs (PU–1, PU–2 and PU–3) differing in the PEG content (0%, 7%, 14% wt.) were examined.

24.2.1.2 SCAFFOLD PREPARATION

The porous scaffolds were obtained using the solvent casting/particulate leaching (SC/PL) method [10]. Solutions with a concentration of 20% wt. of PU in an organic solvent were prepared. To do this, the PU solids were ground to a fraction of about 5 mm, and then dissolved in DMSO at about 100°C. The mixture was poured into a 20 x 20 x 20 mm silicone mold. Subsequently, NaCl crystals were gradually added to the PU solution until the solution was saturated. The whole set was left for 5 days in a refrigerator at a temperature of about –20°C. After this time, the samples were taken out of the mold and placed in distilled water at room temperature. The water was changed several times to wash out the salt and remove the solvent from the scaffold. This stage lasted seven days. After finishing the salt leaching process, the samples were allowed to dry at room temperature for 2 weeks.

The porosity (P) of the PU scaffolds was calculated according to below equation, where ρ is the density of the polymer (g/cm^3), V is the volume of the sample (cm^3), and m is the mass of the sample (g):

$$P = \left(1 - \frac{m}{\rho \cdot V}\right) \cdot 100\%$$

24.2.1.3   MECHANICAL STRENGTH

The compressive strength of polyurethane scaffolds was tested using an Zwick/Roell materials testing machine (Model Z020). The sample dimensions were 20 mm x 20 mm x 20 mm, and the crosshead speed was 20 mm/min with 2.5 kN load of cell.

24.3   RESULTS AND DISCUSSION

In Figure 24.1, the total porosity of PU scaffolds is presented.

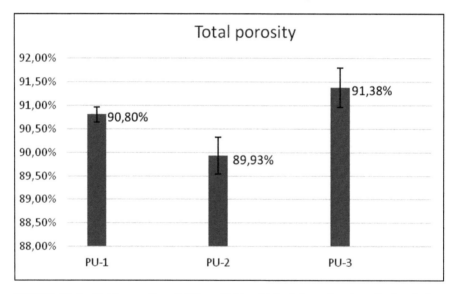

FIGURE 24.1   Total porosity of PU scaffolds obtained by solvent casting/particulate leaching (SC/PL) method.

24.3.1   MICROSCOPIC OBSERVATION OF PORE MORPHOLOGY

PU–1 (0% PEG) was characterized by the most regular distribution of pores in the range of 140–280 μm, with an average pore size of 0.195 μm The scaffold's pores had a cubic shape that corresponds to the shape of the sodium chloride crystals. PU–1 had mostly macropores with a closed structure. No micropores were observed.

PU–2 (7% PEG) was characterized by smaller pore structure. Cross sections of PU–2 scaffolds showed inhomogeneous pore spacing in some places. The inner part of the PU–2 scaffold had small amount of macropores. However, on the outside surface of the scaffolding were present regularly shaped pores.

The PU–3 scaffold (14% PEG) had cubic-shaped pores corresponding to the shape of the porogen agent. However, the pores were smaller in size compared to PU–1. On the PU–3 cross-section, macropores of relatively large size can be seen in the central part of the scaffold. The difference in pore sizes in the central and outer scaffolding may be due to the longer leaching time for sodium chloride crystals inside the scaffold.

Comparing the polymers with different PEG content, an increase in total porosity was observed with increasing PEG content. Scaffolds with 14% wt. PEG (PU–3) had the highest porosity (>90%). The scaffolds with a content of 7% and 14% wt. PEG exhibited very good total porosity, which meets the requirements for scaffold in bone tissue engineering [10].

The mechanical properties for scaffolds studied are shown in Figure 24.2.

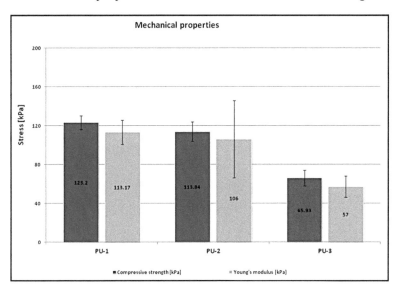

FIGURE 24.2   Mechanical properties of PU scaffolds obtained by solvent casting/particulate leaching (SC/PL) method.

The analysis of the above sets allows to conclude that the PEG content affects the mechanical properties of the obtained scaffolds. With increase of the PEG content, compressive strength and Young's modulus decrease. The scaffold without PEG has the highest compressive strength: 123.20 kPa (PU–1). The scaffold with 7% wt. PEG (PU–2) is characterized by poorer mechanical properties with respect to non-PEG scaffold. Increase in PEG content to 14% by weight caused significant deterioration of mechanical properties to 65.93 kPa for PU–3. Increase in PEG content with 7% wt. up to 14% wt. caused a relatively small change in the total porosity (by about 1%) with a simultaneous large decrease in mechanical properties (reduced compressive strength in the range from 40% to over 50%).

24.4   CONCLUSIONS

The use of the solvent casting method combined with particle leaching (SC/PL) allows obtaining scaffolds with appropriate porosity. The structure of pores, total porosity, and mechanical properties depend on the content of polyethylene glycol. An increase in the polyethylene glycol content caused an increase in porosity. The scaffolds with 14% wt. PEG were characterized by total porosity of over 90%, which meets the requirements for scaffolds applicable in bone tissue engineering. These materials also exhibited an open pore structure. An increase in the polyethylene glycol content caused a deterioration of mechanical properties. The same dependence was found in studies carried out by Gorna and Gogolewski [11]. Scaffolds with a content of 7% by weight poly (ethylene glycol) had a much better compressive strength compared to scaffolding with a content of 14% by weight, while having a total porosity at the 90% limit. For this reason, the scaffolding with a content of 7% wt. of PEG was chosen for farther in vitro studies.

ACKNOWLEDGMENTS

The paper was partly granted by Gdansk University of Technology (DS 2013–2015). Michalina Marzec and is acknowledged for a technical help in the synthesis of PU.

KEYWORDS

- 1,4-butanediol
- hexamethylene diisocyanate (HDI)
- poly(ethylene glycol) (PEG)
- polyurethane
- α,ω-dihydroxy(ethylene-butylene adipate)

REFERENCES

1. Affrossman, S., Barbenel, J. C., Forbes, C. D., MacAllister, J. M. R., Jin, M., Pethrick, R. A., & Scott, R. A., (1991). Surface structure and biocompatibility of polyurethanes, *Clinical Materials 8*, 25–31.
2. Hofmann, A., Ritz, U., Verrier, S., Eglin, D., Alini, M., Fuchs, S., Kirkpatrick, C. J., & Rommensa, P. M., (2008). The effect of human osteoblasts on proliferation and neo-vessel formation of human umbilical vein endothelial cells in a long-term 3D co-culture on polyurethane scaffolds, *Biomaterials, 29*, 4217–422634.
3. Marchant, R. E., Zhao, Q., Anderson, J. M., & Hiltner, A., (1987). Degradation of a poly(ether urethane urea) elastomer: Infrared and XPS studies, *Polymer, 28*, 2032–2039.
4. Gorna, K., & Gogolewski, S., (2006). Biodegradable porous polyurethane scaffolds for tissue repair and regeneration, *J. Biomed. Mater. Res., A., 79*(1), 128–138.
5. Gogolewski, S., Gorna, K., Zaczynska, E., & Czarny, A., (2008). Structure-property relations and cytotoxicity of isosorbide-based biodegradable polyurethane scaffolds for tissue repair and regeneration, *J. Biomed. Mater. Res., A., 85*(2), 456–465.
6. Shokrolahi, S., Mirxadeh, H., Yeganeh, H., & Daliri, M., (2011). Fabrication of poly(urethane-urea)-based scaffolds for bone tissue engineering by a combined strategy of using compression molding and particulate leaching methods, *Iranian Polymer Journal, 20*(8), 645–658.
7. Du, J., Zou, Q., Zuo, Y., & Li, Y., (2014). Cytocompatibility and osteogenesis evaluation of HA/GCPU composite as scaffolds for bone tissue engineering, *International Journal of Surgery, 12*, 404–407.
8. Kucinska-Lipka, J., Marzec, M., Gubanska, I., & Janik, H., (2017). Porosity and swelling properties of novel polyurethane–ascorbic acid scaffolds prepared by different procedures for potential use in bone tissue engineering. *J. of Elastomers & Plastics, 49*(5), 440–456.
9. Kucinska-Lipka, J., Gubanska, I., & Janik, H., (2013). Gelatin-modified polyurethanes for soft tissue scaffold. *Sci. Word. J.:* 450132. doi: 10.1155/2013/450132.
10. Janik, H., & Marzec, M., (2008). A review: Fabrication of porous polyurethane scaffolds. *Materials Science & Engineering C-Materials for Biological Applications, 48*, 586–5918.
11. Gorna, K., & Gogolewski, S., (2003). Preparation, degradation, and calcification of biodegradable polyurethane foams for bone graft substitutes, *J. Biomed. Mater. Res., A., 67*(3), 813–827.

CHAPTER 25

Fractal Dimensions Analysis and Morphological Investigation of Nanomedicine by Machine-Learning Methods

M. GHAMAMI[1] and S. GHAMMAMY[2]

[1]*Department of Mechanical Engineering, Isfahan University of Technology, Isfahan, Iran, E-mail: mghamami@yahoo.com*

[2]*Faculty of Science, Chemistry Department, Imam Khomeini International University, Qazvin, Iran*

ABSTRACT

Nanomedicine is the medical application of nanotechnology. Medications can be more efficiently delivered to the site of action using nanotechnology. Nanotechnology can also reduce the frequency with which we have to take our medications, resulting in improved outcomes with less medication. On a nanoscale of medication, the self-similarity behavior of particles can be seen. Self-similarity is one of the prominent properties in the field of fractal science. The fractal geometry is one of the branches of mathematics, which has shown flexibility and unique ability to interpret and simulate different forms of nature. Using fractal geometry, a clear horizon was devoted for mathematicians and researchers to explaining the behavior of functions, chaotic, and apparently uneven collections. Analysis of fractal dimensions by using of image processing method is a powerful tool for obtaining the morphological information of nanomedicine. Using the Scanning Electron Microscope (SEM) images were provided for the morphological studies of nanomedicine. The fractal dimension was calculated using the image processing technique, fractal dimension and other image properties were analyzed. The goal of this study is to prove the fractal behavior of considered nanomedicine, and some of the hypothesis in the field of measurement of fractal dimension have been investigated. The results show that nanomedicine have a fractalian behavior.

25.1 INTRODUCTION

Recently, researchers have come to the conclusion that some of the properties of medication, such as drug delivery capability, as well as the degree of their stability in the body, can be justified and interpreted using fractal dimension. Fractals can be seen in many natural phenomena, and they can explain and predict many natural phenomena [1–4].

In this research, the morphological properties of SEM images has been investigated by machine learning and image processing technique that calculates fractal dimension. Statistics has been applied for extracting the data from the SEM image and obtained the statistical results such as fractal dimension. The statistical properties will help in identification and improving the performance of nanomedicine properties [5–9].

The fractal dimension is a significant factor that can be used to approximate the surface roughness and the texture segmentation. In this research, first, by using of planetary ball mill drugs that produced by the pharmaceutical companies was prepared at the nanoscale, then by using of scanning electron microscope (SEM), images for morphological studies of the nanomedicine was provided. Using the MATLAB software, linear equation will be extracted for nanomedicine. This equation will guide us to the fractal dimension estimation. At the end, fractal dimension has been calculated

The primary goal of this study is to prove the fractal behavior of nanomedicine. Finally, some of the hypothesis in the field of measurement of fractal dimension were investigated. In some articles, fractal dimension is used as a tool for the classification of tumors, while various tests in this research showed that the fractal dimensions are correlated with the magnification of the images, the binary degree defined in software and other factors. In pathologies laboratory, as well as in the creation of a database on the fractal dimensions of each substance, it is imperative that all test conditions be standardized, since the error value is reached to the minimum.

In this study, two types of medication (Carbidopa and Piracetam) have been considered to study their properties. These medications are used in patients with Alzheimer and Parkinson. Because the way to prevent such diseases is not known, medications are of great importance. In the next section, firstly, the diseases are described and then the medications are introduced.

25.1.1   ALZHEIMER

Alzheimer's disease is a progressive disease that destroys memory and other important mental functions. At first, someone with Alzheimer's disease may notice mild confusion and difficulty remembering. Eventually, people with the disease may even forget important people in their lives and undergo dramatic personality changes. Figure 25.1(a), represent a cross-section of the brain as seen from the front. The cross-section on the left represents a normal brain, and the one on the right represents a brain with Alzheimer's disease. Changes caused by the disease has been displayed. For more information refer to Ref. [2].

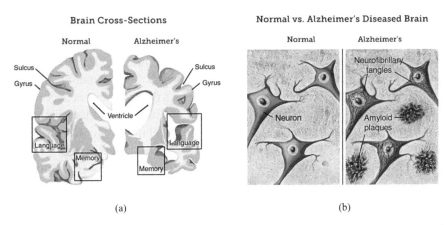

FIGURE 25.1   (See color insert.) Alzheimer disease effect (a) [10], Amyloid plaques and neurofibrillary tangles (b) [10].

In Alzheimer's disease, the brain cells degenerate and die, causing a steady decline in memory and mental function. Current Alzheimer's disease medications and management strategies may temporarily improve symptoms. This can sometimes help people with Alzheimer's disease maximize function and maintain independence for a little while longer. But there's no cure for Alzheimer's disease.

Increasing age is the greatest known risk factor for Alzheimer's disease. Alzheimer's disease is not a part of normal aging, but your risk increases greatly after you reach age 65. The rate of dementia doubles every decade after age 60. In some cases, people with rare genetic changes linked to early-onset Alzheimer's disease begin experiencing symptoms as early as their 30s. The likelihood of having Alzheimer's disease increases substantially after the age of 70 and may affect around 50% of persons over the age of 85. Nonetheless, Alzheimer's disease is not a normal part of aging and is not something that inevitably happens in later life. For example, many people live to over 100 years of age and never develop Alzheimer's disease [11].

Alzheimer's disease is an irreversible, progressive brain disorder that slowly destroys memory and thinking skills and, eventually, the ability to carry out the simplest tasks. It is the most common cause of dementia in older adults. While dementia is more common as people grow older.

25.1.2 PARKINSON

Parkinson's disease is a progressive disorder of the nervous system that affects movement. It develops gradually, sometimes starting with a barely noticeable tremor in just one hand. But while a tremor may be the most well-known sign of Parkinson's disease, the disorder also commonly causes stiffness or slowing of movement. In Parkinson's disease, certain nerve cells (neurons) in the brain gradually break down or die. Many of the symptoms are due to a loss of neurons that produce a chemical messenger in your brain called dopamine. When dopamine levels decrease, it causes abnormal brain activity, leading to signs of Parkinson's disease. Figure 25.2 (a), represent a side-section of the brain. The side-section on the left picture shows the involved region of brain in Parkinson's disease, and the right side picture represents a brain neurons with defective transfer of dopamine in a person with Parkinson's disease.

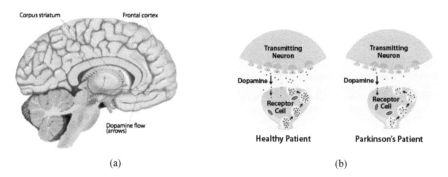

FIGURE 25.2 (See color insert.) Involved region in Parkinson's disease (a) [12], dopamine transfer in brain neurons (b) [11].

Although Parkinson's disease can't be cured, medications may markedly improve the symptoms. In occasional cases, doctor may suggest surgery to regulate certain regions of brain and improve the symptoms. Parkinson's disease symptoms and signs may vary from person to person. Early signs may be mild and may go unnoticed. Symptoms often begin on one side of your body and usually remain worse on that side, even after symptoms begin to affect both sides. Parkinson's signs and symptoms may include: Tremor, slowed movement (bradykinesia), rigid muscles, impaired posture and balance, loss of automatic movements, speech or writing changes [11].

Parkinson's disease is often accompanied by these additional problems, which may be treatable: Thinking difficulties, depression, and emotional changes, swallowing problems, sleep problems and sleep disorders, bladder problems and constipation. Parkinson's disease typically occurs in people over the age of 60, of which about one percent is affected. Males are more often affected than females. When it is seen in people before the age of 50, it is called young-onset PD. The average life expectancy following diagnosis is between 7 and 14 years [11].

Scientists sometimes refer to Parkinson's disease as a synucleinopathy (due to an abnormal accumulation of alpha-synuclein protein in the brain) to distinguish it from other neurodegenerative diseases, such as Alzheimer's disease where the brain accumulates tau protein [13]. Considerable clinical and pathological overlap exists between tauopathies and synucleinopathies. In contrast to Parkinson's disease, Alzheimer's disease presents most commonly with memory loss, and the cardinal signs of Parkinson's disease (slowness, tremor, stiffness, and postural instability) are not normal features of Alzheimer's. In the next section, the common medicines are introduced for above diseases.

25.1.3   CARBIDOPA

Carbidopa is a drug given to people with Parkinson's disease in order to inhibit peripheral metabolism of levodopa. This property is significant in that it allows a greater proportion of peripheral levodopa to cross the blood-brain barrier for central nervous system effect. Figure 25.3 shows the Carbidopa's molecule.

Carbidopa, an inhibitor of aromatic amino acid decarboxylation, is a white, crystalline compound, slightly soluble in water, with a molecular weight of 244.3. It increases the plasma half-life of levodopa from 50 minutes to 1½ hours.

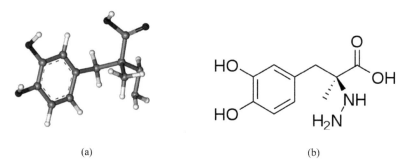

(a)   (b)

FIGURE 25.3   Carbidopa molecule, ball, and stick model (a) [11], Carbidopa molecule structure (b) [11].

25.1.4   PIRACETAM

Since its discovery and usefulness detection, piracetam has been used as a nootropic agent for the enhancement of memory in human beings with Alzheimer's disease. Piracetam is an exceptional nootropic with low contagiousness and insolubility levels and incurs a small number of side effects. Some sources suggest that piracetam overall effect on lowering depression and anxiety is higher than on improving memory [14]. However, depression is reported to be an occasional adverse effect of piracetam. It has been found to increase blood flow and oxygen consumption in parts of the brain, but this may be a side effect of increased brain activity rather than a primary effect or mechanism of action for the medication [15]. Furthermore, it is also used for the reduction of symptoms associated with anxiety, withdrawal from alcohol, and clinical depression. Figure 25.4 shows the piracetam molecule.

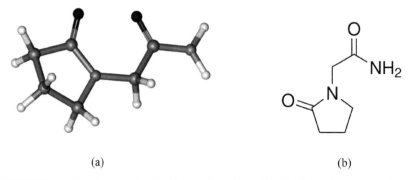

(a)   (b)

FIGURE 25.4   Piracetam molecule, ball, and stick model (a) [11], piracetam molecule structure (b) [11].

From recent and past clinical studies and research, Piracetam has been determined to support memory and aid individuals experiencing memory loss and difficulties in the ability to retain knowledge. Piracetam may enhance, elevate, and improve cognitive functions and abilities linked and associated to the central nervous system, memory development, and memory processes. Piracetam may benefit with senile dementia, dyslexia, alcoholism, and withdrawal syndrome, acute ischemic stroke, circulatory disorders, heart disease, Raynaud's, Alzheimer disease, cortical myoclonus, obsessive-compulsive disorders (OCD), and traumatic brain injury (TBI).

As mentioned before, described medications have been studied on a nanoscale. Materials reduced to the nanoscale can show different properties compared to what they exhibit on a macroscale, enabling unique applications. For instance, opaque substances can become transparent (copper), stable materials can turn combustible (aluminum), insoluble materials may become soluble (gold). A material such as gold, which is chemically inert at normal scales, can serve as a potent chemical catalyst at nanoscales. Much of the fascination with nanotechnology stems from these quantum and surface phenomena that matter exhibits at the nanoscale. In order to better understand, it is necessary to become more familiar with the nanofield. A branch of science that is studying on nanoscale is nanotechnology, and the next section explains more about it.

25.2 NANOTECHNOLOGY

The concepts that seeded nanotechnology were first discussed in 1959 by renowned physicist Richard Feynman. Nanotechnology is manipulation of matter on an atomic, molecular, and supramolecular scale. A more generalized description of nanotechnology was subsequently established, which defines nanotechnology as the manipulation of matter with at least one dimension sized from 1 to 100 nanometers. This definition reflects the fact that quantum mechanical effects are important at this quantum-realm scale, and so the definition shifted from a particular technological goal to a research category inclusive of all types of research and technologies that deal with the special properties of matter which occur below the given size threshold. It is therefore common to see the plural form "nanotechnologies" as well as "nanoscale technologies" to refer to the broad range of research and applications whose common trait is size.

Nanotechnology may be able to create many new materials and devices with a vast range of applications, such as in nanomedicine, nanoelectronics, biomaterials energy production, and consumer products. On the other hand, nanotechnology raises many of the same issues as any new technology, including concerns about the toxicity and environmental impact of nanomaterials. [9]

In the 1980s, a major breakthrough sparked the growth of nanotechnology in modern era. The invention of the scanning tunneling microscope in 1981 which provided unprecedented visualization of individual atoms and bonds, and was successfully used to manipulate individual atoms in 1989.

Nanotechnology is the engineering of functional systems at the molecular scale. This covers both current work and concepts that are more advanced. In its original sense, nanotechnology refers to the projected ability to construct items from the bottom up, using techniques and tools being developed today to make complete, high-performance products. The upper limit is more or less arbitrary but is around the size below which phenomena not observed in larger structures start to become apparent and can be made use of in the nanodevice [16]. Figure 25.5 illustrates the scale of nanotechnology and introduce the Famous structures in this field.

Quantum effects can become significant when the nanometer size range is reached, typically at distances of 100 nanometers or less, the so-called quantum realm. Additionally, a number of physical (mechanical, electrical, optical, etc.) properties change when compared to macroscopic systems. One example is the increase in surface area to volume ratio altering mechanical, thermal, and catalytic properties of materials. The catalytic activity of nanomaterials also opens potential risks in their interaction with biomaterials. Given the presence of quantum effects and self-similarity property in nanoscale materials, these properties need to be further investigated. In the next section, a field of science is introduced that deals with these properties.

25.3 FRACTALS

A branch of mathematical science dealing with endless patterns is fractal science. A fractal is a never-ending pattern. Fractals are infinitely complex patterns that are self-similar across different scales. They are created by repeating a simple process over and over in an ongoing feedback loop. Geometrically, they exist in between our familiar dimensions. Fractal patterns are extremely familiar, since nature is full of fractals [18–21]. For instance: trees, rivers, architecture, coastlines, mountains, clouds, galaxies, seashells, hurricanes, and etc. Figure 25.6 shows examples of fractal patterns in nature.

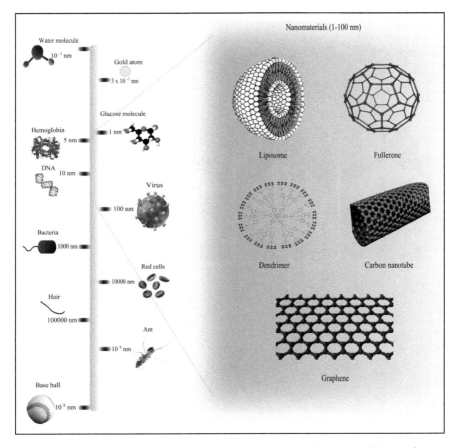

FIGURE 25.5 **(See color insert.)** Comparison of the sizes of nanomaterials and famous structures in nanotechnology [17]. Reprinted from Ref. [17]. Open access article distributed under the Creative Commons Attribution License (CC BY 3.0).

Abstract fractals such as the Mandelbrot Set can be generated by a computer calculating a simple equation over and over. Figure 25.7 shows examples of fractal patterns in mathematics world.

An important (defining) property of a fractal is self-similarity, which refers to an infinite nesting of structure on all scales. Strict self-similarity refers to a characteristic of a form exhibited when a substructure resembles a superstructure in the same form. But if a fractal's one-dimensional lengths are all doubled, the spatial content of the fractal scales by a power. This power is called the fractal dimension of the fractal, and it usually exceeds the fractal's topological dimension [22]. To understand the fractal behavior we need to understand the fractal dimension.

FIGURE 25.6 (See color insert.) Fractal pattern in leaves of plants or trees (a), snowflakes (b), cauliflower (c), romanesco broccoli (d), Architectural patterns (e), historical buildings (f) [11].

25.4 FRACTAL DIMENSION

In mathematics, a fractal dimension is a ratio providing a statistical index of complexity comparing how detail in a pattern (a fractal pattern) changes with the scale at which it is measured. In fractal field, as the measurement unit is scaled smaller and smaller, the amount of the measurement increases. It has also been characterized as a measure of the space-filling capacity of a pattern that tells how a fractal scales differently from the space it is embedded in. A fractal dimension does not have to be an integer [23–25].

Unlike topological dimensions, the fractal index can take non-integer values [26], indicating that a set fills its space qualitatively and quantitatively differently from how an ordinary geometrical set does [23–25]. For instance, a curve with fractal dimension very near to 1, say 1.10, behaves quite like an ordinary line, but a curve with fractal dimension 1.9 winds convolutedly through space very nearly like a surface. Similarly, a surface with fractal dimension of 2.1 fills space very much like an ordinary surface, but one with a fractal dimension of 2.9 folds and flows to fill space rather nearly like a volume [27]. This general relationship can be seen in the two images of fractal curves in Figure 25.8 (a), the 32-segment quadratic fractal has a fractal dimension of 1.67, compared to the perceptibly less complex Koch curve (b), which has a fractal dimension of 1.26.

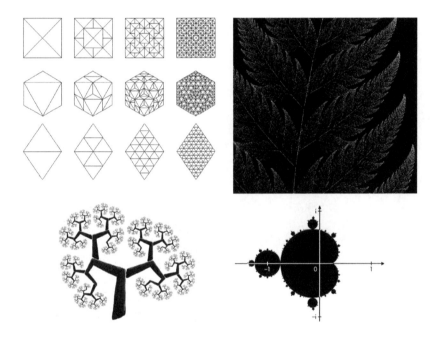

FIGURE 25.7   Examples of fractal patterns in mathematics world.

Fractal complexity may not always be resolvable into easily grasped units of detail and scale without complex analytic methods, but it is still quantifiable through fractal dimensions [28].

In practice, fractal dimensions can be determined using techniques that approximate scaling and detail from limits estimated from regression lines over log vs. log plots of size vs. scale. Several formal mathematical definitions of different types of fractal dimension are listed below. Although for some classic fractals all these dimensions coincide, in general, they are not equivalent:

- Box-counting dimension: D is estimated as the exponent of a power law.
- Information dimension: D considers how the average information needed to identify an occupied box scales with box size, p is a probability.
- Correlation dimension D is based on M as the number of points used to generate a representation of a fractal and g_ε, the number of pairs of points closer than ε to each other.
- Generalized or Rényi dimensions.

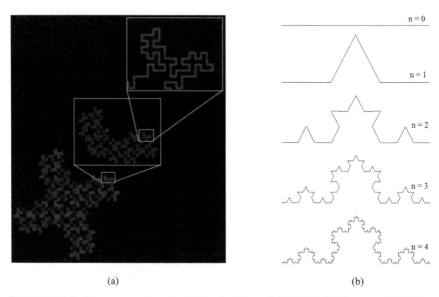

(a)   (b)

FIGURE 25.8 Piracetam molecule, ball, and stick model (a) [11], Koch curve (b) [11].

In this research, box-counting dimension method has been used. To calculate the box-counting dimension, the picture is placed on a grid. The x-axis of the grid is r where r = 1/(width of the grid). For example, if the grid is 240 blocks tall by 120 blocks wide, r = 1/120. Then, count the number of blocks that the picture touches. Label this number N, then, resize the grid and repeat the process. Plot the values found on a graph where the x-axis is the log (r), and the y-axis is the log (N). Draw in the line of best fit and find the slope. The box-counting dimension measure is equal to the slope of that line. Figure 25.9 (a) shows application of Box-counting dimension method on Koch curve and Figure 25.9 (b) illustrate linear diagram of fractal dimension for Koch curve.

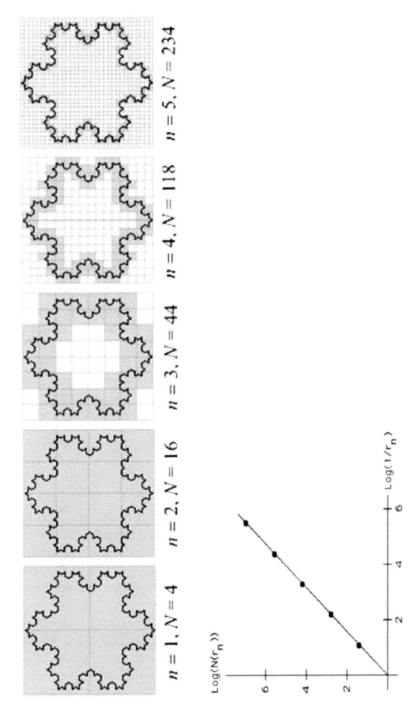

FIGURE 25.9   Box-counting dimension method on Koch curve (a), linear diagram of Koch curve (b) [11].

The Box-counting dimension is much more widely used than the self-similarity dimension since the box-counting dimension can measure pictures that are not self-similar (and most real-life applications are not self-similar). The box-counting method is analogous to the perimeter measuring method. By shrinking the size of the grid repeatedly, we end up more accurately capturing the structure of the pattern.

Using the box-counting method, fractal dimension is again the slope of the line when we plot the value of log (N) on the Y-axis against the value of log (r) on the X-axis. The same equation is used to define the fractal dimension, D. As mentioned, N is the number of boxes that cover the pattern, and r is the magnification, or the inverse of the box size.

$$D = \frac{\log (N)}{\log (r)}$$

The slope of the line equals the dimension, and it is defined as the amount of change along the Y-axis, divided by the amount of change along the X-axis. Slopes (and dimensions) range between 1 and 2 for this kind of analysis, which corresponds to the range between a line that is straight (dimension = 1) and a line that is so wiggly it completely fills up a 2-dimensional plane. A steeper slope means that the object is more "fractally," that is, it gains in complexity as the box size decreases. A flatter, lower-valued slope means that the object is closer to a straight line, less "fractally," and that the amount of detail does not grow as quickly with increasing magnification.

25.5 EXPERIMENTAL

By reducing the size of medicine particle and study of nanoscale materials, different properties become visible that is depends on particle size. Scanning electron microscopy is used to view material particle size. The resulting images can be defined as RGB image. Each RGB image can be defined as a three-dimensional matrix. Note that a RGB image does not necessarily colorful, but can be colorful.

In recent studies on the processing of digital images, edge identification techniques are used to analyze and extract photo attributes, and it's used in applications such as image segmentation, object identification, object restoration. Edge identification methods use localized gray level data to determine the points that are edges. For example, the gradient operator, the

Laplacian operator, and other operators are used to distinguish gray levels at high computational speeds. These algorithms are only sensitive to changes in gray levels, and their performance depends on factors such as the quality of light and noise. In recent years, attention has been focused on tissue properties, and tissue features are used to segment images and categorize images, identify patterns, and etc. In the segmentation of the texture, first, the extraction of the characteristics of the feature done and then these features are classified. There are common methods for extracting the texture of a picture that can be divided into four groups:

- statistical methods;
- geometric methods;
- model-based methods; and
- signal processing methods.

Images in MATLAB are defined as two, three, or four-dimensional matrices. Some types of images are as follows: Indexed images, Binary images, Intensity images, RGB images. Figure 25.10 show an example of a RGB and an indexed image structure.

An RGB or true color image is an image that three numbers are stored per pixel of it in the computer memory, the numbers are between 0 and 255. These numbers represent the intensity of each of the red, green, and blue colors. Therefore, for each pixel of the image, more than 16 million (256 * 256 * 256) different color mode will be possible. Obviously, a RGB image will occupy computer memory three times more than of an intensity image in the same size and at the same proportion requires more processing time.

Each RGB image is defined as a three-dimensional matrix. In the third dimension of matrix, the values of the color of each point (Red, Green, and Blue) are stored. The Figure 25.10 (b) illustrates this concept better. The brightness of each point of the indexed image was determined. By using of Machine learning method and image processing, fractal dimension computing is possible. The RGB numerical value can also be used to determine the bright points of the images. In all images, dark spots are displayed with numbers smaller than 0.5 and a bright spot with numbers larger than 0.5.

25.6 SAMPLES PREPARATION

A planetary ball mill was used to prepare samples because the medication particles shrink and its particle transform to the nanoscale. Particles on the

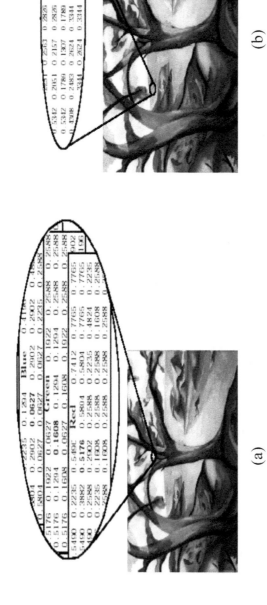

FIGURE 25.10   An example of RGB images (a), an example of an indexed image (b).

nanoscale are easily visible by scanning electron microscopy. One of the tools used in nanoscience is scanning electron microscopy [29–33]. With the electron bombardment, pictures of 10-nanoscale little objects for study. Table 25.1 presents some of the properties of the desired medicines.

TABLE 25.1 Formulation and Molar Mass of Amiodarone

	Carbidopa	Piracetam
Formulation	$C_{10}H_{14}N_2O_4$	$C_6H_{10}N_2O_2$
Molar mass	226.229 g/mol	142.16 g/mol
Melting point (ave)	203°C	152°C
Boiling Point (ave)	384°C	408°C
Density (25°C)	1.41 g/cm³	1.36 g/cm³
Water Solubility (ave)	1.95 mol/L	6.23e–01 mol/L
Surface Tension (ave)	74.1 dyn/cm	51.5 dyn/cm

In Figures 25.11 and 25.12, the SEM image of nanomedicines have been shown.

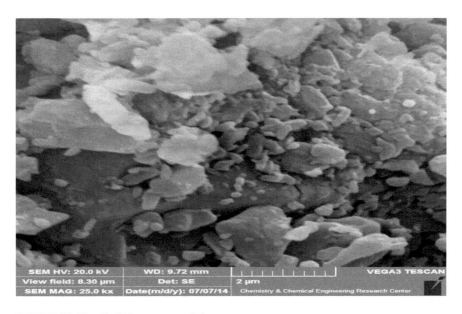

FIGURE 25.11 Carbidopa nanoparticles.

SEM HV: 20.0 kV	WD: 9.69 mm	⊥⊥⊥⊥⊥⊥⊥⊥⊥⊥⊥⊥	VEGA3 TESCAN
View field: 9.03 μm	Det: SE	2 μm	
SEM MAG: 23.0 kx	Date(m/d/y): 07/07/14	Chemistry & Chemical Engineering Research Center	

FIGURE 25.12 Piracetam nanoparticles. The result of box-counting dimension method for nanomedicine have been shown in Figure 25.13. The spatial value of any point on this chart obtained using Matlab software and enter it to the table. In this chart, the independent variable is X. This factor represents 1/r or the box reverse length. The dependent variable or Y is the number of boxes. Achieve the spatial value of these points shows the relationship between the length of the measurement and the number of boxes.

The software has used the same ruler length for all images, difference for each image is in the log (boxnum), which results in the fractal pattern found in each image. Images despite chaotic appearance, referring to these factors and charts, they show a certain standard and pattern.

The table data was analyzed using regression coefficient of linear equations. The slope of the line was calculated. Slope of the fitted line is equal to the fractal dimension.

$$\text{Log (boxnum)} = -1.5178 * \text{Log(res)} + 13.167 \Rightarrow$$
$$\text{FD (Fractal Dimension)} = 1.5178$$

Histograms plot show the symmetry and uniformity between image points. Histograms plots are used in order to determine the differences between image points, especially in nanocomposites field. The maximum and the minimum of numbers and the uncertainty of the data are obtained.

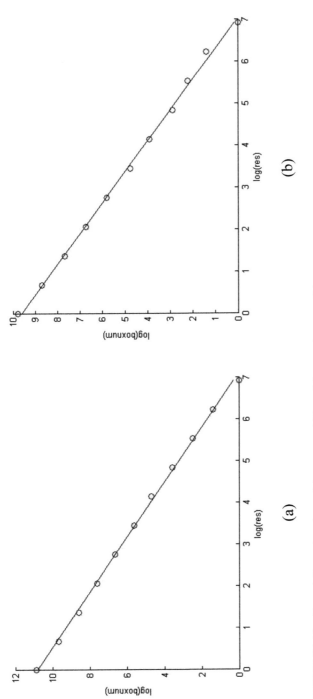

FIGURE 25.13 Fractal dimension chart of the nano–carbidopa (a), nanopiracetam (b).

In a statistical population, the normality of data can be achieved using p-p plot charts. The dispersion of data around a line is indicative of the normality of this statistical society. In order to achieve the dispersion of points in each image, histogram graph can be accessed by Matlab software.

Nano-Carbidopa FD = 1.5178, Nano-piracetam FD = 1.3648

25.7   CONCLUSIONS

In this research, we used and reported new methods for the purification, nanonization, and structural morphological investigations of nanomedicine. Also, the morphological properties of SEM-images nanomedicine has been investigated by an image processing technique that calculates fractal dimensions by using of a computer software. Statistics data extracted from the SEM image and obtained the statistics results such as fractal dimension and histogram plot. The statistical properties will help in identification and improving the performance of this medicine. The fractal dimension is a significant factor that can be used to approximate the surface roughness, the texture segmentation, and an image of the studied compounds. The fractal laws increasingly used in describing cancerous tumors structure, diagnosis of heart and bone disease, understanding, and explaining of the earth natural phenomena, control the scale and distribution of fluids in chemical engineering and many other phenomena in science.

Fractal dimensional analysis using image processing is a powerful tool to obtain nanomedicine morphological information. During the research, first, by using of planetary ball mill, medicines that produced by the pharmaceutical companies, was prepared at the nanoscale, then by using of SEM, images for morphological studies of the nanomedicine was provided. Using programs written in MATLAB software, images fractal dimension has been calculated, and histogram charts and other features of the images have been discussed. At the end, the correlation between independent and dependent variables on the fractal dimension graph will be studied. Linear equation will be extracted for nanomedicine.

In some articles, fractal dimension is used as a tool for the classification of tumors, while various tests in this research showed that the fractal dimensions are correlated with the magnification of the images, the binary degree defined in MATLAB software and other factors. Therefore, in pathologies laboratory, as well as in the creation of a database on the fractal dimensions of each substance, it is imperative that all test conditions be standardized. The amount of error in judgment based on the fractal dimension reach as little as possible.

ACKNOWLEDGMENTS

The support of the Materials Engineering Department of Imam Khomeini International University (IKIU) is gratefully acknowledged.

KEYWORDS

- **fractal dimension**
- **image processing**
- **machine learning**
- **nanomedicine**
- **scanning electron microscope**

REFERENCES

1. Esgiar. A. N., Naguib, R. N. G., Sharif, B. S., Bennett, M. K., & Murray, A., (2002). Fractal analysis in the detection of colonic cancer images, *IEEE Journal of Biomedical and Health Informatics, 6*(1), 54–58.
2. Schroeder, M., Freeman, W., & Fractals, H., (1991). *Chaos, Power Laws* (p. 39). New York Press.
3. Mandelbrot, B., (1983). *The Fractal Geometry of Nature* (1st edn). W. H. Freeman and Company.
4. Krasowska, M., Grzywna, Z. J., Mycielska, M. E., Djamgoz, M. B., & Eur Biophys, J., (2009). Fractal analysis and ionic dependence of endocytotic membrane activity of human breast cancer cells, *European Biophysics Journal, 38*(8), 1115–1125.
5. Gonzalez, R. C., & Woods, R. E., (2002). *Digital Image Processing* (2nd edn), Prentice-Hall Inc.
6. Jacobs, I. S., & Bean, C. P., (1963). In: Rado, G. T., & Suhl, H., (eds.), *Fine Particles, Thin Films, and Exchange Anisotropy, in Magnetism* (Vol. III, pp. 271–350), New York: Academic.
7. Ng, W. W., Panu, U. S., & Lennox, W. C., (2007). Chaos-based analytical techniques for daily extreme hydrological observations. *Journal of Hydrology, 342,* 17–41.
8. Murr, L., (1984). *Electron and Ion Microscopy and Microanalysis,* McGraw-Hill, New York.
9. Goldstein, J., & Yakowitz, H., (1975). *Practical Scanning Electron Microscopy: Electron and Ion Microprobe Analysis*, Plenum Press, New York.
10. https://www.brightfocus.org/alzheimers (accessed on 11 January 2019).
11. https://en.wikipedia.org (accessed on 11 January 2019).
12. https://futurism.com/were-about-to-enter-a-new-era-in-parkinsons-disease-treatments (accessed on 11 January 2019).

13. Galpern, W. R., & Lang, A. E., (2006). Interface between tauopathies and synucleinopathies: A tale of two proteins, *Annals of Neurology, 59*(3), 449–58.
14. Malykh, A. G., & Sadaie, M. R., (2010). Piracetam and piracetam-like drugs: From basic science to novel clinical applications to CNS disorders, *Drugs, 70*(3), 287–312.
15. Jordaan, B., Oliver, D. W., Dormehl, I. C., & Hugo, N., (1996). Cerebral blood flow effects of piracetam, pentifylline, and nicotinic acid in the baboon model compared with the known effect of acetazolamide, *Arzneimittel-Forschung, 46*(9), 844–7.
16. Allhoff, F., Lin, P., & Moore, D., (2010). *What is Nanotechnology and Why Does it Matter?: From Science to Ethics* (pp. 3–5). John Wiley and Sons Press.
17. Panneerselvam, S., & Choi, S., (2014). Nanoinformatics: Emerging databases and available tools, *International Journal of Molecular Sciences, 15*, 7158–7182.
18. Tan, C. O., Cohen, M. A., Eckberg, D. L., & Taylor, A. J., (2009). Fractal properties of human heart period variability: Physiological and methodological implications, *The Journal of Physiology, 587*(15).
19. Liu, J. Z., Zhang, L. D., & Yue, G. H., (2003). Fractal dimension in human cerebellum measured by magnetic resonance imaging, *Biophysical Journal, 85*(6), 4041–4046.
20. Karperien, A. L., Jelinek, H. F., & Buchan, A. M., (2008). Box-counting analysis of microglia form in schizophrenia, Alzheimer's disease and affective disorder, *Fractals, 16*(2).
21. Jelinek, H. F., Karperien, A., Cornforth, D., Cesar, R., & Leandro, J. G., (2002). MicroMod-an L-systems approach to neural modeling, *Workshop Proceedings: The Sixth Australia-Japan Joint Workshop on Intelligent and Evolutionary Systems*. University of New South Wales.
22. Audrey, B. K., (2004). *Defining Microglial Morphology: Form, Function, and Fractal Dimension* (p. 86). Charles Sturt University Press.
23. Falconer, K., (2003). *Fractal Geometry* (p. 308). New York, Wiley Press.
24. Sagan, H., (1994). *Space-Filling Curves* (p. 156). Berlin, Springer-Verlag Press.
25. Vicsek, T., (1992). *Fractal Growth Phenomena* (p. 10). World Scientific Press.
26. Sharifi-Viand, A., Mahjani, M. G., & Jafarian, M., (2012). Investigation of anomalous diffusion and multifractal dimensions in polypyrrole film, *Journal of Electroanalytical Chemistry, 671*, 51–57.
27. Mandelbrot, B., (2004). *Fractals and Chaos* (p. 38), Springer Press.
28. Mandelbrot, B., (1983). *The Fractal Geometry of Nature*, Macmillan.
29. Reed, S., (1975). *Electron Microprobe Analysis*, Cambridge University Press.
30. Wells, O., Boyde, A., Lifshin, E., & Rezanowich, A., (1974). *Scanning Electron Microscopy*, McGraw-Hill Inc.
31. Russ, J., (1990). *Computer-Assistant Microscopy: The Measurement and Analysis of Images*, Springer Plenum Press.
32. Armstrong, M., & Dowd, P. A., (1993). *Geostatistical Simulations,* Kluwer Academic Publishers.
33. Oneil, M. J., (2001). *The Merck Index, an Encyclopedia of Chemicals, Drugs, and Biologicals* (13th edn., p. 1342). Whitehouse Station, Merck Co. Inc.

CHAPTER 26

Effect of Fullerene on Physicochemical Properties and Biological Activity of the Poly(Methyl Methacrylate)/Fullerene Composite Films

O. ALEKSEEVA and A. NOSKOV

G.A. Krestov Institute of Solution Chemistry, Russian Academy of Sciences, Akademicheskaya str., 1, Ivanovo, 153045, Russia, E-mail: avn@isc-ras.ru

ABSTRACT

Composite polymer materials are promising field of advanced material sciences with scope of using in biology, medicine, electronics, etc. Special attention is paid to fullerene-containing polymers which have unique features of both fullerenes and polymers. Modification by carbon nanoparticles results to occurrence of new properties of the polymer, for example, biological activity. Poly(methyl methacrylate) (PMMA) is one of the polymers capable to content nanocarbonic particles. In the present study, we report results of the tests that have been performed to compare physicochemical properties and biological activity of the PMMA films and PMMA/fullerene composite films. A solvent casting of perspective components (PMMA and fullerenes C_{60}) from solutions was employed for film fabrication. The structural characteristics of the PMMA films modified with fullerenes were researched by X-ray diffraction (XRD) technique. Thermal behavior of the PMMA/C_{60} composites was studied by the DSC and TG techniques. It was found the DSC curve mode to be depended on the composite composition. The thermograms of pure PMMA and PMMA/C_{60} composite have presented that these film materials decomposed in three stages. According to the UV and IR spectroscopy data, it was suggested a non-covalent interaction of the PMMA donor macromolecules with the fullerene acceptor molecule. To assess bioactivity of pure PMMA and filled composite with fullerene the laboratory tests were conducted in biologic fluids (blood serum). We studied the influence of polymer materials researched on free-radical processes in blood serum in vitro. Lipid peroxidation was evaluated by induced chemiluminescence. We supposed that fullerene-containing films can influence peroxidation processes. Antimicrobial activity of PMMA/fullerene composites was tested against *Escherichia coli* and *Candida albicans*. The test results showed absolute death of the microorganisms under the modified film.

26.1   INTRODUCTION

The development of polymeric composites with controllable structure and properties is one of priority lines of the modern chemistry and materials science. Harsh conditions of polymer operation in power engineering and in chemical, petroleum, and pulp-and-paper industries impose stringent requirements upon the properties of the polymers (hardness, strength, and electrophysical parameters). A number of publications deal with various types of fillers for polymers and various methods of composites production [1–12]. Insertion of carbon fillers such as fullerenes, carbon nanotubes, graphene can lead to creation of materials with improved physical and chemical properties and the main service characteristics [13–15].

The discovery of soccer-ball-shaped Buckminsterfullerene in 1985 [16] was an exciting and unexpected discovery that established an entirely new branch of chemistry. Since a new direction of fullerene science has been developed, and the related studies in this direction have concerned the preparation of fullerene-containing polymers, which combine the unique characteristics of fullerene with the useful properties of matrix polymers [17].

There are two acknowledged procedures for the preparation of two different types of fullerene-containing polymers and related products. It is shown that fullerene doped polymeric materials can be produced by covalent bonding of fullerene molecules with polymeric circuits or as a result of formation of polymer-fullerene complexes due to donor-acceptor interactions [18]. Meanwhile, noncovalent interactions of polymer with fullerene, in opinion of authors of Ref. [19], can provide uniform distribution of nano-carbonic particles in a polymeric matrix. It's suggested that significant importance in the non-covalent linkage of fullerenes in a composite is connected with aromatic rings presence in the structure of a macromolecule [19] which increases fullerene-connecting ability of polymers. Moreover, owing to their structural features, fullerenes are capable of multipoint noncovalent interactions. This phenomenon is of special significance for interaction of fullerene molecules with numerous electron-donor macromolecular fragments [15].

Scientific interest in the doping of polymers with fullerene is likely related to the simplicity of fullerene incorporation, either in its native form or as solutions in organic solvents, and to the use of minor amounts of modifying agents. Furthermore, less significant changes in the electron structure and, hence, less dramatic changes in the characteristics of fullerene molecules in the absence of any covalent interaction between fullerene molecules and polymer chain fragments were expected. These modifiers, when introduced in small amounts (up to several percent), can be nucleating agents and can affect the degree of crystallinity of the polymer [20].

Modification by carbon nanoparticles results to occurrence of new properties of the polymer, for example, biological activity. Biological activity of fullerenes are due to, firstly, lipophilic properties, so that they can penetrate into the cell membrane, secondly, electron deficit, promoting to react with free radicals, and, thirdly, capacity of excited C_{60} to generate active oxygen species [21].

However, biological activity of polymer/fullerene composites studied not enough. It can be assumed that the insertion of fullerenes into a polymer matrix will result in creation of biocomposites, which may be used as agents for drug delivery, antiseptic preparations [21].

Among the high-molecular compounds, which are widely used as matrices for fullerene composites one can mark out poly(methyl methacrylate) (PMMA). The main advantages of this polymer are its accessibility, sustainability to environmental impact, ease, low cost. PMMA is a transparent thermoplastic material that allows it to be successfully used in many industries: aviation, instrumentation, electronics, building sector, food industry, as well as in the production of medical supplies [22–25].

But the conditions of processing and use in these areas impose certain requirements on polymer thermal characteristics, such as glass transition temperature, temperature limits of thermal destruction, and thermal oxidative degradation. PMMA is a polymer with low glass transition temperature and heat resistance that is substantially inferior to that of many other materials. Its thermal stability may be increased by doping with inorganic fillers.

PMMA is well-known film-forming polymer often used for different modifications with low molecular compounds of special properties, including fullerenes. PMMA is well dissolved in benzene, toluene, o-xylene which are also solvents for fullerenes. It is widespread procedure of polymer-fullerene composite formation that consists in preparation of the base-polymer solution and fullerene solution in the same organic solvent and mixing them with following evaporation of the solvent.

In the current chapter, we report on study of the structure, thermal, electrical properties, and biological activity of both PMMA films and PMMA/fullerene composite films. We also discuss on intermolecular interactions of PMMA and fullerene in polymer/C_{60} composite films. Findings have very important practical significance for materials science because PMMA used often for various modifications with low molecular compounds, including fullerenes.

26.2   EXPERIMENTAL

26.2.1   MATERIALS

PMMA purchased from Aldrich (US) with molecular mass of 120,000 was used as a polymer matrix. Fullerene C_{60} (NeoTechProduct Ltd, Russia) was used as filler agent. For fabrication of films, a solvent casting of perspective components from solutions was employed. Preliminary purification of organic solvent (toluene) was made by standard technique [26].

To produce the PMMA films, a polymer batch was dissolved in toluene (17 wt.% of PMMA), and the solution was stirred for about 1 day. After casting onto a glass substrate, the solvent was slowly evaporated at room temperature over several days until the thin film formation.

PMMA/C_{60} composition films were fabricated as follows. Fullerene batch was dissolved in toluene at required concentration. Then PMMA batch was dissolved in obtained solution (17 wt.% of PMMA), and the mixed solution was stirred for about 1 day before being cast into thin film. After casting the solvent was slowly evaporated at room temperature over several days to produce the composite film. By this technique, we prepared some of composites samples in the form of films that contain fullerene in the required concentrations.

The samples obtained have been examined by optical microscope "Boetius" (Germany). We found both pure PMMA films and PMMA/fullerene-composite films transparent, that is, the films are homogeneous on the optical level. Unmodified PMMA samples were colorless, whereas the PMMA/fullerene-composite films were light purple. The thickness of the films was in the range of 0.71~0.85 mm.

26.2.2   METHODS

26.2.2.1   X-RAY DIFFRACTION (XRD)

Cristal structure of both pure PMMA films and PMMA/C$_{60}$ composite films was evaluated by the X-ray diffraction (XRD) measurements on the base of Debye-Sherrer method. XRD patterns of film samples were obtained by X-ray diffractometer DRON-UM1 (Russia) equipped with MoK$_a$ radiation that monochromate by the Zr-filter, λ = 0.071 nm. X-ray diffractometer was modernized for substances in condensed and polycrystalline state. The voltage and the current of the X-ray tubes were 40 kV and 40 mA, respectively. A scan rate of 0.04 degrees was used. We investigated structure of pure PMMA and fullerene-containing composites by the XRD technique in wide angles from 2 to 40 degrees.

26.2.2.2   DIELECTRIC MEASUREMENTS

To determine the dielectric characteristics of composites (capacitance, *C*, and dielectric loss tangent, *tanδ*) we used Solartron 1255 frequency response analyzer (UK). The numerical values of the parameters were obtained at room temperature using a two-electrode cell with round clamp electrodes 19.8 mm in diameter. The dielectric constant (ε') of the film substance was calculated using the flat-plate capacitor formula:

$$C = \frac{\varepsilon' \varepsilon_0 S}{d}, \tag{1}$$

where *S* is the electrode area, *d* is the film thickness, and $\varepsilon_o = 8.854 \times 10^{-12}$ F·m^{-1} is the electric constant. The ac resistivity was determined by the relationship

$$\rho = (2\pi f \varepsilon' \varepsilon_0 \tan \delta)^{-1} \tag{2}$$

where *f* is the frequency of alternating current.

26.2.2.3   DIFFERENTIAL SCANNING CALORIMETRY (DSC)

DSC measurements were performed using DSC 204 F 1 apparatus (Netzsch, Germany) in argon atmosphere (15 cm^3·min^{-1}). For this purpose, a stack of films with a diameter of 4 mm and a mass of 4~5 mg was placed in a press-fitted aluminum crucible covered the pierced lid. The samples were undergone first heating up to 423 K with a scan rate of 10 K·min^{-1} to remove volatile substances from the polymer and cooled down to 283 K by means of liquid nitrogen. Second heating of the samples was carried out up to 423 K with a scan rate of 10 K·min⁻¹. The glass transition temperatures were determined from data of the second heating. The reference aluminum crucible was empty. All measurements were performed relative to the baseline obtained with two empty crucibles. Three measurements required for each composite (baseline, sample, and standard) were carried out on the same day.

26.2.2.4   THERMAL GRAVIMETRY

The *thermogravimetric analysis was* performed by the TG 209 F1 thermal analyzer (Netzsch, Germany) using platinum crucibles in argon atmosphere (30 ml/min). The samples of pure PMMA and PMMA/C_{60} composite films with a mass of 3~5 mg were heated from 298 to 773 K at a heating rate of 10 K/min, and weight loss was measured during test. The accuracy in sample mass measurement was 1×10^{-6} g, and accuracy in temperature measurement was 0.1 K. Five measurements were performed for each film sample and granule. The results were represented as mean values ± standard deviations.

26.2.2.5   ULTRAVIOLET (UV) AND INFRARED (IR) SPECTROSCOPY

Optical absorption spectra for fullerene in toluene solution were recorded in the wavelength range 280–800 nm in quartz cell (size is equal to 1.0 cm) by the Spectrophotometer U–2001 (HITACHI, Japan) with a working range of 190–1100 nm. A deuterium lamp was as light source for UV zone; photometric accuracy is ±0.002 Abs. Optical absorption spectra of PMMA and the PMMA/C_{60} composite films were recorded by the Spectrophotometer U–2001 (HITACHI, Japan) with a reflection attachment SMART MULTI-BOUNCE HATR, crystal is ZnSe 45°.

Infrared spectra (IR) spectra of films were recorded by Avatar 360 FT-IR ESP spectrometer (Thermo Nicolet, US) in the wave number range of 4000–400 cm⁻¹.

26.2.2.6   CHEMILUMINESCENCE (CHL) ANALYSIS

Induced chemiluminescence was used to evaluate effect of PMMA/C_{60} film composites on free-radical oxidation of lipids in biologic fluid (blood serum). For this purpose, the ChL tests were performed by BChL–07 luminometer

(Medozons, Russia). Subject of research was native blood serum of 10 patients managed in V.N. Gorodkov Research Institute of Maternity and Childhood (Ivanovo, Russia). Specimen of pure PMMA or composite film containing 3 wt.% of fullerene was put into test tube with blood serum (1 ml). System was incubated for 1 hour at 277 K. Then, the film specimen was removed from tube. The parameters of lipid peroxidation in serum after exposure of the film nanomaterials were determined by chemiluminescent analysis. Hydrogen peroxide and ferric sulfate have been used as inductors of ChL. 0.1 mL of serum, 0.4 mL of phosphate buffer (pH 7.5), 0.4 mL of 0.01M ferric sulfate and 0.2 mL of 2% hydrogen peroxide were put into cuvette. Luminescence was registered for 40 s.

26.2.2.7 MICROBIOLOGICAL TESTS

Antibacterial activity of the PMMA films and PMMA/C_{60} composite films was screened against gram-negative bacteria *Escherichia coli 1257* that were purchased from All-Russian collection of microorganisms – VKM (Moscow, Russia).

Study of bacteriostatic effect was performed by the agar diffusion method (Kirby-Bauer method) that was as follows. We have inoculated the aqueous suspension bacterial spores into the nutrient medium. Concentration of bacterial spores was equal to 1000·CFU/ml. Endo agar was used as a nutrient medium. Most gram-negative organisms grow well in this medium, while growth of gram-positive organisms is inhibited.

The bacterial mixture was thoroughly blended, poured into Petri dishes, and dried in the desiccator for 20 min. Then we put the test film specimens on the mixture surface and placed the Petri dishes into the thermostat at a temperature of 310 K for 24 hours. By the end of test, we took the samples and assessed stability against bacteria.

The pure PMMA films and the PMMA films containing fullerene were screened for their antifungal activity against *Candida albicans* (NCTC 885–653). It is an opportunistic fungus which is a form of *yeast*. Clinical strains were purchased from Russian collection of pathogenic fungi (St. Petersburg, Russia).

An antifungal potency of fabricated films was estimated using the agar diffusion method too. For this, we inoculated *Candida* into the nutrient medium (Sabouraud medium). Concentration of fungi was equal to 1000·CFU/ml. Then the mixture was thoroughly blended, poured into Petri dishes, and dried in the desiccator for 20 min. Thereafter we put the test PMMA and PMMA/ composites specimens on the mixture surface and placed the Petri dishes into the thermostat at a temperature of 310 K for 24 hours. By the end of test, we took the samples and assessed stability against fungi.

26.3   RESULTS AND DISCUSSION

26.3.1   CRYSTAL STRUCTURE OF THE PMMA/FULLERENE COMPOSITE FILMS

XRD patterns of both pure PMMA and PMMA/C_{60} composite films with 3 wt.% of C_{60} are given in Figure 26.1.

It can be seen that diffraction pattern does not change under insertion of fullerene into the polymer matrix. Both pure PMMA and PMMA/C_{60} composite films exhibit broad diffraction peak (halo). Its location does not depend on the film composition. Abscissa of maximum, $2\theta_m$, is equal to 11–12 degrees.

XRD of solid C_{60} was reported by Krätschmer et al. [27]. The authors found that the strongest fullerene reflex (111) appeared near 11 degree. That is, it overlaps with the PMMA halo. Therefore the reflex associated with fullerene is absent in the XRD pattern of composite film (Figure 26.1). Apparently, this is due to the concentration of C_{60} in the composite film under study is inadequate for the occurrence of reflex.

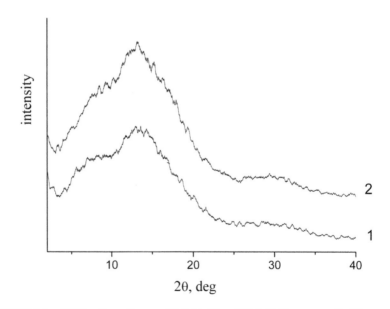

FIGURE 26.1   XRD patterns for pure PMMA (1) and PMMA/C_{60} composite (2).

This finding correlates to the data of Ref. [28], in which there is a comparison of wide-angle diffraction patterns of pure PMMA and PMMA/ C_{60} composites containing 1 and 10 wt.% of filler. It is found in ref. [28] that diffraction pattern does not change under insertion of C_{60} (1 wt.%) into the polymer matrix, but in diffraction pattern of PMMA/C_{60} composite with 10 wt.% of filler, there are additional peaks associated with fullerene aggregates.

26.3.2 DIELECTRIC PARAMETERS OF THE PMMA/FULLERENE COMPOSITE FILMS

Figure 26.2 shows the frequency dependences of the capacitance that are recorded for samples of pure PMMA film and PMMA/C_{60} composite film. It can be seen for both samples under study, capacity monotonically decreases with increasing frequency up to 10 Hz (for pure polymer) and up to 1000 Hz (for composite). In case of modified polymer, $C(f)$ curve is above than the curve for pure PMMA.

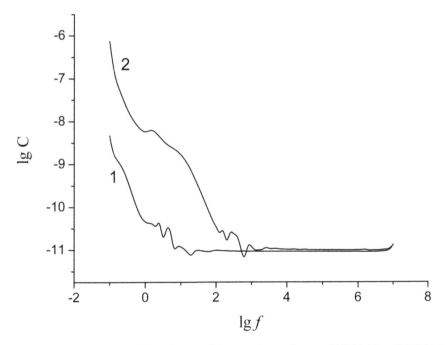

FIGURE 26.2   Frequency dependences of the capacitance for pure PMMA (1) and PMMA/ C_{60} composite with 3 wt.% of filler (2).

With further increase in the frequency, the curves practically coincide, and the capacity of samples does not change. Such frequency dependence of the capacitance is characteristic of the polar dielectrics which includes PMMA.

Table 26.1 shows the dielectric constant values calculated from the capacitance at a frequency of 1000 Hz using Eq. (1). For unmodified polymer, the ε value is equal to 2.5; that is, close to the literature data. Modification of PMMA with fullerene results to increase in the dielectric constant to 3.97.

TABLE 26.1 Dielectric Parameters of the PMMA/C$_{60}$ Composite Films as Functions of C$_{60}$ Content

C$_{60}$ content in film, wt.%	tanδ	ε	$\rho \times 10^{-8}$, $\cdot$Ohm$\cdot$m
0	0.02	2.50	3.14
3	0.28	3.97	0.16

Also, Table 26.1 shows the values of dielectric loss tangent measured at 1000 Hz. It can be seen that doping of PMMA with fullerene results to increase in the tanδ value of more than 10 times. The specific resistance, ρ, calculated using Eq. (2) decreases with growth in the fullerene concentration that also indicates the conductivity increases.

In materials with higher conductivity, reduction of static charge is more probable. Consequently, doping PMMA with fullerenes prevents the static electricity accumulation in polymer.

26.3.3 SECOND-ORDER PHASE TRANSITIONS IN PMMA/C$_{60}$ COMPOSITES

Differential scanning calorimetry (DSC) was used to research the phase transitions from the glassy state to elastic one in PMMA/C$_{60}$ composites. Phase transition was characterized by the following parameters:

T_1 is the extrapolated temperature of the phase transition onset;
T_2 is the extrapolated temperature of the phase transition end;
T_g is the temperature of DSC curve inflection taken as the glass transition point.

Figure 26.3 shows the DSC curves for the pure polymer film and composite filled with 3 wt.% of fullerene. It can be is seen that for pure PMMA there is a reversible phase transition from the glassy state to elastic one, which manifests itself as a step of heat flow in endothermic direction. This is the second-order phase transition.

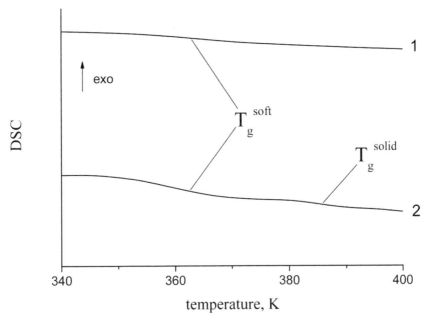

FIGURE 26.3 DSC curves of the PMMA/C$_{60}$ composite films with various filler concentration, wt.%: a) 0 (1); 3 (2).

But the thermal behavior of composite containing 3 wt.% of C$_{60}$ is more complex. Namely, two steps are observed in DSC curve (Figure 26.3, curve 2). According to [29], the thermograms of this type are characteristic for filled polymers with two glass transition temperatures, which correspond to transitions in the "soft phase" and "hard phase." Glass transition temperature of "solid phase," T_g^{solid}, is usually above the glass transition of temperature "soft phase," T_g^{soft}, which is associated with lower mobility segments. For PMMA/C$_{60}$ composites examined in this study, the difference is about 20 K (Figure 26.3).

Table 26.2 shows the average values of the characteristic temperatures of phase transition for the "soft phase" obtained from the thermograms analysis of the materials studied. It can be seen for composite containing 3 wt.% of C$_{60}$, the glass transition temperature of "soft phase" exceeds the value of T_g^{soft} for pure PMMA indicating deceleration of polymer chains mobility due to interaction of them with carbon nanoparticles [30].

TABLE 26.2   Parameters of Phase Transition for "Soft Phase" in Pure PMMA and PMMA/ Fullerene Composite Film

C_{60} content in film, wt.%	T_1, K	T_g^{soft}, K	T_2, K
0.0	350.3	362.6	374.3
3	354.7	363.5	370.4

As noted above, for composites film containing 3 wt.% of fullerenes, two steps are observed in the DSC curve (Figure 26.3, curve 2) which correspond to phase transitions in "soft phase" and "solid phase." Occurrence of the second glass transition temperature, T_g^{solid}, was also found in Refs. [31, 32], dedicated to the study of the thermomechanical properties of some polymer composites. To explain this effect, an idea concerning an interfacial layer on the surface modifier particles was developed. In this layer, the mobility of the polymer chains is reduced. Furthermore, it was shown in Refs. [31, 33] that the such features of thermal behavior are characteristic only for composites with nanometer-sized filler particles (7 nm) and are not observed in the case of micron particles (<44 microns), as well as unmodified polymers. Thus, occurrence of the second glass transition temperature for the composites containing C_{60} nanoparticles is consistent with the findings of other researchers.

26.3.4   THERMAL STABILITY OF PMMA/C_{60} COMPOSITES

Figures 26.4(a, b) show the TG and DTG curves for pure PMMA and PMMA film filled with 3 wt.% of fullerene. It can be seen there are three steps of degradation. The first step is observed in the temperature range from 414 K to 485 K. Probably, this mass loss (10~12 wt.%) is caused by the solvent evaporation out of the film.

The second and third stages in the TG and DTG curves for all film samples under study (Figures 26.4(a, b)) are attributed to the actual degradation of substance. The second step of degradation (about 540~580 K) is due to depolymerization initiated at unsaturated chain ends. In the third stage (about 630~700 K), the weight loss is associated with random scission of the backbone chains. The temperature ranges corresponding to each of the destruction stages and the temperatures corresponding to the DTG peaks $T_{1,2,3}^m$ are listed in Table 26.3.

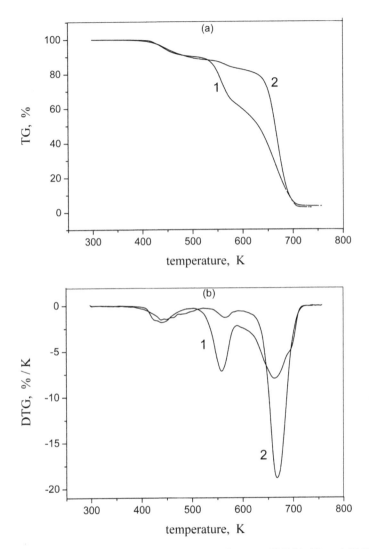

FIGURE 26.4 TG (a) and DTG (b) thermograms for pure PMMA (1) and PMMA/C$_{60}$ composite with 3 wt.% of fullerene (2).

TABLE 26.3 DTG Peaks and Temperature Ranges of the Separate Destruction Stages for the Samples Under Study

C$_{60}$ content, wt.%	First stage		Second stage		Third stage	
	Range, K	T$_1^m$, K	Range, K	T$_2^m$ K	Range, K	T$_3^m$ K
0	416.7~463.6	440.6±2.0	540.4~571.9	558.6±2.1	635.9~696.5	662.5±0.9
3	414.9~484.7	439.51.5	551.9~577.1	565.8±1.5	649.1~689.1	667.3±2.1

One can be seen in Figure 26.4(a, b) that for composite film containing 3 wt.% of C_{60} the TG and DTG curves are slightly shifted towards higher temperatures in comparison with pure PMMA. It indicates the improvement in thermal stability of polymer due to fullerene loading.

Table 26.4 shows the mass losses of samples in each of the destruction stages as the functions of film composition. These values are denoted as Δm_1, Δm_2, and Δm_3, respectively. It can be seen that the fullerene content almost does not affect the weight loss of the sample in the first stage caused by removal of solvent out of it. But C_{60} dramatically influences the actual degradation of polymer (the second and third stages). Pure PMMA film loses about 30% of its mass under heating in the temperature range of 540~580 K (the second stage of destruction), whereas insertion of filler leads to inhibition of decomposition under like conditions. So PMMA/C_{60} composites lose the major portion (80~90%) of mass at higher temperatures (630~700 K) that are assigned to the third stage of destruction.

TABLE 26.4   Effect of the Fullerene Content on the Mass Losses of Samples in the Separate Destruction Stages

C_{60} content, wt.%	Δm_1, %	Δm_2, %	Δm_3, %
0	10.0±0.5	29.3±1.2	58.0±0.9
3	11.9±1.1	5.4±1.0	79.3±2.4

Authors of Refs. [34–35] have reported on acceptor property of fullerene. So we can assume that fullerene effectively captures the active macroradicals and allyl radicals (-$COOCH_3$) formed at chain ends scission during the second stage of thermal degradation. Hence fullerene retards the process of polymer thermal degradation in the temperature range of 540~580 K thereby prominently increasing Δm_3 magnitude. Therefore for PMMA/C_{60} composites, Δm_3 values are much higher than that of Δm_2 whereas for pure polymer these values are comparable (Table 26.2). So, the introduction of fullerenes into PMMA leads to change in the mass loss distribution over the stages of destruction.

26.3.5   INTERMOLECULAR INTERACTIONS IN THE PMMA/C_{60} COMPOSITE FILMS

The interaction of fullerene with PMMA was studied by UV and IR spectroscopy.

The electronic absorption spectrum of the diluted solution of fullerene in toluene in the range of 250–750 nm displays two sets of absorption bands denoted as ρ (λ = 334 and 407 nm) and β bands (λ = 553 and 595 nm) (Figure 26.5). The absorption intensity of these bands regularly grows with a decrease in wavelength. The intense absorption band is observed at λ = 334 nm corresponding to allowed π-π* electronic transitions. In the visible part of the spectrum, there are weak bands in region of 550–600 nm. The latter bands are attributed to the forbidden n−π* excited electronic transitions.

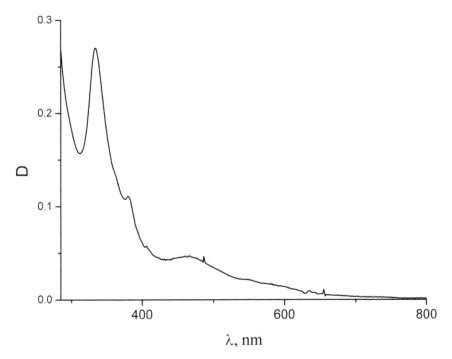

FIGURE 26.5   Electronic absorption spectrum of C$_{60}$ in toluene.

In the 250–300 nm region, the UV spectrum of the pure PMMA film (Figure 26.6, spectrum 1) contains a wide band, probably due to the *n-π* * transition in the carbonyl group of polymer [36]. It should be noted that the PMMA has not absorption bands in Vis region of spectrum. The UV spectrum of the PMMA/C$_{60}$ composite film in the 250–300 nm region (Figure 26.6, spectra 2–5) is a set of overlapping absorption bands of the polymer carbonyl group and the fullerene absorption band at λ = 260 nm.

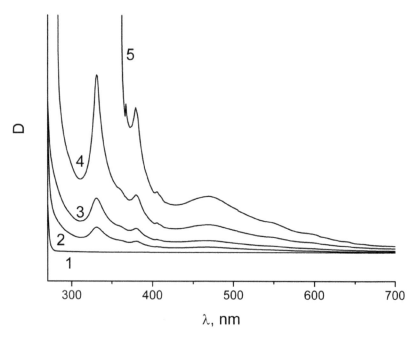

FIGURE 26.6 Electronic absorption spectra of pure poly(methyl methacrylate) (1) and PMMA/C$_{60}$ composite films with various contents of fullerene, wt.%: 0.05 (2), 0.1 (3), 0.5 (4), 1 (5).

Figure 26.6 shows that in the electronic spectrum of PMMA/C$_{60}$ composite film, compared to the PMMA spectrum, fullerene absorption bands appear at λ = 331, 405 and 469 nm. The maximum at λ = 331 nm is shifted by 3 nm to the short-wavelength region compared to the C$_{60}$ spectrum in toluene. Intensity of this bands increases with the C$_{60}$ concentration (Table 26.5).

Based on the analysis of the electronic spectra it can be assumed that in composites the intermolecular complexes form containing the electron-donor functional groups of PMMA (C = O, C-O-C) and π-electron system of the fullerene molecule.

Additional information about interaction of fullerene with PMMA was obtained from the analysis of the vibrational spectra of the pure polymer and PMMA modified with C$_{60}$. Figure 26.7 shows the IR spectra of pure PMMA and PMMA/fullerene composites containing 3 wt.% of filler. The IR spectrum of PMMA (Figure 26.7, spectrum *1*) indicates the details of functional groups. The broad peak at 3436 cm^{-1} and the sharp, intense peak at 1732 cm^{-1} appear due to the presence of carbonyl group (C = O) stretching vibration.

TABLE 26.5 Optical Properties of Fullerene in Various Systems

C_{60} in toluene solution		PMMA/C_{60} composites						
		0.05 wt.% of C_{60}		0.1 wt.% of C_{60}		0.5 wt.% of C_{60}		1 wt.% of C_{60}
λ, nm	D	λ, nm	D	λ, nm	D	λ, nm	D	Λ, nm	D
334	0.270	331	0.511	331	1.056	331	>3	331	>3
380	0.111	379	0.248	379	0.488	379	1.116	379	2.761
406	0.058	405	0.146	405	0.282	405	0.592	405	1.179
466	0.047	469	0.139	469	0.261	469	0.559	469	1.094

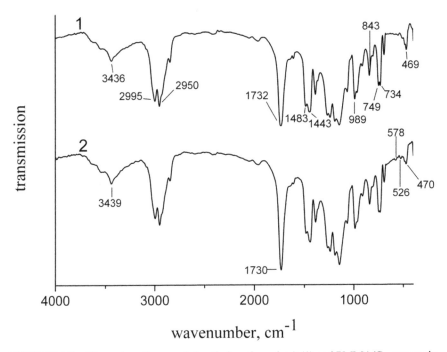

FIGURE 26.7 IR spectra of pure poly(methyl methacrylate) (1) and PMMA/C_{60} composite films with 3 wt.% of fullerene (2).

The band with two maxima at 2995 cm⁻¹ and 2950 cm⁻¹ refers to the stretching vibrations of C-H bond in the O-CH₃, CH₃ and CH groups. The deformation vibrations of CH₂ and CH₃ manifest band with two peaks (1443 and 1483 cm⁻¹). The broad peak ranging from 1260–1000 cm⁻¹ can be explained owing to the C-O (ester bond) stretching vibration. The broadband from 950–650 cm⁻¹ is due to the bending of C-H. The band at 989 cm⁻¹ belongs to the v (C-O-C), mixed with the γ (CH₃-O). The band at 843 cm⁻¹

corresponds to CH$_2$ rocking vibration bands. Doublet at 734 cm^{-1} and 749 cm^{-1} corresponds to (CH$_2$) vibrations mixed with (C-C). The band at 469 cm^{-1} belongs to the vibration v (C-O-C) group [37].

Figure 26.8 shows the IR spectrum of the fullerene molecule. It can be seen active four vibrations with absorption bands at 527, 578, 1180 and 1427 cm^{-1} [35].

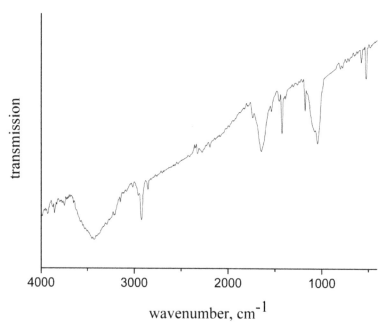

FIGURE 26.8   IR spectrum of the fullerene.

In the spectrum of PMMA/C$_{60}$ composite containing 3 wt.% C$_{60}$, there are significant changes in comparison with the spectrum of pure PMMA (Figure 26.7, spectrum 2). In the spectrum of the composite, the absorption band at 3436 cm^{-1} is shifted by 3 cm^{-1} to higher frequencies region and the band at 1732 cm^{-1} is shifted by 2 cm^{-1} to low frequencies region. It is noted the appearance of the new bands in the spectrum at 578 and 526 cm^{-1} for composites with 3 wt.% of C$_{60}$. This indicates the presence of C$_{60}$ in the composite in unbound form.

Taking into account the combination of changes in the electronic and vibrational spectra of PMMA/C$_{60}$ in comparison with pure PMMA, it can be assumed a non-covalent interaction of the PMMA donor macromolecules with the fullerene acceptor molecule.

26.3.6 ANTIOXIDANT ACTIVITY OF THE PMMA/C$_{60}$ COMPOSITE FILMS

It is considered that modification of polymer matrix with fullerene can produce biocomposites which have medical potential as drug transporters, antiseptics, and antioxidants. Today regulation of free-radical processes is adjusted by both natural and synthetic pharmaceutical compositions [38]. As any other medicine, some antioxidants may produce adverse events. So finding of safe preparations with high antioxidant activity is still actual.

In present study, we researched influence of PMMA/C$_{60}$ film composites on free-radical oxidation of lipids in biologic fluid (blood serum) in vitro. For this purpose, induced chemiluminescence was used. Kinetics of chemiluminescence in serum was researched after exposure of pure PMMA and composite films.

Induction of chemiluminescence by hydrogen peroxide and iron sulfate is based on Fenton reaction: at mixing the components, the catalytic decomposition of hydrogen peroxide takes place by divalent iron ions. Thus formed free radicals oxidize the lipoproteins of blood serum in the test samples, leading to the formation of new free radicals. At the recombination of radicals, unstable products are formed and decomposed with the release of photons.

Generally, a chain reaction in which formation of radicals leads to chemiluminescence may be represented by the scheme [39]:

$$\longrightarrow R\bullet \xrightarrow{\ k\ } P* \xrightarrow{\ k_e\ } P + photon$$

where $R\bullet$ – the free radicals; $P*$ – the radical decay products in excited state; P – ones in ground state.

The intensity of chemiluminescence, I, is proportional to the rate reaction, v:

$$I = \eta v, \tag{3}$$

where η – quantum efficiency of the chemiluminescent reaction.

For the process, the ChL curve dips because of antioxidant agents. The decay rate constant of free radicals, k, is defined by the dip rate of ChL curve. Therefore, the main indicator of the antioxidant activity of the system is tangent of maximum slope angle of ChL curve towards time axis, tanα.

Also, to estimate the intensity of lipid peroxidation, we used following parameters:

- I_m is maximum intensity of ChL during the experiment. Value of I_m quantifies the level of free radicals, i.e., gives an idea of the potential ability of the blood serum to free radical lipid peroxidation;
- A is an area covered by intensity curve or total light sum. Value of A is inversely proportional to the antioxidant activity of the sample;
- $A_n = A/I_m$ is normalized light sum. The value of A_n evaluates antioxidant activity more correctly than the A value because the total area covered by ChL curve depends on value of I_m.

The mean values of ChL indices in native serum without film adding were used as controls (100%). 6–8 measurements required for each film were carried out on the same day. The results have been expressed as percentages relative to controls and presented as values corresponding to the first, second (median) and third quartiles in the non-normal population. A p-value of 0.05 was chosen as the significance limit.

Table 26.6 shows the main parameters obtaining from analysis of ChL curves. Statistical processing of the results gave for pure PMMA film that the ChL indices were approximate to controls ($p > 0.05$). In case of PMMA/C_{60} composites for values of I_m, A, A_n, and tanα, there are significant differences compared to control because $p < 0.05$. This indicates a change in the concentration of free radicals and in antioxidant activity of system.

TABLE 26.6 ChL Indices in Blood Serum After Exposure of Pure PMMA Film and PMMA/Fullerene Composite Film[*]

ChL indices	PMMA	PMMA/C_{60}
I_m,%	(99, 106, 115)	(78, 82, 90)[**]
A,%	(100, 106, 119)	(74, 80, 83)[**]
A_n %	(96, 101, 104)	(94, 97,98)[**]
tan α,%	(94, 99, 117)	(82, 91, 92)[**]

[*]results have been presented as values corresponding to the first, second, and third quartiles in the distribution.

[**]significant differences compared to control ($p < 0.05$).

Thus, changing the nanocarbon particles concentration in the composite film allows specifically regulate processes of free radical oxidation of lipids in biological fluids.

We suggest the fullerene molecules can cross the external cellular membrane and they localize to the mitochondrions. Formation of oxygen free radicals occur due to the electrons leakage in the mitochondrial electron transport chain. Therefore, the localization of fullerenes near mitochondrions may contribute to their antioxidant action. Note similar assumption has been proven for fullerene derivative $C_{61}(CO_2H)_2$ [40]. The authors of Ref. [40] researched distribution of the [^{14}C] radioactivity in various cellular compartments and concluded $C_{61}(CO_2H)_2$ molecules localize preferentially to the mitochondrions.

26.3.7 ANTIMICROBIAL ACTION ACTIVITY OF THE PMMA/C_{60} COMPOSITE FILMS

It was noticed above that modification by nanoparticles can result in occurrence of new properties of polymer, for example, biological activity. Below we describe data of the tests that were performed to compare bacteriostatic effect and fungistatic effect of the pure PMMA films and PMMA filled with fullerene. For these purposes, spores of bacterial and fungal cultures were incubated under conditions that are optimal for their growth and development, and effect of the produced films was researched.

As a result of testing, it was found that the pure polymer does not suppress *Escherichia coli* growth whereas composites containing 1 and 3 wt.% of fullerene inhibit the growth of the test culture. Figure 26.9 shows the overall death of bacteria under the samples of fullerene-containing composites. Probably one of the reasons for the bacteria death is the adhesion of bacterial cells on the surface of the modified polymer. As a result, the cell membrane is damaged, the cell wall is destroyed which leads to its death. It should be noted that the dynamics of inactivation of *Escherichia coli* persists for a month.

Figure 26.10 shows the fungicidal activity of polymer materials under study against *Candida albicans*. It can be seen that the pure PMMA film does not suppress the growth of test culture, while film materials containing 1 and 3 wt.% of C_{60} inhibit the growth of them.

It can be assumed that the fungicidal effect of polymeric composites containing 3 wt.% of fullerene upon introduction of them into the nutrient medium is associated with the formation of active oxygen-containing radicals (HO^*; O_2^*), which oxidize proteins, nucleic acids, and lipids of the microorganism cell, that can lead to its death [41].

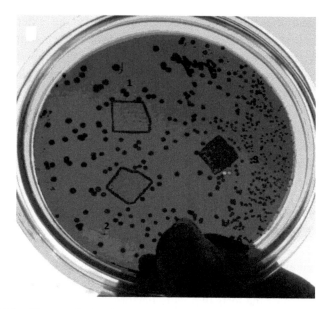

FIGURE 26.9 Photograph showing effect of film to development of *Escherichia coli* on nutrient medium (Endo agar): *1* – pure PMMA; *2* – PMMA/C$_{60}$ composite (1 wt.% of filler); *3* – PMMA/C$_{60}$ composite (3 wt.% of filler).

FIGURE 26.10 Photograph showing effect of film to development of *Candida albicans* on nutrient medium (Sabouraud medium): *1* – pure PMMA; *2* – PMMA/C$_{60}$ composite (1 wt.% of filler); *3* – PMMA/C$_{60}$ composite (3 wt.% of filler).

26.4 CONCLUSIONS

The experimental results presented in current study have shown that fullerene to be quite effective modifiers for PMMA. Many researchers concluded that an improvement in physical and chemical properties of PMMA/C_{60} composites compared to ones for pure polymer is related to structural changes under the effect of introduced fullerene. However, the mechanism of these improvements is still far from clear. Therefore the effect of fullerene doping on phase transition from the glassy state to elastic one, thermal stability, dielectric parameters, antioxidant, antimicrobial activity, etc., requires complex analysis. This study is very important, since the use of PMMA/fullerene composites will allow expansion of the applications borders of PMMA traditionally used in construction industry, home appliances, medicine, etc., and will allow them to be replaced with fullerene-containing materials.

KEYWORDS

- **biological activity**
- **DSC**
- **fullerene**
- **IR spectroscopy**
- **poly(methyl methacrylate)**
- **x-ray diffraction dielectric**

REFERENCES

1. Ayatollahi, M. R., Shadlou, S., & Shokrieh, M. M., (2011). *Compos. Struct., 93*(9), 2250.
2. Burnside, S. D., & Giannelis, E. P., (2000). *J. Polym. Sci. Polym. Phys., 38*(12), 1595.
3. Chen, K., Wilkie, C. A., & Vyazovkin, S., (2007). *J. Phys. Chem., 111*(44), 12685.
4. Dittrich, B., Wartig, K. A., Hofmann, D., Mülhaupt, R., & Schartel, B., (2013). *Polym. Degrad Stab., 98*(8), 1495.
5. Fonseca, M. A., Abreu, B., Gonçalves, F. A. M. M., Ferreira, A. G. M., Moreira, R. A. S., & Oliveira, M. S. A., (2013). *Compos. Struct., 99*, 105.
6. Hattab, Y., & Benharrats, N. A., (2015). *J. Chem., 8*(3), 285.
7. Lazzeri, A., Zebarjad, S. M., Pracella, M., Cavalier, K., & Rosa, R., (2005). *Polymer, 46*, 827.

8. Meenakshi, K. S., & Sudhan, E. P. J. A., (2016). *J. Chem., 9*(1), 79.

9. Alekseeva, O. V., Bagrovskaya, N. A., & Noskov, A. V., (2018). *Arab. J. Chem., 11*(7), 1160. http://dx.doi.org/10.1016/j.arabjc.09.008.

10. Alekseeva, O. V., Rodionova, A. N., Bagrovskaya, N. A., Agafonov, A. V., & Noskov, A. V., (2017). *Cellulose, 24*(4), 1825.

11. Alekseeva, O. V., Rodionova, A. N., Bagrovskaya, N. A., Agafonov, A. V., & Noskov, A. V., (2017). *Journal of Chemistry*, Article ID 1603937.

12. Alekseeva, O. V., Rudin, V. N., Melikhov, I. V., Bagrovskaya, N. A., Kuzmin, S. M., & Noskov, A. V., (2008). *Dokl. Phys. Chem., 422*(2), 275.

13. Uğur, Ş., Yargi, Ö., & Pekcan, Ö. C., (2010). *J. Chem., 88*(3), 267.

14. Rana, S., Bhattacharyya, A., Parveen, S., Fangueiro, R., Alagirusamy, R., & Joshi, M., (2013). *J. Polym. Res., 20*(12), Article ID 314.

15. Badamshina, E. R., & Gafurova, M. P., (2008). *Polym. Sci. B, 50*(7 & 8), 215.

16. Kroto, H. W., Heat, J. R., O'Brien, S. C., Curl, R. F., & Smalley, R. E., (1985). *Nature, 318*, 162.

17. Verner, R. F., & Benvegnu, C., (2012). *Handbook on Fullerene: Synthesis, Properties, and Applications* (p. 548). Nova Science Publishers Inc., New York.

18. Sibileva, M. A., Tarasova, E. V., & Matveyeva, N. I., (2004). *Rus. J. Phys. Chem. A, 78*(4), 526.

19. Krakovyak, M. G., Nekrasova, T. N., Arsenyeva, T. D., & Anufriyeva, E. V., (2002). *Polym. Sci. B, 44*(9 & 10), 271.

20. Potalitsin, M. G., Babenko, A. A., Alekhin, O. S., Alekseev, N. I., Arapov, O. V., Charykov, N. A., et al., (2006). *Rus. J. Appl. Chem., 79*(2), 306.

21. Da Ros, T., (2008). Twenty years of promises: Fullerene in medicinal chemistry. In: Cataldo, F., & Da Ros, T., (eds.), *Carbon Materials: Chemistry and Physics* (Vol. 1, p. 1). *Medicinal Chemistry and Pharmacological Potential of Fullerenes and Carbon Nanotubes*, Springer.

22. Arshad, M., Masud, K., Arif, M., Rehman, S., Arif, M., Zaidi, J. H., et al., (2009). *J. Therm. Anal. Calorim., 96*(3), 873.

23. Harper, C. A., (2000). *Modern Plastics Handbook* (p. 1298). McGraw-Hill, New York.

24. Andrade, C. K. Z., Matos, R. A. F., Oliveira, V. B., Duraes, J. A., & Sales, M. J. A., (2010). *J. Therm. Anal. Calorim., 99*(2), 539.

25. Thangamani, R., Chinnaswamy, T. V., Palanichamy, S., Bojja, S., & Charles, A. W., (2010). *J. Therm. Anal. Calorim., 100*(2), 651.

26. Coetzee, J. F., (1982). *Recommended Methods for Purification of Solvents and Tests for Impurities* (p. 59). Pergamon Press, Oxford.

27. Krätschmer, W., Lamb, L. D., Fostiropoulos, K., & Huffman, D. R., (1990). *Nature, 347*, 354.

28. Ginzburg, B. M., Pozdnyakov, A. O., Shepelevskij, A. A., Melenevskaya, E. Y., Novoselova, A. V., Shibaev, L. A., et al., (2004). *Polym. Sci. A, 46*(2), 169.

29. Tugov, L. I., & Kostrykina, G. L., (1989). *Khimiya i Fizika Polimerov (Polymer Chemistry and Physics)* (p. 432). Khimiya, Moscow, (in Russian).

30. Zanotto, A., Spinella, A., Nasillo, G., Caponetti, E., & Luyt, A. S., (2012). *Express Polym. Lett., 6*(5), 410.

31. Tsagaropoulos, G., & Eisenberg, A., (1995). *Macromolecules, 28*(1), 396.

32. Tsagaropoulos, G., & Eisenberg, A., (1995). *Macromolecules, 28*(18), 6067.

33. Robertson, C. G., & Rackaitis, M., (2011). *Macromolecules, 4*(5), 1177.

34. Krusic, P. J., Wasserman, E., Parkinson, B. A., Malone, B., Holler, E. R., Keizer, P. N., et al., (1991). *J. Am. Chem. Soc., 113*(16), 6274.
35. Konarev, D.V., (1999). *Russ. Chem. Rev., 68*(1), 19.
36. Kurmaz, S. V., & Ozhiganov, V. V., (2011). *Polym. Sci. A, 53*(3), 232.
37. Dechant, J., Danz, R., Kimmer, W., & Schmolke, R., (1972). *Ultrarotspektroskopische Untersuchungen an Polymeren* (p. 472). Akademie-Verlag, Berlin. (in German).
38. Okovitiy, S. V., (2003). *FARMindex: Praktik, 5*, 85 (in Russian).
39. Vladimirov, Yu. A., & Proskurnina, E. V., (2009). *Biochemistry (Moscow), 74*(13), 1545.
40. Foley, S., Crowley, C., Smaihi, M., Bonfils, C., Erlanger, B. F., Seta, P., et al., (2002). *Biochem. Biophys. Res. Commun., 294*(1), 116.
41. Diezmann, S., (2014). *Fungal. Biol. Rev., 28*(4), 126.

CHAPTER 27

Development of *In Vitro* Prostate Cancer Biomarkers on the Basis of Gelatin Matrix Incorporated Gold Nanoparticles Functionalized with Fluorescence Dye and Prostate Specific Membrane Antigen

K. CHUBINIDZE[1, 2], B. PARTSVANIA[2], A. KHUSKIVADZE[3], G. PETRIASHVILI[2], and M. CHUBINIDZE[3]

[1]*Tbilisi State University, 1 Ilia Chavchavadze Ave., Tbilisi 0179, Georgia*

[2]*Georgian Technical University, Institute of Cybernetics, S. Euli 5, Tbilisi 0186, Georgia*

[3]*Tbilisi State Medical University, 7 Mikheil Asatiani St, Tbilisi 0186, Georgia, E-mail: Chubinidzeketino@yahoo.com*

ABSTRACT

It's known that prostate-specific membrane antigen (PSMA) is one of the most well established and highly specific prostate epithelial cell membrane antigens. PSMA is a type II transmembrane zinc metallopeptidase, belonging to the M28 peptidase family. It possesses hydrolyzing enzyme activities and is also known as FOLH1 (foliate hydrolase 1). The combination of a nanoparticle platform with targeting ligands for tumor cell-surface biomarkers is a promising architecture for achieving selective drug delivery and uptake into target cells. Gold nanoparticles (GNPs) are prone to be attached to many biological probes such as antibodies, enzymes, lectins, glycans, nucleic acids, and receptors. In this study as the prostate cancer (CaP) biomarker, we propose GNPs functionalized with PSMA and fluorescent dyes. We have investigated a possibility to obtain an increased fluorescence signal, gained from GNPs conjugated with fluorescent dye and PSMA. The electric charge on the GNPs, the distance between GNPs and luminescent dye molecules has a significant effect on the luminescence intensity, and this enhancement highly depends upon the excitation wavelength of pumping laser source. Charged antigen, such as PSMA, can absorb on GNPs via electrostatic interaction. As the *in vitro* platform, we have used a biotissue mimicking phantoms based on gelatin matrix. Gelatin-based materials are attractive due to their stable mechanical properties and ease of fabrication. Proposed method that is specific and reliable for detecting cancers at early stages and is easily accessible so that it can function as the first-line guidance is of utter importance. Further, the unique physical properties of developed nanoscale materials can be utilized to produce novel and effective sensors for cancer diagnosis, agents for tumor imaging, and therapeutics for treatment of cancer.

27.1   INTRODUCTION

Cancer is a major public health problem in the worldwide. More than 11 million people are diagnosed with cancer every year. It is estimated that there will be 16 million new cases every year by 2020. Lung, colon, prostate, and breast cancers continue to be the most common causes of cancer death, accounting for almost half of the total cancer deaths among men and women. CaP is the second most common cancer diagnosed in men globally. Survival of a cancer patient depends heavily on early detection and thus developing technologies applicable for sensitive and specific methods to detect cancer is an inevitable task for cancer researchers. Biomarkers have an important role in today's diagnostics. A biomarker is a molecule that is up or down-regulated depending on the physical state of the body. This makes biomarkers interesting as deviations in biomarker levels that can reveal information about a patient's health condition [1, 2]. It's known that PSMA is one of the most well-established and highly specific prostate epithelial cell membrane antigen. PSMA is a type II transmembrane zinc metallopeptidase, belonging to the M28 peptidase family. It possesses hydrolyzing enzyme activities and is also known as FOLH1 (foliate hydrolase 1) (Figure 27.1).

Pathological studies have indicated that virtually all CaP expresses PSMA. PSMA is an excellent target for next reasons: (1) PSMA is mainly expressed in the prostate; (2) PSMA is highly expressed at all stages of the disease and is expressed on the cell surface as an integral membrane protein and not released into the circulation. PSMA is expressed in other tissues, including normal (benign) prostate epithelium, the small intestine, renal tubular cells, and salivary gland. This "nontarget" expression is fortunately 100- to 1,000-fold less than baseline expression in CaP. The combination of a nanoparticle platform with targeting ligands for tumor cell-surface biomarkers is a promising architecture for achieving selective drug delivery and uptake into target cells (Figure 27.2) [3]. GNPs are prone to be attached to many biological probes such as antibodies, enzymes, lectins, glycans, nucleic acids, and receptors.

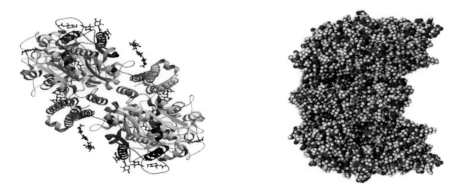

FIGURE 27.1 **(See color insert.)** Chemical structure of prostate-specific membrane antigen (PSMA, Glutamate carboxypeptidase II).

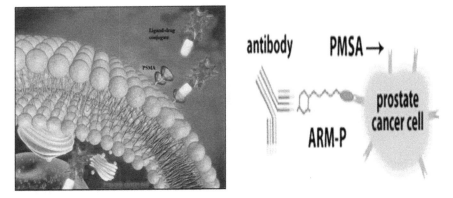

FIGURE 27.2 **(See color insert.)** Transporter molecules carrying therapeutic drugs to PSMA targets on a prostate cancer cell.

27.2 EXPERIMENTAL

27.2.1 MATERIALS

In this investigation as the CaP biomarker, we propose GNPs functionalized with PSMA and fluorescent dye Nile Blue (Nb). All experimental materials: GNPs, Nb fluorescent dyes, PSMA, are certified and commercially available (GNPs and Nb were purchased from Sigma-Aldrich and PSMA – from thermolab respectively) (Figure 27.3). The aim of this work was to investigate a possibility to obtain an increased fluorescence signal, gained from GNPs conjugated with Nb fluorescent dye and PSMA. GNPs are easily bioconjugated, and their special optical properties are unique in comparison with other optical probes. Several methods have been developed for conjugation of biomolecules to GNPs, including direct ligand exchange, covalent coupling, electrostatic adsorption, and surface coating. In our case as a covalent coupling, small bifunctional molecules such as Nb fluorescent dye were used for further bioconjugation with GNPs and PSMA. Charged antigen, such as PSMA, can absorb on GNPs via electrostatic interaction.

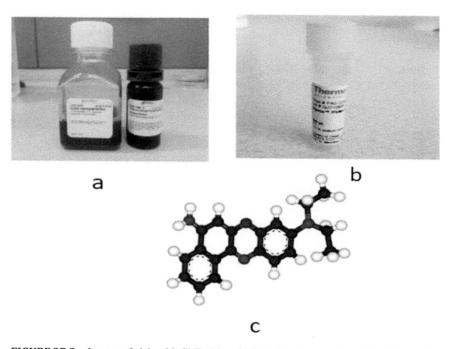

FIGURE 27.3 Images of vials with GNPs (a) and PSMA (b), structural formula of Nb dye (c).

In the description of the nature of the energy transfer from an organic luminescent dye to a GNP, the distance dependence between the luminescent dye and the surface of a GNP plays a crucial role. The altered electromagnetic field around the GNPs changes the properties of a dye that is placed in the vicinity. It can cause two enhancement effects: the first is an increase in the quantum efficiency of the dye and the second is an increase in the excitation rate of the dye. The induced collective electron oscillations associated with the surface Plasmon resonance give rise to induced local electric fields near the nanoparticle surface. Energy transfer from luminescent organic dyes to GNPs is generally considered to be the major process leading to the excited-state activation/deactivation of the dyes [4, 5]. The most familiar mechanism is that of energy transfer via dipole-dipole interactions, i.e., FRET, from an energy donor to an energy acceptor (Figure 27.4).

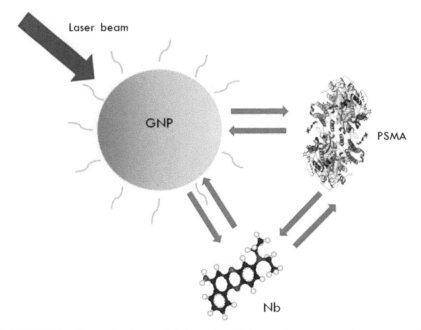

FIGURE 27.4   (See color insert.) Schematic of light to enhancement and energy transfer in GNPs/Nb /PSMA nanocomposite.

27.2.2   *MEASUREMENT*

Absorbance, transmittance, and luminescence spectra of the samples were recorded by multi-fiber optical spectrometer (Avaspec–2048, "Avantes").

Photoexcitation of the nanocomposites were performed at different wave-length, using corresponding laser light sources. Some specific experiments we have carried out in the LiCryL/Cemif. Cal department at the University of Calabria (Italy). This gave us an opportunity to utilize in our research such modern and precise scientific devices, an atomic force microscope (AFM, Aotoprobe CP VEECO), scanning electron microscope (SEM), and confocal laser scanning microscope (CLSM) (Figure 27.5). Purchased GNPs were stabilized in the citrate buffer, and a surface *plasmon* resonance of GNPs was measured using a spectrometer (Figure 27.6).

FIGURE 27.5   Laboratory-based set-up of the confocal laser scanning microscope.

27.3   RESULTS AND DISCUSSION

As the *in vitro* platform, we have used a biotissue mimicking phantoms based on gelatin matrix. Traditionally, phantoms that closely mimic the physical properties of various human tissues have been invaluable for the development and testing of medical imaging modalities. Gelatin-based materials are attractive due to their stable mechanical properties and ease of fabrication. To carry

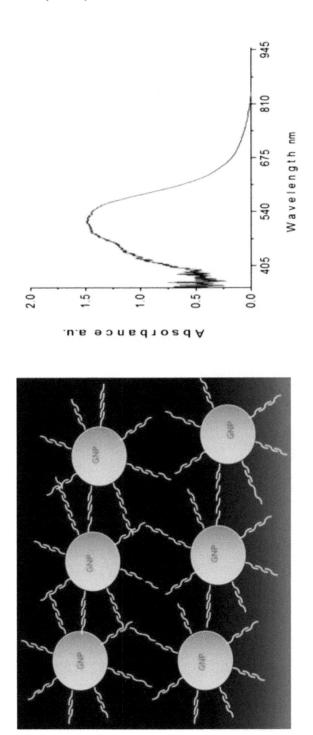

FIGURE 27.6 GNPs stabilized suspension in citrate buffer (a), The surface *plasmon* resonance of GNPs (b).

out a gelatin phantom based experiments, we used two samples prepared: (1) Gelatin matrix doped with Nb fluorescent dye and PSMA, and (2) Gelatin matrix doped with Nb fluorescent dyes, PSMA, and GNPs. The gelatin-based solution was prepared as follows: 5 gram of gelatin powder was doped and dissolved in 100 mL of deionized water embedded in laboratory flask and was left for 2 hours at rooms temperature. After that, the solution was heated to 35°C and vigorously stirred for 20 minutes to create an uniform substance. Prepared solution was divided into two parts using Petri Dishes. One was filled with a gelatin solution doped Nb/PMSA substance with next concentration: 50 mL/water-gelatin/2×10^{-5} g NB: 2.5 mL water/5.0 mg PSMA, and another one with 50 mL/water-gelatin/2×10^{-5} g NB: 2.5 mL water/5.0 mg PSMA/$7.15 \times 10^{10} N$ GNPs/ml, dispersed in an aqueous buffer (0.02 mg/ml). Dishes were stored in a humidified atmosphere at 37°C, for 24 h in order to achieve a desired incubation rate of Nb and NB/GNPs/ PSMA composites on gelatin matrix. Then solutions were cooled to about at room temperature and deposed by drop coating to the glass surfaces treated with deionized water. The coated films on substrates were stored for 24 hours at room temperature. Each sample was examined thoroughly, in order to estimate and calibrate such important parameters, as thicknesses and optimal concentrations of Nb fluorescence dyes, PSMA, GNPs (Figure 27.7).

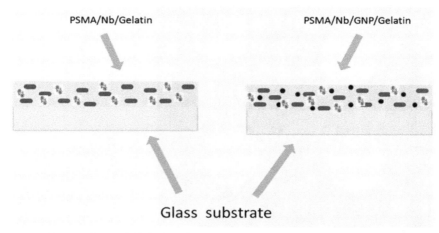

FIGURE 27.7   PSMA/Nb/Gelatin and PSMA/Nb/GNPs/Gelatin nanocomposites coated on the glass substrates.

After all procedures and solvent evacuation films were detached gently from the glass substrates, ready for examination. Prepared films thicknesses vary between 200–300 mm (Figure 27.8).

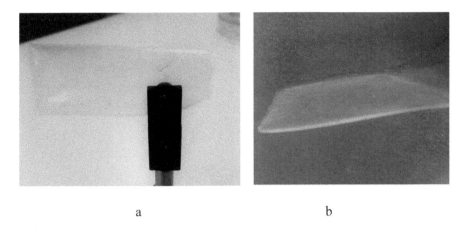

a   b

FIGURE 27.8   PSMA/Nb/Gelatin and PSMA/Nb/GNPs/Gelatin films.

For the demonstration of fluorescence enhancement, we assembled a setup. Laser light with λ = 532 nm was directed toward the gelatin films (Figure 27.9). In order to compare the intensities of the output lights from the Gelatin/PSMA/Nb and Gelatin/PSMA/GNPs/Nb composites, we recorded the fluorescence spectra using a spectrometer (Figure 27.10).

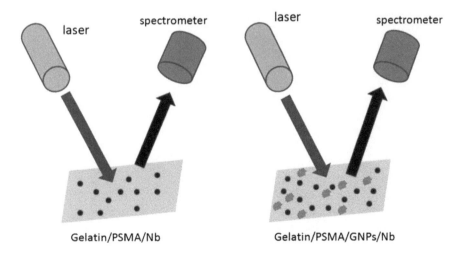

Gelatin/PSMA/Nb   Gelatin/PSMA/GNPs/Nb

FIGURE 27.9   An optical set-up for the demonstration of fluorescence enhancement in GNPs/Nb/PSMA and GNPs/Nb nanocomposites.

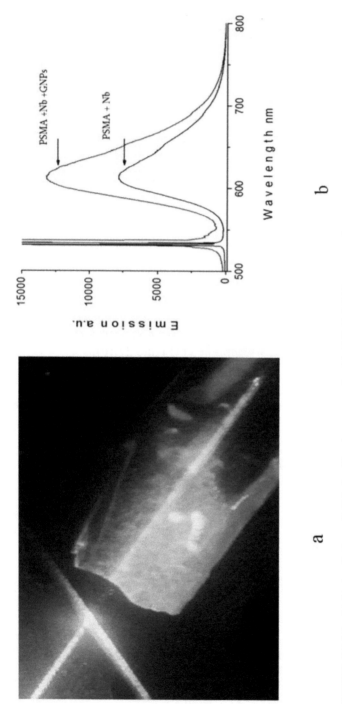

FIGURE 27.10 Light irradiation of Gelatin film incorporated with PSMA/Nb/GNPs (a) fluorescent enhancement in PSMA/Nb/GNPs nanocomposite stimulated by GNPs (b).

To visualize the distribution of GNPs/Nb pairs in the Gelatin matrix, we used a confocal microscope and an AFM. During the experiments, we found that some amounts of GNPs/Nb are prone to aggregate and form the clusters, which are the condensed quantity of GNPs and Nb dyes. The vivid "halos" surrounding the clusters demonstrate the enhancement of light brightness, which confirms the strong FRET effect between GNPs and Nb (Figures 27.11, 27.12).

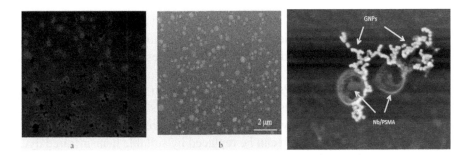

FIGURE 27.11 Distribution of GNP/Nb/PSMA nanocomposite inside the gelatin matrix (a) under a confocal laser scanning microscope (b) under atomic force microscope.

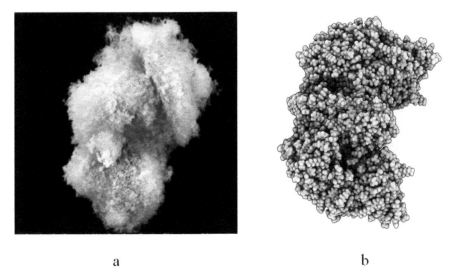

FIGURE 27.12 SEM image of PSMA functionalized with GNPs/Nb nanocomposites (a) Structural image of PSMA (b).

27.4   CONCLUSIONS

To summarize, we have developed highly reliable and sensitive screening test diagnostic tools for the detection of prostate cancer (CaP) in the early stage of its development.

In our study, we have shown that the fluorescence enhancement strongly depends on:

- size, shape, and concentration of GNPs.
- distances between GNRs, fluorescent dye Nb and PSMA.
- relative spectral positions between the Plasmon absorptions of GNPs and the absorption and emission of the fluorescent dyes.

Furthermore, while GNPs have the potential to improve contrast with structural imaging modalities, functionalized GNPs could be useful in the field of photothermal therapy that uses light to destroy cancer cells by heat. Besides, a fluorescent dye Nb can be replaced by suitable near-infrared dye, because the light irradiation in this region can penetrate deeper inside the tissues, increase the spatial resolution and cause less photodamage than UV/ blue part, as well as to avoid an overlapping of the signal with autofluorescence of biological samples.

KEYWORDS

- **biomarker**
- **cancer**
- **fluorescent dye**
- **GNPs**
- **prostate**
- **PSMA**

REFERENCES

1. Bhatt, N., Mathur, R., Farooque, A., Verma, A., & Dwarakanath, B. S., (2010). Cancer biomarkers - Current perspectives. *Indian J. Med. Res., 132,* 129–149.

2. Choi, Y. E., & Kwak, J. W., (2010). *Park Nanotechnology for Early Cancer Detection, 10*, 428–455.

3. https://www.sciencedaily.com/releases/2009/07/090706161306.htm (accessed on 11 January 2019).

4. Thomas, K. G., & Kamat, P. V., (2003). "Chromophore functionalized gold nanoparticles," *Acc. Chem. Res., 36*, 888–898.

5. Ketevan, C., Besarion, P., Tamaz, S., Aleksandre, K., Elene, D., & Nana, K., (2014). "*Luminescence Enhancement in Nanocomposite Consisting of Polyvinyl Alcohol Incorporated Gold Nanoparticles and Nile Blue 690 Perchlorate*" (Vol. 53, No. 31, 7177–7181). Applied Optics.

CHAPTER 28

Antibacterial Activity of Hyperbranched Poly(Acrylic Acid-Co-3-Hydroxypropionate) Hydrogels

E. ÇATIKER[1], T. FILIK[1], and E. ÇIL[2]

[1]*Faculty of Art and Science, Department of Chemistry, Ordu University, 52200, Ordu, Turkey, E-mail: ecatiker@gmail.com*

[2]*Faculty of Education, Department of Math and Science, Ordu University, 52200, Ordu, Turkey*

ABSTRACT

In this research, the antibacterial activity of poly(acrylic acid-co-3-hydroxypropionate) (PAcHP) was studied by disc diffusion, broth macrodilution minimum inhibitory concentration (MIC), and minimum bactericidal concentration (MBC) methods. For this purpose, three Gram-positive (*B. subtilis* NRRL B-209, *M. luteus* NRRL B-1018 and *S. aureus* ATCC 6538) and three Gram-negative (*E. coli* ATCC 25922, *P. aeroginosa* NRRL B-2679, *P. vulgaris* NRRL B-123) bacteria were selected. Firstly, antibacterial activity study was carried out by disc diffusion method using 100 microgram polymer. No bacterial growth was observed in *B. subtilis* plate, while the 29.63 mm inhibition zone diameter was obtained in the *M. luteus* plate. No inhibition zone diameters were observed in *P. vulgaris*, *S. aureus*, *E. coli*, and *P. aeruginosa* plates. The minimum inhibitory concentration (MIC) Broth Macrodilution and MBC method were then applied to four different polymer concentrations (25–50–75–100 mg/2 mL) in *E. coli*, *S. aureus* and *M. luteus* bacteria. There was no growth of *S. aureus* and *M. luteus* plaques in samples with 100 mg PAcHP /2 mL concentration. For this reason, it was concluded that the PAcHP polymer was bactericidal against three Gram-positive bacteria in this study.

28.1   INTRODUCTION

Human health is threatened by many microorganisms, such as bacteria, fungi, and parasites that cause numerous infectious diseases resulting in a large number of deaths. Many infectious diseases are easily expanded via contact with infected individuals and environmental sources such as air, drinking water, and any infected materials. Sanitation by use of traditional disinfecting agents is mostly inadequate to inhibit the spread of infectious diseases. Moreover, routine cleaning procedures are generally costly and offer short-term protection. Hence, use of antimicrobial materials in daily life is a developing approach to prevent from infectious diseases. Antimicrobial polymers [1–4] are promising materials in combating with pathogens considering the wide use as commodity materials, relatively low cost, structural diversity and ease of their structural modification. However, the approach has more applications in medical devices [5–7], water purification systems [8, 9], textiles [10, 11], and food packaging industry [12, 13] comparing to commodity materials.

The concept of antimicrobial polymers was propounded by Cornell and Dunraruma [14] for the first time in 1965. Although many antimicrobial polymers were introduced up to now, they may be basically classified into three main groups [15, 16] considering to working principles of polymeric systems; (i) *biocidal polymers* which are inherently antimicrobial, (ii) *polymeric biocides* which involve covalently bonded non-deliverable biocides, (iii) *biocide-releasing* polymer composites. Antimicrobial polymer may also be classified as *solution-based polymers* and *surface-bound polymers* [15] according to application way. The former is effective when dissolved in a solvent. However, the latter is effective on the microorganisms on its surface. Although the working principles or the application styles are completely different with each other, all of the polymeric systems either kills directly or inhibit proliferation of target cells through several definite mechanisms. One of the most common mechanisms is based on a strong electrostatic interaction between positive charged sites on the biocide and cell wall of the target cell, damage of cytoplasmic membrane and then death of the cells [17]. Another common mechanism is to deliver a biocidal molecule into the target cells and then to cause disorderliness in cell functions [18, 19]. Besides, a new approach is inhibition of biofilm formation through repelling mechanisms which are mainly based on electrostatic repulsion between the surface of material and the target cell [20]. Recent studies [21, 22] revealed that poly(acrylic acid) (PAA) copolymers exhibit antimicrobial effect via the latter mechanism. Gratzl et al. [21] were reported that PAA copolymers of polystyrene and poly(methyl methacrylate) against *S. Aureus*, *E. coli*, and *P. Aeruginosa*. Gratzl et al. [21] evaluated the dependence of antibacterial activity on the degree of acrylic acid units in the block copolymers. Ping et al. [22] evaluated the *E. coli* resistance of PET-g-PAA films.

Recent studies [20–22] showed that PAA copolymers also have biocidal effect on some pathogenic bacteria. The promising results motivated us to investigate antimicrobial properties of hyperbranched copolymer of acrylic acid and 3-hydroxy propionate (PAcHP) on the both Gram-positive and negative pathogenic bacteria.

28.2 EXPERIMENTAL

28.2.1 MATERIALS

Hyperbranched poly(acrylic acid-co-3-hydroxypropionate) (PAcHP) was synthesized through base-catalyzed hydrogen transfer polymerization of acrylic acid as reported previously by Catiker and Filik [23]. Structural characterization of the polymer pointed out the chemical structure given in Figure 28.1.

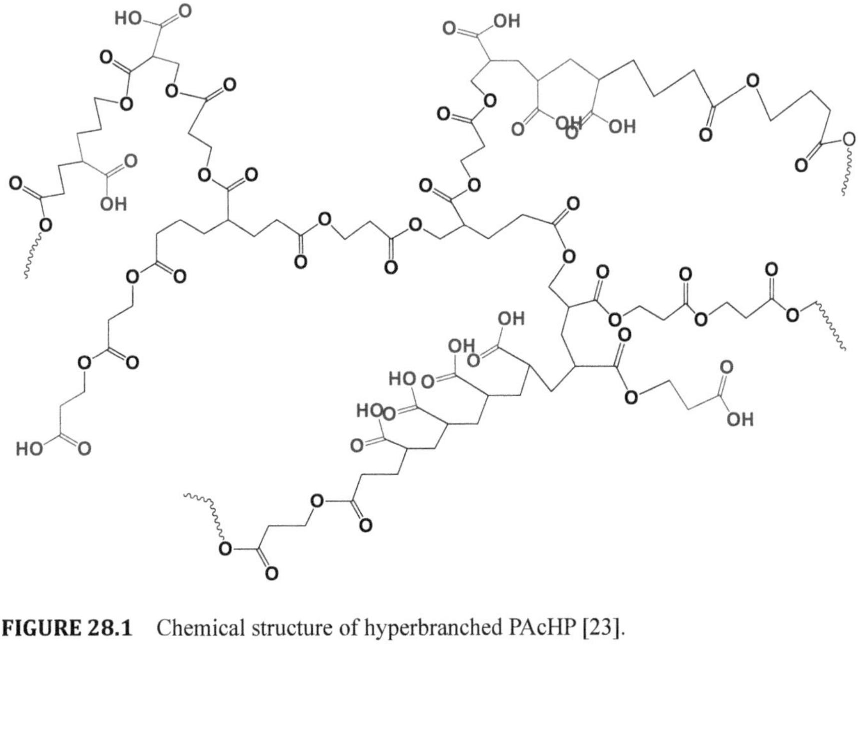

FIGURE 28.1 Chemical structure of hyperbranched PAcHP [23].

28.2.1.1 MICROORGANISMS AND CULTURE CONDITIONS

Gram positive and Gram negative total six bacteria strains used in the study, two of them were obtained from ATCC (American Type Culture Collection, Rockville, Maryland) and four of them were obtained from NRRL (Agricultural Research Service, United States of America). *B. subtilis* NRRL B-209, *M. luteus* NRRL B-1018, *P. aeroginosa* NRRL B-2679, *P. vulgaris* NRRL B-123, *S. aureus* ATCC 6538 and *E. coli* ATCC 25922.

Mueller Hinton agar (MHA) is considered the best medium to use for routine susceptibility testing of nonfastidious bacteria for supporting satisfactory growth of most nonfastidious pathogens. And also MHA is had low in sulfonamide, trimethoprim, and tetracycline inhibitors and is showed acceptable batch-to-batch reproducibility for susceptibility testing [24]. All bacterial strains were grown in cation-adjusted Mueller Hinton Broth (MHB) medium (Merck) for 24 h, at 37°C. 20 mL sterile cation-adjusted MHA medium (Merck) were poured into each 90 x 17 mm disposable Petri dishes (Isolab, Turkey). 24 well plate Corning® Costar® TC-treated multiple well plates were used for MIC Broth macrodilution method.

28.2.2 METHODS

For disc diffusion method, discs were prepared according to the methods described by Kirby-Bauer with slight modification. To prepare the discs, 100 microgram polymer was loaded in 10 mm round punch mold and pressed by hand stamp maker press machine. Bacterial suspension turbidity 0.5 McFarland standard was prepared in sterile ringer solution. Freshly prepared 100 µl ringer solution with bacteria (approximately 10^8) was placed over the agar and dispersed. Then, 10 mm diameter 100 microgram polymer discs were placed on MHA plates and incubated for 24 h, at 37°C. [24]. All the assays were performed in triplicate.

Minimum inhibitory concentrations (MICs) are defined as the lowest concentration of an antimicrobial that will inhibit the visible growth of a microorganism after incubation which are under optimum time and media. Because of the high concentration of polymer (100–75–50–25 mg) should be preferred to use the broth macrodilution method for determination of MIC, according to the Clinical and Laboratory Standards Institute standard procedures [25, 26]. For this purpose, each concentrations of (100–75–50–25 mg) the polymer in 1 mL sterile MHB and 1 mL 0.5 McFarland microorganism, total 2 mL suspension were poured in the well of 24 well plate. All multiple well plates were incubated for 24 h, at 37°C and 250 rpm in orbital shaker. Positive control (bacteria and growth media without polymer) and negative control (growth media) were used for each test [27]. All the assays were performed in triplicate.

Minimum bactericidal concentrations (MBCs) as the lowest concentration of antimicrobial that prevents the growth of a microorganism after subculture on to antibiotic-free media [28]. After 24h from bacteria inoculation, 100 μl the remaining bacteria with polymer suspension, was separated from each well and dispersed onto nutrient agar media. Plates were incubated for 24 h, at 37°C. After that, each plates was evaluated positive (no bacterial growth) or negative (killed the entire inoculum).

28.3 RESULTS AND DISCUSSION

A 100 microgram polymer discs were used to screen in vitro antibacterial activity by disc diffusion method. There was no bacterial growth was observed in the five aforementioned test strain's disc diffusion plates, but the 29.63 mm inhibition zone diameter was obtained in the *M. luteus* plate (Figures 28.2–28.5). So we planned to use higher concentrations of the polymer for MIC and choosed macrodilution for this purpose. All concentrations of (100–75–50–25 mg) the polymer, was effective to *M. luteus* but only one concentration (100 mg) was effective to *S. aureus* at antibacterial activity assay. A 25 mg and higher dosage polymer application to *M. luteus,* 100 mg and higher dosage polymer application to *S. aureus* were observed no bacterial growth. It can be concluded that PAcHP have a bactericidal effect on *M. luteus* and *S. Aureus* which are Gram-positive bacteria. Similar results were previously reported for block copolymers of acrylic acid/styrene [21], and PAA grafted PET films [22].

FIGURE 28.2 Antimicrobial effect of 10 mm PAcHP disc on *Micrococcus luteus.*

FIGURE 28.3   No antimicrobial effect of 10 mm PAcHP disc on *E. coli.*

FIGURE 28.4   No antimicrobial effect of 10 mm PAcHP disc on *Bacillus subtilis.*

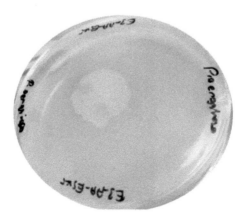

FIGURE 28.5   No antimicrobial effect of 10 mm PAcHP disc on *Pseudomas aeroginosa.*

KEYWORDS

- **disc diffusion method**
- **MIC broth macrodilution method**
- **PAcHP**

REFERENCES

1. Kenawy, E. R., Worley, S. D., & Broughton, R., (2007). *Biomacromolecules., 8*(5), 1359.
2. Grace, J. L., Huang, J. X., Cheah, S. E., Truong, N. P., Cooper, M. A., Li, J., Davis, T. P., Quinn, J. F., Velkov, T., & Whittaker, M. R., (2016). *RSC Adv., 6*(19), 15469.
3. Jain, A., Duvvuri, L. S., Farah, S., Beyth, N., Domb, A. J., & Khan, W., (2014). *Adv. Healthcare Mater.* doi: 10.1002/adhm.201400418.
4. Palza, H., (2015). *Int. J. Mol. Sci., 16,* 2099.
5. Kuroda, K., & Caputo, G. A., (2013). *Wiley Interdiscip. Rev. Nanomed. Nanobiotechnol., 5,* 49.
6. Gomes, A. P., Mano, J. F., Queiroz, J. A., & Gouveia, I. C., (2015). *Carbohydr. Polym., 127,* 451.
7. Yao, Y., Ohko, Y., Sekiguchi, Y., Fujishima, A., & Kubota, Y., (2008). *J. Biomed. Mater. Res. B. Appl. Biomater., 85*(2), 453.
8. Tyag, M., & Singh, H., (2000). *J. Appl. Polym. Sci., 76,* 1109.
9. Sun, Y., & Sun, G., (2002). *Macromolecules*, 35(23), 8909.
10. Simoncic, B., & Tomsic, B., (2010). *Text. Res. J., 80*(16), 1721.
11. Periolatto, M., Ferrero, F., Vineis, C., Varesano, A., & Gozzelino, G., (2017). In: Dr. Ranjith, K., (eds.), *Novel Antimicrobial Agents and Processes for Textile Applications, Antibacterial Agents*, InTech, DOI: 10.5772/intechopen.68423. Available from: https://www.intechopen.com/books/antibacterial-agents/novel-antimicrobial-agents-and-processes-for-textile-applications.
12. Gharsallaoui, A., Joly, C., Oulahal, N., & Degraeve, P., (2016). *Crit. Rev. Food Sci. Nutr., 56,* 1275.
13. Appendini, P., & Hotchkiss, J. H., (2002). *Innov. Food. Sci. Emerg., 3*(2), 113.
14. Cornell, R. J., & Donaruma, L. G., (1965). *J. Med. Chem., 8*(3), 388.
15. Huang, K. S., Yang, C. H., Huang S. L., Chen, C. Y., Lu, Y. Y., & Lin, Y. S., (2016). *Int. J. Mol. Sci., 17,* 1.
16. Santos, M. R. E., Fonseca, A. C., Mendonça, P. V., Branco, R., Serra, A. C., Morais, P. V., & Coelho, J. F., (2016). *J. Materials., 9,* 1.
17. Xue, Y., Xiao, H., & Zhang, Y., (2015). *Int. J. Mol. Sci., 16,* 3626.
18. Jamsa, S., Mahlberg, R., Holopainen, U., Ropponen, J., Savolainen, A., & Ritschkoff, A. C., (2013). *Prog. Org. Coat., 76,* 269.
19. Petersen, R. C., (2016). *AIMS Mol. Sci., 3*(1), 88.
20. Banerjee, I., Pangule, R. C., & Kane, R. S., (2011). *Adv. Mater., 23,* 690.

21. Gratzl, F., Paulik, C., Hild, S., Gugenbichler, J. P., & Lackner, M., (2014). *Mater. Sci. Eng. C., 38*, 94.

22. Ping, X., Wang, M., & Xuewu, G., (2011). *Radiat. Phys. Chem., 80*, 567.

23. Çatıker, E., & Filik, T., (2015). *Int. J. Polym. Sci.,* vol. 2015, Article ID 231059, 7 pages. https://doi.org/10.1155/2015/231059.

24. Hudzicki, J., (2009). *Kirby-Bauer Disk Diffusion Susceptibility Test Protocol.* http://www.asmscience.org/content/education/protocol/protocol.3189.

25. Clinical and Laboratory Standards Institute (CLSI). http://www.clsi.org.

26. Wiegand, I., Hilpert, K., & Hancock, R. E. W., (2008). *Nat. Protoc., 3*(2), 163.

27. Çatıker, E., Çil, E., & Filik, T., (2016). *Ordu Üniversitesi Bilim ve Teknoloji Dergisi., 6*(2), 117.

28. Andrews, J. M. J., (2001). *Antimicrob, Chemoth., 48*, 5.

CHAPTER 29

The Isolation of Glycosaminoglycans from Fish Eyeballs and Their Potential Application

B. KACZMAREK and A. SIONKOWSKA

Department of Chemistry of Biomaterials and Cosmetics,
Faculty of Chemistry, Nicolaus Copernicus University, Gagarin 7,
87–100 Torun, Poland, E-mail: beatakaczmarek8@gmail.com

ABSTRACT

Biomaterials are substances designed for biomedical purposes. They have to be biocompatible, biodegradable, and non-toxic. Biomaterials can be obtained from natural and synthetic polymers. Natural high molecular weight compounds have to be isolated from natural sources. The management of food industry wastes is important because they are biologically active and can be hazardous for human and environment. Fish eyeballs can be a potential source of glycosaminoglycans (GAGs) as for instance hyaluronic acid (HA) or chondroitin sulfate (CS). In the experimental studies, GAGs were isolated from vitreous of *Cyprinus Carpio* and *Salmonidae* eyeballs. The concentration of HA and CS was determined by the spectrophotometric method at 520 and 525 nm, respectively, with the use of a standard curve. The results showed higher content of HA and CS in the *Salmonidae* than in *Cyprinus Carpio.* However, in each isolated GAGs mixture, both polymers were detected. It can be assumed that fish vitreous eyeballs can be managed as the natural source of GAGs.

29.1 INTRODUCTION

Demographic studies indicate an aging population, resulting in forecasted that a group of people with dysfunction in bone or cartilage tissue will become more numerous. Reconstructive medicine is developing rapidly because of the increased progress in implantation possibilities [1]. Tissue engineering is science, the aim of which is the use of medical and material engineering knowledge to obtain functional substitutes for damaged human tissues. It refers to the growth of a new tissue using living cells basing on the structure, which is called scaffold. The scaffold materials have to be compatible with growing living cells. Most of them are made from natural polymers. Materials for orthopedic use should be formed as porous scaffolds, what provides space for cells and allows reconstruction of living tissues. Tissue engineering includes studies from cells to organ biology and physiology as well as the interactions with biomaterials.

Biomaterial can be defined as any material which is used to obtain devices for replacing parts or functions in the body. The goal of using biomaterials is to improve human health by restoring the function of natural living tissues and organs in the body. Nowadays, bone-cartilage implants are produced from different types of material – ceramic, metallic, and polymeric. Materials based on polymers are used to obtain implants for fulfill of small bone or cartilage tissue cavities [2]. They are characterized by defined porosity, whereby cells can penetrate deep into material, and then proliferate and as a result, tissue is reconstructed [3]. Due to the vulnerability of biodegradable polymers based on these materials are slowly degraded, while the reconstruction process occurs [4].

The use of natural polymers to obtain these kinds of materials eliminates the necessity of removing implant after the tissue reconstruction. It is fundamental for using polymeric biomaterials as implants. Materials based on natural polymers should be biocompatible; cells should adhere to the material, what cause the incorporation deep into implant [5]. Biodegradability and biocompatibility are two main properties of material, which make the possibility for use of material to produce implants. It is essential to observe the angiogenesis process after the implantation of biomaterial because blood vessels provide the nutrients which are necessary for the cells proliferation [6]. There are numerous known protein and polysaccharide polymers which found application to produce biomaterials. Such compounds are isolated from the natural sources have several advantages. They are biocompatible, biodegradable, and do not occur any immunological reactions after incorporating them into the human body [7]. Nowadays, polymeric implants are produced mainly on the base of collagen and chitosan.

Chitosan is a natural cationic polyelectrolyte copolymer derived from chitin, which is obtained in a deacetylation process. Chitin is a natural homopolymer of 2-acetamido–2-deoxy-β-D-glucopyranose units. It is the second most abundant natural polymer in nature, and it is present in a wide range of natural sources (crustaceans, fungi, insects, annelids, mollusks, coelenterate, etc.). However, chitosan is mainly manufactured from crustaceans (crab and crayfish), because a large amount of the crustacean exoskeleton is available as a by-product of food processing. The degree of deacetylation of chitosan (DA) is related to the balance between two kinds of residues. Chitin has to be deacetylated at least in 50% (mostly 70–90%) to be called chitosan.

Collagen is one of the most abundant protein in human body (one-third of all total proteins). It has very complicated structure, which nowadays is not artificially synthesized. Collagen is isolated from the natural parts of animal body. Main sources of collagen are cows derma or muscle and rat tail tendons. Moreover, there are scientific works in which collagen is isolated from fish skin or scales and marine sponges. Collagen is a main protein of connective tissue. It is responsible for strength of skin, durability, and soothe healthy look. Moreover, it provides the structural and mechanical support by providing scaffold for tissues, cells, and organs. Materials from chitosan and collagen were tested for their mechanical, thermal [8], and biological [9] properties. Nevertheless, collagen and chitosan materials have several disadvantages as well. Firstly, they are not elastic, and as a result, they do not return to their previous volume after the mechanical force effect. The second main disadvantage is their poor stability in aqueous environment (they swell and then dissolve). It is, therefore, necessary to modify materials by the addition of different natural or synthetic polymers, or by the cross-linking process. In the case of biomaterials, it is unwanted to change such properties as biocompatibility and biodegradability. Neverthe-less, it is advisable to modify the stability in aqueous environment and the mechanical properties as elasticity to not change the shape or structure after the force effect [10].

Nowadays polymeric compounds for biomedical application are proposed from proteins as well as polysaccharides groups as gelatin, hyaluronic acid (HA), elastin or for instance sodium alginate. The examples of natural compounds from polysaccharides group are also glycosaminoglycans (GAGs). GAGs have long chains consisting of repeating units as an amino sugar with uronic sugar or galactose. GAGs can be classified to four groups based on the repeating unit type:

- hyaluronic acid;
- keratin sulfate;

- heparin sulfate;
- chondroitin sulfate (CS)/dermatan sulfate.

They have appropriate biological properties as biocompatibility and biodegradability. Moreover, they do not cause any immunological reactions after implantation. The increasing interest in their application in regenerative medicine is currently observed. Natural polymers are used to obtain biomaterials in various forms. Porous structures called scaffolds can be obtained by porogen addition or by lyophilization process. Thin solid films can be obtained for instance by the solvent evaporation. Hydrogels are the products of polymeric materials holding large amounts of water in their three-dimensional networks.

GAGs are polymers which are isolated from natural sources. For many years the main GAGs source were rooster combs [11], fish eyeballs [12], or fish skin [13]. An innovative idea is the isolation of GAGs from food industry wastes as vitreous of eyeballs and skin of popular in Poland fish species eaten in high amounts as *Cyprinus Carpio* (17.5 thousands of tons in total per year) and *Salmonidae* (12 thousands of tons in total per year).

The management of wastes is very important. The need for biological treatment of biodegradable wastes results from the high biological activity of this type of wastes that can be hazardous to human health and life directly and indirectly. Fish heads and skin are the by-products in the food industry, and therefore they can be used as a cheap and readily available potential natural source of GAGs.

The aim of the study was to isolate the GAGs from vitreous of fish eyeballs of *Cyprinus Carpio* and *Salmonidae*. The HA and CS content was determined by spectrophotometric method with the use of standard curve.

29.2　EXPERIMENTAL

29.2.1　MATERIALS

The procedure of GAGs isolation from vitreous of fish eyeballs was according to the procedure reported elsewhere [14]. The eyeballs of the *Cyprinus Carpio* and *Salmonidae* (from Acipol, Konin, Poland) were mechanically isolated from the fish heads. Then they were cut and defatted in acetone (85 g of tissue per 150 g acetone) for 48h. Obtained precipitate was dried at 60°C for 24 h. 100 mM sodium acetate buffer (pH = 5.5) containing 5 mM EDTA and 5mM cysteine (20 mL of solution per 1g of precipitate) was added to the obtain precipitate, and then papain was added (100 mg per 1g of obtained precipitate), and incubated for 1 h in 60C. The solution was boiling for 10 min at 100°C, cooled, and centrifuged at 10,000 x g for 15 min. Three volumes of saturated ethanol with sodium acetate were added, and mixture was incubated in 4°C for 24 h. The precipitate was centrifuged at 10,000 x g for 15 min and dried in 60°C. Obtained precipitate was dissolved in distilled water to obtain solution with 1% concentration.

29.2.2   MEASUREMENT

The solution of 1,9-dimethylmethylene blue (DMB) was prepared by dissolving 1.6g of DMB and mixing with 0.5ml of ethanol, 1.0g sodium formate and 1.0 mL of formic acid and then filled to 0.5 l with distilled water [15]. The GAGs, such as HA and CS were identified by spectrophotometric method. The 2.5 mL of DMB dye solution was added to 250 μl of isolated from the eyeballs GAGs mixture. The wavenumber for the maximum absorbance was determined (HA at 520 cm^{-1}, CS at 525 cm^{-1}) and the standard curves for purchased HA and CS were made in their solutions concentration range 0.25–2%.

The morphology of the samples was studied using Scanning Electron Microscope (SEM) (LEO Electron Microscopy Ltd, England). Scaffolds were frozen in liquid nitrogen for 3 min and gently cut with a razor scalpel for the interior structure observation. Samples were covered by gold and SEM images were made with resolution 200 μm.

29.3   RESULTS AND DISCUSSION

29.3.1   GLYCOSAMINOGLYCANS IDENTIFICATION

The isolated GAGs mixture was dissolved in the distilled water at 1% concentration. The pure purchased HA and CS were dissolved at 0.25–2% concentration. For the prepared mixture, the standard curve was obtained for the maximum absorbance at 525 (CS) and 520 nm (HA). The GAGs as HA and CS were identified in the isolated mixture by calculation with the use of obtained equation form standard curves (Figure 29.1) [15]. Obtained equations were then used for the calculation of HA and CS content in the isolated GAGs mixture from eyeballs of *Cyprinus Carpio* and *Salmonidae* fish (Table 29.1).

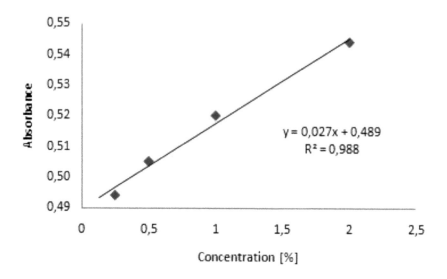

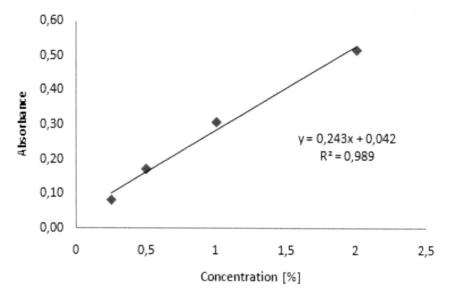

FIGURE 29.1   The standard curve of (a) chondroitin sulfate and (b) hyaluronic acid.

The GAGs mixture was isolated from the fish eyeballs of two species: *Cyprinus Carpio* and *Salmonidae*. Each mixture contains HA as well as CS. Higher content of those polymers was noticed in *Salmonidae*; however, the differences are not high. It is possible that polysaccharides content is higher than detected. The naturally occurred polysaccharides are covalently bonded to proteins, but such complexes cannot be detected by the spectrophotometric method.

TABLE 29.1 The Concentration of Pure Chondroitin Sulfate and Hyaluronic Acid in GAGs Mixture

Compound	Wavelength [nm]	Absorbance	Concentration [%]
Cyprinus Carpio			
HA	520	0.121	0.33
CS	525	0.509	0.77
Salmonidae			
HA	520	0.151	0.45
CS	525	0.521	1.19

The GAGs mixture was also isolated from fish skin of *Salmonidae* [13]. The content of HA was 1.26%, CS 2.03% and was higher than for the GAGs mixture isolated from vitreous of fish eyeballs. Moreover, the isolation from fish skin was characterized by higher efficiency. The need for eyeballs collection takes some period of time because they are smaller than the pieces of skin. Nevertheless, potentially skin, as well as vitreous of eyeballs, can be managed as GAGs natural source.

GAGs obtained in this study can be used for modification of biopolymeric materials, for example, collagen and chitosan. The SEM images of scaffolds based on chitosan/collagen with the 2 and 5% addition of GAGs mixture are shown in Figure 29.2. It can be observed that scaffolds have porous structure with interconnected pores. It is necessary for the application in tissue engineering science.

The influence of isolated GAGs on the scaffolds properties was noticed [16]. Scaffolds were biocompatible with human osteosarcoma cells. The presence of GAGs enhances the cells proliferation which can then affect the improvement of tissue regeneration process. The influence of GAGs addition to the chitosan/collagen mixture was also detected. From the polymeric mixture, scaffolds were obtained by lyophilization process [13]. The scaffolds supplemented with GAGs demonstrated higher biocompatibility compared to the porous materials without GAGs. Moreover, the cells attachment to the scaffold surface was detected by SEM observation [17]. It can be assumed that the addition of isolated GAGs to the polymeric mixture improves their biological properties and as a results can be proposed for biomedical application.

FIGURE 29.2   The scanning electron microscope images of chitosan/collagen scaffolds with (a) 2% and (b) 5% of glycosaminoglycans addition.

29.4 CONCLUSIONS

The GAGs mixture was isolated from vitreous of fish eyeballs of two species: *Cyprinus Carpio* and *Salmonidae.* The presence of HA and CS were detected by spectrophotometric method with the use of standard curve. In both isolated precipitation HA as well as CS were noticed; however, their concentration was different. From the *Salmonidae,* the isolated HA and CS content was higher than from *Cyprinus Carpio.* In the previous studies, GAGs were isolated from *Salmonidae* skin, and the detected polymers content was higher than from eyeballs. Isolated mixture of GAGs added to the polymeric matrixes enhances their biocompatibility. Moreover, the fish eyeballs are the by-products of food industry, and their management is ecologically beneficial. It can be assumed that fish eyeballs are the potential source of GAGs mixture which then can be used to obtain biocompatible materials.

ACKNOWLEDGMENTS

Financial support from the National Science Centre (NCN, Poland) Grant No UMO–2015/19/N/ST8/02176 is gratefully acknowledged.

KEYWORDS

- **glycosaminoglycans**
- **natural polymers**
- **polysaccharides**
- **spectrophotometry**

REFERENCES

1. Agrawal, C. M., (1998). *Emerging Technol., 50,* 31–35.
2. Ma, L., Gao, C., Mao, Z., Zhou, J., Shen, J., Hu, X., & Han, C., (2003). *Biomaterials, 24,* 4833–4841.
3. Yang, S., Leong, K. F., Du, Z., & Chua, C. K., (2004). *Tissue Eng., 7,* 679–689.
4. Dang, J. M., & Leong, K. W., (2006). *Adv. Drug Del. Rev., 58,* 487–499.
5. Nair, L. S., & Laurencin, C. T., (2007). *Prog. Polym. Sci., 32,* 762–798.
6. Naderi, H., Martin, M. M., & Bahrami, A. R., (2011). *J. Biomater. App., 26,* 383–417.

7. Karageorgiou, V., & Kaplan, D., (2005). *Biomaterials, 26,* 5474–5491.

8. Sionkowska, A., Wiśniewski, M., Skopińska, J., Poggi, G. F., Marsano, E., Maxwell, C. A., & Wess, T., (2006). *J. Polym. Deg. Stab., 91,* 3026–3032.

9. Sionkowska, A., Kaczmarek, B., Stalinska, J., & Osyczka, A. M., (2014). *Key Eng. Mater., 587,* 205–210.

10. Yang, L., Korom, S., Welti, M., Hoerstrup, S. P., Zund, G., Jung, F. J., Neuenschwander, P., & Weder, W., (2003). *Eur. J. Cardio. Surg., 24,* 201–207.

11. Nakano, T., & Sim, J. S., (1989). *Poult. Sci., 68,* 1303–1306.

12. Vazques, J. A., Rodriguez-Amado, I., Montemayor, M. I., Fraguas, J., Del Pilar Gonzalez, M., & Murado, M. A., (2013). *Mar. Drugs, 11,* 747–774.

13. Kaczmarek, B., Sionkowska, A., Łukowicz, K., & Osyczka, A. M., (2017). *Mater. Lett., 206,* 166–168.

14. Sadhasivam, G., Muthuvel, A., Pachaiyappan, A., & Thangavel, B., (2013). *Int. J. Biol. Macromol., 54,* 84–89.

15. Farndale, R. W., Sayers, C. A., & Barrett, A. J., (1982). *Connect Tissue Res., 9,* 247–248.

16. Kaczmarek, B., Sionkowska, A., & Osyczka, A. M., (2017). *Polym. Test., 62,* 132–136.

17. Kaczmarek, B., Sionkowska, A., & Osyczka, A. M., (2018). *Polym. Test., 65,* 163–168.

CHAPTER 30

Biopolymer Films Based on the Blends of Silk Fibroin and Collagen for Applications in Hair Care Cosmetics

S. GRABSKA and A. SIONKOWSKA

Department of Chemistry of Biomaterials and Cosmetics, Faculty of Chemistry, Nicolaus Copernicus University in Torun, Gagarin 7, 87–100 Torun, Poland, E-mail: sylwiagrabska91@gmail.com

ABSTRACT

In the present study, thin films based on the blend of silk fibroin (SF) and collagen were prepared by evaporation of the solvent and their properties were characterized by FTIR spectroscopy and thermogravimetric analysis (TGA). Additionally, the swelling ability was measured. It was found that SF film has higher thermal stability than collagen film. Materials made from SF and collagen was wettable. The addition of SF to collagen decreases the swelling ability of the material. The hair protection possibility of SF/collagen was studied using scanning electron microscopy imaging (SEM) and the mechanical testing of hair coated by the blends.

30.1   INTRODUCTION

Biopolymers are widely used in the cosmetic industry, particularly in hair care products [1–4]. Healthy and beautiful hair is desired by most of the population. For most people, grooming, and maintenance of hair and skin is a daily process [5]. Therefore, the need for products that improve the look and feel of hair surface has created a huge industry for hair care [6]. Combing, permanent wave treatment, chemical dyeing and weather conditions contribute a large amount of chemical and mechanical damage to the fibers. It leads to degradation of structure and mechanical properties. As a result, the fibers become weak and more susceptible to breakage after time, which is undesirable for healthy hair [7]. Everybody wants to have beautiful and healthy hair.

The conditioner is used to coat the hair with a thin film in order to protect it and provides desirable look and feel. Conditioners repair hair damage and make the hair easier to comb, prevent flyaway, add feel, shine, and softness. Conditioners, which typically repair the hair surface, respectively, have a distinct effect on its mechanical properties as well. To meet the needs of consumers, many ingredients, such as synthetic and natural polymers are added to conditioners [5, 6].

Conventional and commercial cosmetics contain synthetic polymers, such as polyethylene, polypropylene, polystyrene, which are not biodegradable. The microparticles of these cosmetic products, after application and final wash, go in the wastewater stream and eventually enter the aquatic environment. This can be dangerous for the environment [8]. Biopolymers are compounds which occur naturally in living organisms or are produced by them [9]. They represent a viable alternative to not biodegradable polymers. Ingredients in preparations for cosmetic applications should be non-toxic, because these materials are brought into contact with the skin or body. Natural polymers are biodegradable, bioresorbable, biocompatible, bioactive, and non-toxic. They could replace conventional polymers in cosmetics and personal care products [9].

Silk fibroin (SF) and collagen (Coll) are macromolecular compounds, which belong to the group of protein [9, 10]. Those biopolymers are widely used in cosmetic industry because they have fiber-forming and film-forming properties. Collagen is a major structural protein of extracellular matrix, and it is one of the most excellent material used in tissue engineering and cosmetic industry. However, materials containing collagen exhibit poor mechanical properties. Therefore, it is necessary to search for new materials that can be potentially used in cosmetic field. Significantly better mechanical properties have been demonstrated for SF [11]. SF is a protein composed of raw silk, where it performs structural functions [12]. For the production of materials with better properties, mixtures of two or more biopolymers should be used. Two or even more natural polymers can be mixed to form new materials with unique parameters and better properties in comparison with those consisting of a single component [13–15].

The contact angle measurements, mechanical properties, and atomic force microscopy imaging of films made of SF and collagen were studied by us previously [16]. We obtained materials which mechanical properties such as tensile strength and Young's modulus were much better for Col/SF blend films than for pure SF films. The results of contact angle and the surface free energy reveal that collagen films were more polar than SF films. In the case of AFM studies, surface roughness increased with the addition of collagen into SF matrix [16].

The purpose of the present work was the characterize SF/Coll films by FTIR spectroscopy, thermogravimetric analysis (TGA) and swelling ability measurements. We also studied properties of hair covered by thin films made of the blend of SF and collagen. The fresh aspect of our experiment is to investigate the influence of SF/Coll films to human hair for potential cosmetic applications in hair care products.

30.2 EXPERIMENTAL

30.2.1 MATERIALS

SF was prepared in our laboratory form *Bombyx mori* cocoons following the method described by Kim et al. [17] with slight modifications. *Bombyx mori* cocoons were supplied thanks to kindness of President of "Jedwab Polski Sp. z o.o." company. Empty cocoons were boiled two times in 0.5% Na_2CO_3 for 1 hour. After the removing of solution, SF was washed for 5 min in deionized water and boiled in 5% alkaline soap solutions and 20 min in deionized water. This procedure was repeated three times. The degummed silk was dissolved in a $CaCl_2/H_2O/CH_3CH_2OH$ (molar ratio: 1:8:2) at 80°C for 4 h. Then the fibroin solution was filtered and dialyzed against distilled water for 3 days to yield a fibroin aqueous solution [18]. The final fibroin concentration was 1% (it was determined by weighing the remaining residue after drying). Collagen was obtained from rat-tail tendons also in our laboratory. Tendons were washed in deionized water and dissolved in 0.1M acetic acid for 3 days in 4°C. Undissolved parts were centrifuged for 10 min at 10,000 rpm [19, 20]. The obtained solution was frozen at −18°C and lyophilized at −55°C and 5 Pa for 48 h (ALPHA 1–2 LD plus, CHRIST, Germany). Collagen was dissolved in 0.1M acetic acid to obtain 1% weight solution.

SF and collagen were mixed together in the volume ratios 90/10, 80/20, 70/30, 60/40, 50/50, 40/60, 30/70, 20/80, 10/90. SF and collagen were left as a control samples. Samples were dried at room temperature and humidity until solvent evaporated. Biopolymer films were obtained.

The film-forming properties of the blends on hair surface were measured by immersing the straight blond human hair of a 25-year-old volunteer in polymer solution for 1 h and drying for 24 h at room temperature and humidity [21].

30.2.2   MEASUREMENT

Following characteristics were studied for the biopolymer films based on the blends of SF and collagen: FTIR spectroscopy, TGA and swelling ability measurements. Mechanical properties analysis and scanning electron microscopy imaging (SEM) were studied for hair covered with biopolymeric blends.

30.2.2.1   FTIR SPECTROSCOPY

The interactions between functional groups of polymers were evaluated by attenuated total reflection infrared spectroscopy using Nicolet iS10 equipment. All spectra were recorded by absorption mode at 4 cm^{-1} intervals and 64-times scanning. The absorption values were obtained in the range of 400–4000 cm^{-1}.

30.2.2.2   THERMOGRAVIMETRIC ANALYSIS (TGA)

TGA was carried out using a Thermal Analysis SDT 2960 Simultaneous TGA-DTA analyzer from TA Instruments in the temperature range of 20°C to 650°C at a heating rate of 20°C/min in nitrogen. From the thermogravimetric curves, the characteristic temperature of maximum decomposition rate (T_{max}) was obtained.

30.2.2.3   SWELLING TESTS

Swelling behavior was measured by immersing the composites fragments in phosphate-buffer saline (PBS) solution, pH = 7.4 [22]. After 30 minutes and 1 h of immersion, materials were gently dried by putting them between two sheets of paper and weighted [23]. Swelling ratios were then calculated using the equation:

$$swelling\ ratio = \frac{(m_t - m_0)}{m_0} \times 100\% \quad (1)$$

where m_t is the weight of the material after time of immersion in PBS and m_0 is the weight of a material before immersion.

30.2.2.4   TREATMENT OF HAIR

Mechanical tests were made for hair covered with polymeric blends and without such a treatment. The mechanical properties of hair were tested by using a Zwick and Roell testing machine with 5 mm/min speed starting position and 0.1 N initial force. Analysis was repeated for 10 samples, and the standard deviation was calculated [21].

30.2.2.5   SCANNING ELECTRON MICROSCOPY (SEM)

The surface of human hair was studied using scanning electron microscope (LEO Electron Microscopy Ltd, England). Samples were covered by gold and images were made with the resolution 50 μm. The thickness of hair shafts was calculated for native hair and for hair covered with the polymer film.

30.3   RESULTS AND DISCUSSION

The aim of this research was to prepare and investigate the physicochemical properties of biopolymeric films based on the blends of SF and collagen by means of FTIR spectroscopy, TGA and swelling ability. The aim of our work was also investigate the influence of biopolymeric films for human hair. In this work, the properties of hair covered by thin films made of the blend of SF/collagen were studied.

30.3.1   FTIR SPECTROSCOPY

The position of the main band (cm^{-1}) in IR spectra of SF, collagen, and their blend films are shown in Table 30.1. There is a relationship between the position of the absorption bands and the structure of the molecule. Three basic areas are distinguished on the FTIR spectra: amide I, amide II and amide III.

The bands showing the presence of the β-sheet, occur at the values of wave numbers: 1616–1637 cm^{-1} (amide I), 1515–1525 cm^{-1} (amide II) and 1265 cm^{-1} (amide III). However, those located at the following values: 1638–1660 cm^{-1} (amide I), 1540–1545 cm^{-1} (amide II) and 1235 cm^{-1} (amide III) correspond to the statistical structure of the ball/α-helix [24].

SF and collagen belong to the group of protein polymers; therefore their FTIR spectra are similar. Table 30.1 shows the characteristic wave numbers of bands and the associated vibrations. The spectrum of pure SF film shows absorption bands, corresponding to the values of the wavenumber: 1647 cm^{-1} (amide I), 1518 cm^{-1} (amide II) and 1234 cm^{-1} (amide III), which indicates the presence of confirmation of the statistical ball/α-helix. A band at 1540 cm^{-1} (amide II) was also observed, which is characteristic of β-sheets. This means that fibroin in the form of a pure film simultaneously adopts the structure of SF I and SF II.

On the spectrum of pure collagen film, there are bands corresponding to the wavenumbers: 1647 cm^{-1} (amide I), 1541 cm^{-1} (amide II), 1235 cm^{-1} (amide III). This location indicates that the protein adopts the confirmation of the statistical ball/α-helix. The spectra of films based on fibroin and collagen with different weight ratios do not show significant differences. Data summarized in Table 30.1 clearly indicate that there are no visible shifts between bands. Small changes in the position can be seen only in the case of a50/50Coll/SF film. Then, the amide band III (1244 cm^{-1}) shifts towards higher values, while the amide band II shifts towards the lower values (1537 cm^{-1}). It can be concluded that in this case there are intermolecular interactions between the studied polymer components.

TABLE 30.1   The Position of Main Bands (cm^{-1}) in FTIR Spectra of Silk Fibroin, Collagen, and Their Blend Films

w_{Coll} [%]	w_{SF} [%]	Amid A	Amid B	Amid I	Amid II	Amid III
0	100	3272	3079	1647	1518 1540	1234
10	90	3270	3078	1647	1518 1541	1231
20	80	3271	3079	1647	1518 1541	1227
30	70	3274	3078	1647	1518 1540	1233
40	60	3272	3078	1647	1518 1540	1234
50	50	3273	3067	1648	1537	1244
60	40	3271	3079	1647	1518 1540	1230
70	30	3271	3078	1647	1518 1541	1230
80	20	3271	3079	1647	1518 1541	1233
90	10	3293	3078	1647	1518 1541	1235
100	0	3305	3079	1647	1541	1235

w_{Coll} – Weight Fraction of Collagen, w_{SF} – Weight Fraction of Silk Fibroin

30.3.2   *THERMOGRAVIMETRIC ANALYSIS (TGA)*

Changes in the thermal stability of pure polymers and their blends films measured under nitrogen flows were examined by TGA. The weight loss curves of the films made of SF and collagen as a function of temperature are shown in

Figure 30.1. The weight loss and temperature at maximum decomposition rate (T_{max}) of pure polymers and their blends films are shown in Table 30.2.

The TGA curves presented for the SF, Coll, and their blends samples consist of two stages (Figure 30.1 and Table 30.3). The first stage at 50–66°C was due to the loss of moisture and residual acetic acid, and it showed an approximate 7–13% loss in weight for SF/Coll blends. These results indicate that the binary SF/Coll blends with a small amount of collagen, entrap a lower amount of water in their structure in comparison to the SF/Coll with a big amount of collagen.

The second stage is due to the breakdown of side chains of amino acids and the breakdown of protein peptide bonds. The film of SF in the first stage (T_{max} = 57°C) loses 8.7% of its mass. In the case of a collagen film, a decrease of 12.7% is observed (T_{max} = 59°C). In the second stage, the weight loss is much higher, namely 58.1% (T_{max} = 308°C) for fibroin and 69.2% (T_{max} = 327°C) for collagen. The films of both polymers are not completely degraded at 650°C.

Based on this data, it can be concluded that SF has a higher thermal stability than Coll. In addition, the results of thermal properties of Coll/SF mixtures (Table 30.3) show the relationship between the SF content in the system and the final mass of the film. Increasing the amount of SF in Coll/SF systems improves the thermal stability of materials.

TABLE 30.2 Temperature of the Maximum Rates of the Reaction (T_{max}) and the Weight Loss (Δm) for SF, Coll, and SF/Coll Blends

Sample		T_{max} [°C]		Δm [%]	
w_{Coll} [%]	w_{SF} [%]	I	II	I	II
0	100	57	308	8.7	58.1
10	90	53	301	7.7	63.7
20	80	66	303	8.1	56.8
30	70	64	307	10.5	60.0
40	60	61	301	8.7	59.5
50	50	63	307	7.5	58.6
60	40	65	312	8.3	54.2
70	30	65	321	8.2	54.4
80	20	51	317	9.3	53.2
90	10	63	328	11.1	60.5
100	0	59	327	12.7	69.2

w_{Coll} – Weight Fraction of Collagen, w_{SF} – Weight Fraction of Silk Fibroin

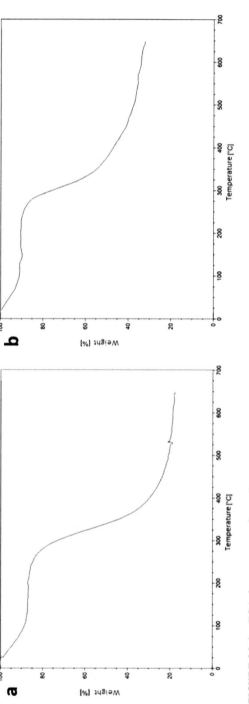

FIGURE 30.1   TGA thermograms of a) collagen; b) silk fibroin.

30.3.3 SWELLING ABILITY

The fast swelling behavior is the characteristic property of materials uses in biomedical and cosmetic industry. Collagen is a hydrophilic polymer with many functional groups, and for this reason, it is easily wettable by polar solvents, for example, PBS (*Phosphate-Buffered Saline*). The percentage of swelling of SF/collagen films after the immersion in PBS for 30 minutes and 1 h is presented in Table 30.3.

TABLE 30.3 The Percentage of Swelling for Films After the Immersion in PBS for 30 min and 1 h

Sample		Swelling [%]	
w_{Coll} [%]	w_{SF} [%]	30 min	1 h
0	100	-	-
10	90	-	-
20	80	129 ± 36	224 ± 29
30	70	156 ± 48	289 ± 36
40	60	352 ± 39	405 ± 42
50	50	209 ± 42	358 ± 38
60	40	232 ± 28	384 ± 39
70	30	250 ± 36	432 ± 48
80	20	310 ± 41	-
90	10	342 ± 44	502 ± 42
100	0	350 ± 39	543 ± 56

During 30 minutes of incubation, the film made of pure collagen absorbed around 350% of PBS. For SF/collagen materials a decrease in the swelling degree was observed. The addition of SF to collagen decreases the swelling ability of the material. However, the swelling ability of films made of SF and collagen is still very big. Binary films absorb around 130–340% PBS. Films made of pure SF and mixtures of SF and collagen in 90/10 weight fraction are not stable in water conditions.

During 1 hour incubation, the films made of pure collagen and SF/collagen mixtures absorbed more PBS than the same samples during the 30 minutes of incubation. During 1-hour incubation films made of pure SF and mixtures of SF and collagen in 90/10 and 20/80 weight fraction are not stable in PBS.

30.3.4 FILM FORMING PROPERTIES ON THE SURFACE OF HAIR

The possibility of protecting hair using SF/collagen thin films was studied using SEM. The mechanical properties of hair after the topical application of the blend were studied. Samples of human hair were cut from the hair shaft without any chemical treatment. Hair were immersed in polymeric mixture for 1 h and then dried at room temperature and humidity for 24 h. The mechanical tests were made. The results are shown in Table 30.4. Mechanical parameters such as Young's modulus (E_{mod}), the breaking force (F_{max}) and elongation at break (d_L) after the treatment of hair by polymer mixture are as follows. The hair without polymer has the lowest Young's modulus. Applying SF/Coll mixtures to the hair leads to an increase of E_{mod}. The highest Young's modulus was observed for the sample covered with film SF/Coll in 70/30 weight ratio. The tensile strength and elongation at break did not change significantly after the application of the SF/Coll films.

TABLE 30.4 Mechanical Parameters of Human Hair With and Without Treatment by the Blends of Biopolymers

	w_{SF} [%]	E_{mod} [GPa]	F_{max} [N]	d_L [%]
Without polymer		4.51 ± 0.97	0.878 ± 0.145	52.3 ± 1.6
0	100	6.31 ± 0.49	1.020 ± 0.010	51.2 ± 1.8
10	90	5.51 ± 0.29	0.929 ± 0.048	51.3 ± 0.9
20	80	5.38 ± 0.58	0.854 ± 0.027	49.9 ± 2.1
30	70	6.00 ± 0.91	0.853 ± 0.056	52.8 ± 1.8
40	60	5.45 ± 0.46	0.724 ± 0.076	51.8 ± 1.0
50	50	5.12 ± 0.11	0.634 ± 0.034	51.2 ± 2.8
60	40	6.02 ± 0.34	0.818 ± 0.068	49.1 ± 2.9
70	30	6.54 ± 1.34	0.777 ± 0.061	50.5 ± 0.8
80	20	5.32 ± 0.29	0.649 ± 0.098	47.2 ± 1.0
90	10	5.89 ± 0.78	0.633 ± 0.043	51.2 ± 2.6
100	0	6.28 ± 0.30	0.976 ± 0.049	50.2 ± 1.8

30.3.5 SCANNING ELECTRON MICROSCOPY IMAGING (SEM)

Figure 30.2 shows SEM micrographs of control sample of hair (without polymer), hair covered by SF, collagen, and SF/collagen mixtures in various weight ratio. The thickness of hair shafts with and without polymer covering was calculated from the SEM images in three places, and it is shown in Table 30.5. A polymer film was formed on the hair by evaporation of the solvent. In the case of hair covered with SF/Coll mixtures in weight ratio: 80/20, 70/30, 50/50, 40/60, 10/90 the surface of the hair is smoothed. Visibly smoothing of hair can be observed. The thickness of a hair shaft with the polymer covering is bigger than without it. The application of each studied SF/Coll mixtures leads to the increase of hair shaft thickness. The polymer mixture is adsorbed on the hair surface due to weak forces of adhesion. The highest thickness was observed for hair covered with SF/Coll mixture in 70/30 weight ratio. It can be concluded that such a composition of the blends leads to the formation of film with very good adhesion to the hair surface.

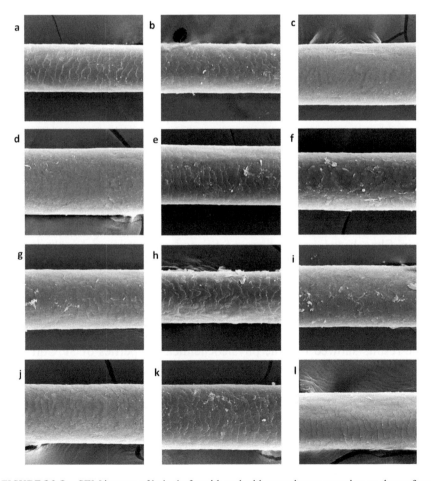

FIGURE 30.2   SEM images of hair shafts with and without polymer covering on the surface (a) without polymer,(b) SF/Coll 90/10, (c) SF/Coll 80/20, (d) SF/Coll 70/30, (e) SF/Coll 60/40, (f) SF/Coll 50/50, (g) SF/Coll 40/60, (h) SF/Coll 30/70, (i) SF/Coll 20/80, (j) SF/Coll 10/90, (k) silk fibroin, (l) collagen.

TABLE 30.5 The Thickness of Hair Shafts Without and With Covering by Biopolymer Mixtures

Hair covering		Thickness [μm]
w_{Coll} [%]	w_{SF} [%]	
Without polymer		55.97
0	100	75.37
10	90	58.21
20	80	71.64
30	70	76.87
40	60	67.91
50	50	67.16
60	40	68.28
70	30	59.70
80	20	70.52
90	10	68.66
100	0	63.06

30.4 CONCLUSIONS

Thin films based on mixtures of SF and collagen can be obtained in different weight ratios. FTIR analysis showed that in all films SF has two structures: silk I and silk II. SF has a higher thermal stability than collagen. The application of SF/collagen mixtures on the hair leads to smoothing of its surface and increases the thickness of the hair. Therefore, these mixtures can be used in cosmetics preparations as ingredients for hair care products. The results showed that the physicochemical properties of SF/Coll films depend on their composition. The application of SF and Coll solutions to the hair improves its elasticity and smoothes its surface. Mixtures of SF and collagen can be considered as promising materials for cosmetic applications.

ACKNOWLEDGMENTS

Grateful acknowledgments to Aleksandra Andrzejczyk for obtaining SF/Coll films and help with preparing the data from analysis.

KEYWORDS

- **biopolymer blends**
- **biopolymer films**
- **collagen**
- **hair care cosmetics**
- **silk fibroin**

REFERENCES

1. Yang, J., (2017). Hair care cosmetics. In: *Cosmetic Science and Technology: Theoretical Principles and Applications* (p. 601). Elsevier, Amsterdam, Netherlands.
2. Lochhead, M. Y., (2017). The use of polymers in cosmetic products. In: *Cosmetic Science and Technology: Theoretical Principles and Applications* (p. 171). Elsevier, Amsterdam, Netherlands.
3. Reich, C., & Su, D. T., (2001). Hair conditioners. In: *Handbook of Cosmetic Science and Technology* (p. 331). Marcel Dekker, New York, USA.
4. Wolfram, L. J., (2001). Hair cosmetics. In: *Handbook of Cosmetic Science and Technology* (p. 581). Marcel Dekker, New York, USA.
5. Bhushan, B., (2008). *Prog. Mater. Sci., 53*, 558.
6. Weia, G., Bhushan, B., & Torgerson, P. M., (2005). *Ultramicroscopy, 105*, 248.
7. Bhushan, B., La Torre, C., & Wei, G., (2004). Structural, nanomechanical and nanotribological characterization of human hair using atomic force microscopy and nanoindentation. In: *Springer Handbook of Nanotechnology* (p. 1223). Springer, New York, USA.
8. Niaounakis, M., (2015). *Biopolymers: Applications and Trends* (p. 407). Elsevier, Waltham, USA.
9. Sionkowska, A., (2011). *Prog. Polym. Sci., 36*, 1254.
10. Lv, Q., Hu, K., Feng, Q., & Cui, F., (2008). *J. Appl. Polym. Sci., 109*, 1577.
11. Joseph, B., & Raj, S. J., (2012). *Front. Life Sci., 6*(3–4), 55.
12. Zafar, M. S., & Al-Samadani, K. H., (2014). *J. Taibah Univ. Sci., 9*(3), 171.
13. Sionkowska, A., Lewandowska, K., Grabska, S., Kaczmarek, B., & Michalska, M., (2016). *Mol. Cryst. Liq. Cryst., 640*, 21.
14. Chung, T. W., & Chang, Y. L., (2010). *J. Mater. Sci-Mater. M, 21*, 1343.
15. Lewandowska, K., Sionkowska, A., & Grabska, S., (2015). *J. Mol. Liq., 212*, 879.
16. Sionkowska, A., Grabska, S., Lewandowska, K., & Andrzejczyk, A., (2016). *Mol. Cryst. Liq. Cryst., 640*, 13.
17. Kim, U. J., Park, J., Kim, H. J., Wada, M., & Kaplan, D. L., (2005). *Biomaterials, 26*, 2775.
18. Sionkowska, A., & Płanecka, A., (2013). *J. Mol. Liq., 186*, 157.
19. Sionkowska, A., Lewandowska, K., Michalska, M., & Walczak, M., (2016). *J. Mol. Liq., 215*, 323.

20. Sionkowska, A., & Kozłowska, J., (2010). *Int. J. Biol. Macromol., 47,* 483.
21. Sionkowska, A., Kaczmarek, B., Michalska, M., Lewandowska, K., & Grabska, S., (2017). *Pure Appl. Chem., 89*(12), 1829.
22. Jayakumar, G. C., Kanth, S. V., Sai, K. P., Chandrasekaran, B., Rao, J. R., & Nair, B. U., (2012). *Carbohyd. Polym., 87,* 1482.
23. Rodrigues, S. C., Salgado, C. L., Sahu, A., Garcia, M. P., Fernandes, M. H., & Monteiro, F. J., (2013). *J. Biomed. Mat. Res. A101A,* 1080.
24. Zhou, J., (2010). *Int. J. Biol. Macromol., 47,* 514.

Index

T - #0806 - 101024 - C474 - 234/156/21 - PB - 9781774634387 - Gloss Lamination